Chemistry of Glasses

Chemistry of Glasses

Second edition

A. PAUL

Professor of Materials Science
Indian Institute of Technology
Kharagpur

London   New York
CHAPMAN AND HALL

c3879380

CHEMISTRY

First published 1982 by
Chapman and Hall Ltd
11 New Fetter Lane, London EC4P 4EE
Published in the USA by
Chapman and Hall
29 West 35th Street, New York, NY 10001

Second edition 1990

© 1982, 1990 A. Paul

Typeset in 10/11 Times by
Macmillan India Ltd., Bangalore
Printed in Great Britain by
St. Edmundsbury Press, Bury St. Edmunds, Suffolk

ISBN 0 412 27820 0

British Library Cataloguing in Publication Data

Paul, A. (Amal)
  Chemistry of glasses. – 2nd. ed.
  1. Glass. Chemical properties
  I. Title
  666'.1042

  ISBN 0–412–27820–0

Library of Congress Cataloging-in-Publication Data

Paul, A. (Amal)
  Chemistry of glasses/A. Paul.—2nd ed.
    p.   cm.
  Includes bibliographical references.
  ISBN 0–412–27820–0
  1. Glass.   I. Title.
  QD 139.G5P38   1990
  666'.1—dc20   89–22364
                                                    CIP

To my teachers
Professor D. Lahiri and Professor R.W. Douglas
who introduced me to glass science

Contents

Preface to the second edition

It is encouraging to note that the first edition of the book was well received by the students and teachers – whom the book is primarily meant for. Two important topics related to chemistry of glasses were missing from the first edition: the chemistry of glass batch reactions, and the sol–gel method of glass making. The present author does not have adequate experience of these two topics. Dr Pavel Hrma of Case Western Reserve University, a well known expert in his field, has kindly volunteered to write on the chemistry of glass batch reactions which is included as Chapter 5 in the present edition; we hope the readers find it interesting and helpful. Dr I. Strawbridge has contributed a new chapter dealing with the sol–gel method. Dr A. K. Varshneya of New York State College of Ceramics, Alfred University has written the section 'Strength of Glasses' – which undoubtedly will increase the usefulness of the present edition.

Finally, the author would like to express his sincere thanks and gratitude to all the reviewers of the first edition of the book for their kind and valuable comments.

Amal Paul
Indian Institute of Technology
Kharagpur

Preface

'The preface . . . either serves for the explanation of the purpose of the book, or for justification and as an answer to critics'.

Lermontov

This book is based mainly on the lectures on the Chemistry of Glasses which I gave at the University of Sheffield to the final year honours and postgraduate students of Glass Technology and Materials Science. Most books reflect the interests and enthusiasm of their authors, and the present one is no exception.

The chemistry of glass is a rapidly developing field because the frontiers of advanced chemistry and advanced physics are merging together and consequently this book will soon require considerable amplification and modification. However, my experience in teaching the chemistry of glasses for more than a decade has shown me that there is much need for a good text-book on the subject. This book is therefore intended to be a stop-gap which, until it receives that new revision, may serve as a useful reference work for students and research workers alike.

I gratefully acknowledge the influence on my thinking of many of those colleagues at Sheffield with whom I have been in contact during the past twenty years or so. In addition to these personal influences, other published works have had considerable influence in modifying my approach, especially Cotton and Wilkinson's *Advanced Inorganic Chemistry*. Dr Peter James helped me in writing Chapter 2, and Professor Peter McMillan not only read the whole manuscript but also made a number of most helpful suggestions.

Finally, I wish to thank Professor Roy Newton, ex-Director of the British Glass Industry Research Association, for his encouragement and those valuable suggestions which helped this book to be less parochial than it would otherwise have been.

Amal Paul

Materials Science Centre
Indian Institute of Technology
Kharagpur
India

CHAPTER 1

Glass Formation

1.1 GENERAL ASPECTS

The term *glass* is commonly used to mean the fusion product of inorganic materials which have been cooled to a rigid condition without crystallizing. This generally means the ordinary silicate glasses which are used for making windows and bottle-ware. Literally hundreds of other glasses, each with its characteristic properties and chemical composition, have been made and these do not necessarily consist of inorganic materials. Examples of two familiar glasses made from cane sugar are lollipops and cotton candy; the former are in the shape of a rigid block; and the latter are flexible fibres. Substances of quite diverse chemical composition have been obtained as glasses and it is becoming widely recognized that the property of glass-formation is not, strictly speaking, an atomic or molecular property but rather one of a state of aggregation. Thus the word glass is a generic term and, instead of speaking of 'glass', one should speak of glasses as we speak of crystals, liquids, gases, etc.

Glasses are characterized by certain well-defined properties which are common to all of them and different from those of liquids and crystalline solids. X-ray and electron diffraction studies show that glasses lack long-range periodic order of the constituent atoms. That they resemble liquids and not crystalline solids in their atomic distribution is illustrated in Figure 1.1, in which the radial distribution function of a hypothetical material in the glassy state is compared with that of the gas, liquid and crystalline state of the same composition.

Unlike crystals, glasses do not have a sharp melting point and do not cleave in preferred directions. Like crystalline solids they show elasticity – a glass fibre can be bent almost double in the hand and, when released, springs back to its original shape; like liquids, they flow under a shear stress but only if it is very high, as in the Vickers Hardness Test.* Thus we see that the glassy form of matter combines the 'short-time' rigidity characteristic of the crystalline state with a little of the 'long-time' fluidity of the liquid state. Glasses, like liquids, are isotropic, a property which is of immense value in their use for a variety of purposes.

A glass is generally obtained by cooling a liquid below its freezing point and this has been considered as part of the definition of the glassy state, although as we shall see later it can also be obtained by compressing a liquid. The classical explanation for the formation of a glass is that, when a liquid is cooled, its fluidity

---

* There has been widespread misunderstanding of this point in the past, statements having been made that old windows have become thicker at the bottom and that glass tubing bends more and more, with time, when stored in horizontal racks. However, neither has been demonstrated beyond doubt.

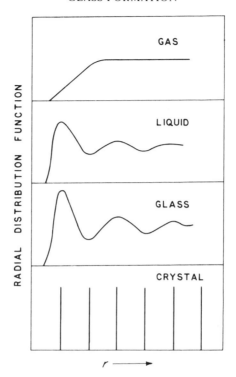

Fig. 1.1 Comparison of the radial distribution function of a glass with that of the gaseous, liquid and crystalline states.

(reciprocal viscosity) decreases and, at a certain temperature below the freezing point, becomes nearly zero. Our liquid becomes 'rigid'.

The relation between crystal, liquid and glass can easily be explained by means of a volume–temperature diagram as shown in Figure 1.2. On cooling a liquid from the initial state A, the volume will decrease steadily along AB. If the rate of cooling is slow, and nuclei are present, crystallization will take place at the freezing temperature T_f. The volume will decrease sharply from B to C; thereafter, the solid will contract with falling temperature along CD.

If the rate of cooling is sufficiently rapid, crystallization does not take place at T_f; the volume of the supercooled liquid decreases along BE, which is a smooth continuation of AB. At a certain temperature T_g, the volume–temperature graph undergoes a significant change in slope and continues almost parallel to the contraction graph CD of the crystalline form. T_g is called the transformation or glass transition temperature. Only below T_g is the material a glass. The location of E, the point corresponding to T_g, varies with the rate of cooling – and thus it is appropriate to call it a transformation *range* rather than a fixed point. At T_g the viscosity of the material is very high – about 10^{13} poise.

If the temperature of the glass is held constant at T, which is a little below T_g, the

2

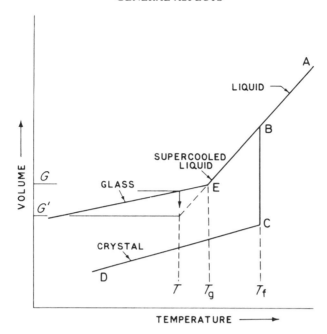

Fig. 1.2 Relationship between the glassy, liquid and solid states.

volume G will continue to decrease slowly. Eventually it reaches the level G' on the dotted line, which is a smooth continuation of the contraction graph BE of the supercooled liquid (undercooled is perhaps a more appropriate word and is used in this book). Other properties of the glass also change with time in the vicinity of T_g. This process by which the glass reaches a more stable condition is known as stabilization. Above T_g no such time-dependence of properties is observed. As a result of the existence of stabilization effects, the properties of a glass depend to a certain extent on the rate at which it has been cooled, particularly through the transformation range.

To understand the glass transition phenomenon let us take an example of a liquid and consider how its different physical properties change on undercooling. Glucose, a familiar substance, is an example of a material which readily undercools to form a glass. It melts at 414 K and, once molten, can be kept below this temperature for a long time without crystallization. The enthalpy, specific heat, specific volume, and thermal expansivity of glucose are shown as functions of temperature in Figure 1.3. We observe that, as the melt is cooled below about 300 K (T_g), its specific heat decreases almost by a factor of two. The specific volume and enthalpy show no analogous change, but they do show a slight discontinuity. There is no volume change or latent heat at this transition but the thermal expansivity decreases by a factor of four. Glucose stays optically transparent and there is no change in refractive index at this temperature, although the temperature coefficient of the refractive index suddenly decreases.

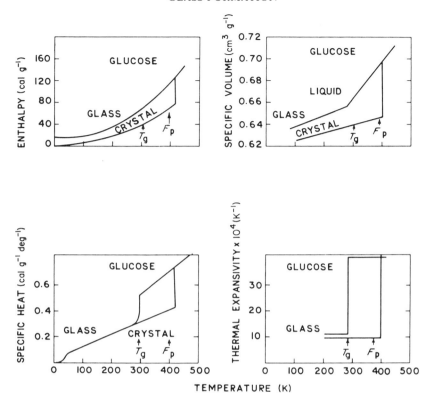

Fig. 1.3 The enthalpy, specific heat, specific volume and thermal expansivity of glucose as a function of temperature. (After Johari [16]).

Liquids can also be transformed into the glassy state by the application of pressure. The melting point of selenium under one atmosphere pressure is 493 K. The volume of selenium at 313 K changes non-linearly with pressure; near 11 kbar there is a discontinuity in the curve which is similar to that seen at E on cooling in Figure 1.2. The compressibility, obtained from the slope of the curve, decreases by about 40 per cent at 11 kbar, in very nearly the same way as the thermal expansivity. At higher pressures the compressibility of liquid selenium is very close to that of the crystalline phase. The pressure at which there is a sudden decrease in compressibility is known as the pressure of glass transition, P_g, and selenium at a pressure above 11 kbar is in the glassy state.

We see from the above that we can now have a phenomenological rather than a generic definition of the glassy state: a glass is a state of matter which maintains the energy, volume and atomic arrangement of a liquid, but for which the changes in energy and volume with temperature and pressure are similar in magnitude to those of a crystalline solid.

As is evident from Figure 1.3, at the glass transition the liquid and glass differ in

4

the second derivative of the free energy, G, with respect to temperature, T, and pressure, P, but not in the free energies themselves, or in their first derivatives. In Figure 1.3 the specific volume of glucose, given by

$$V = \left(\frac{\partial G}{\partial P}\right)_T$$

is unchanged at the transition, but the thermal expansivity

$$\alpha = \frac{1}{V}\left(\frac{\partial V}{\partial T}\right)_P = \frac{1}{V}\left(\frac{\partial^2 G}{\partial P \, \partial T}\right)$$

and the compressibility

$$\beta = -\frac{1}{V}\left(\frac{\partial V}{\partial P}\right)_T = -\frac{1}{V}\left(\frac{\partial^2 G}{\partial P^2}\right)_T$$

undergo an abrupt change. Analogously, the enthalpy

$$H = G - T\left(\frac{\partial G}{\partial T}\right)_P$$

does not change, but the heat capacity

$$C_p = \left(\frac{\partial H}{\partial T}\right)_P = -T\left(\frac{\partial^2 G}{\partial T^2}\right)_P$$

ιges at the transition. These considerations indicate that the glass transition more or less the characteristics specified for a second-order thermodynamic sition. Whether or not it is a true thermodynamic transition is a question that not yet been satisfactorily answered.

1.2 GLASS-FORMERS

ε ability of a substance to form a glass does not depend upon any particular ·mical or physical property. It is now generally agreed that almost any ɔstance, if cooled sufficiently fast, could be obtained in the glassy te – although in practice crystallization intervenes in many substances.

Table 1.1
Maximum undercooling of pure liquids*

ιbstance	(T_m) Melting point (K)	(ΔT) Extent of supercooling (K)	$\Delta T/T_m$
ſercury	234	77	0.33
ın	506	105	0.21
latinum	2043	370	0.18
′arbon fetrachloride	250	50	0.20
ɛenzene	278	70	0.25

After Staveley [14].

Most common liquids, when pure and in the form of a small drop, can be undercooled before spontaneous crystallization. Some typical examples are shown in Table 1.1.

These liquids can be cooled to within 20 per cent of their melting temperature before spontaneously returning to the thermodynamically stable crystalline form. A few liquids, on the other hand, can be undercooled so much that they fail to crystallize and eventually become glass. These glass-forming liquids are often, although not without exception, liquids which are very viscous at the melting point; liquids which do not form glasses have much lower viscosities. Some typical results are shown in Table 1.2.

Table 1.2
Viscosity of various liquids at their melting
temperatures*

Substance	Melting temp. (°C)	Viscosity (poise)
H_2O	0	0.02
LiCl	613	0.02
$CdBr_2$	567	0.03
Na	98	0.01
Zn	420	0.03
Fe	1535	0.07
As_2O_3	309	10^6
B_2O_3	450	10^5
GeO_2	1115	10^7
SiO_2	1710	10^7
BeF_2	540	10^6

* After Mackenzie [15].

However, a high viscosity at the freezing point is not a necessary or sufficient condition for the formation of a glass. Figure 1.4 shows the viscosity at the freezing point of aqueous sucrose solutions. No glasses are formed in this system with less than about 60 wt % sucrose. Although the viscosity of the solution containing 50 wt % sucrose is the same as that with about 80 wt % sucrose, the latter forms a glass and the former does not.

The viscosity in the system TeO_2–PbO at the liquidus temperature is below 1 poise, but it forms a glass; this is to be compared with ordinary silicate glasses where the viscosity at the liquidus temperature is around 10^5 poise. Figure 1.5 shows part of the phase equilibrium diagram for the system TeO_2–PbO; here glass formation appears to cease at the composition $4TeO_2.PbO$. It is to be noted that the primary phase of crystallization also changes from TeO_2 to $4TeO_2$ PbO at this composition ratio. Thus it may tentatively be suggested that the ability of this material to form a glass is in some way related to the fact that there must be difficulty in forming TeO_2 crystals from the liquid, while the formation of $4TeO_2$:PbO is relatively easy. It is important to note that the difficulty in forming crystals may be due to a high viscosity of the melt as in Table 1.2 and to

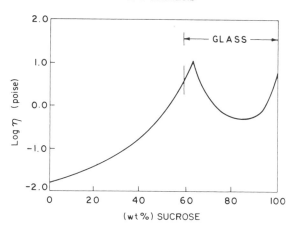

Fig. 1.4 Viscosity at the liquidus temperature in the sucrose–water system.

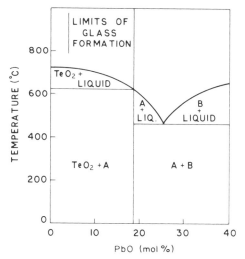

Fig. 1.5 Phase equilibrium diagram for the TeO_2–PbO system
showing limits of glass formation.
Probable compositions: $A - 4TeO_2:PbO$
$B - 3TeO_2:2PbO$

the amount of rearrangement of the atoms (change in configurational entropy) that is necessary in order that the particular crystals may be formed when cooled at a typical rate.

1.2.1 Glass-forming elements

Of all the elements in the Periodic Table, only a few in Groups V and VI can form a glass on their own:

7

Phosphorus: White phosphorus when heated at 250°C under a pressure of more than 7 kbar produces a glass. The same material can also be prepared by heating white phosphorus with mercury (catalyst) in an evacuated sealed tube at 380°C.

Oxygen: Oxygen has been claimed to be prepared in the glassy form by cooling liquid oxygen, but this is controversial, since the material may be the cubic γ-phase of crystalline oxygen.

Sulphur and selenium: Sulphur and selenium form glasses easily with different ring and chain equilibria.

Tellurium: On the basis of irregular volume changes when molten tellurium solidifies, it has been suggested that tellurium may form a glass, however this has not yet been proved beyond doubt.

1.2.2 Glass-forming oxides

B_2O_3, SiO_2, GeO_2 and P_2O_5 readily form glasses on their own and are commonly known as 'glass-formers' for they provide the backbone in other mixed-oxide glasses. As_2O_3 and Sb_2O_3 also produce glass when cooled very rapidly. TeO_2, SeO_2, MoO_3, WO_3, Bi_2O_3, Al_2O_3, Ga_2O_3 and V_2O_5 will not form glass on their own, but each will do so when melted with a suitable quantity of a second oxide. TeO_2, as discussed before, will not form a glass, but a melt of composition $9TeO_2$:PbO will produce on cooling a glass even though PbO is not a glass-former either. Figure 1.6 shows a section of the periodic table, the ringed elements having simple glass-forming oxides and the boxed elements having the second type of oxides, 'conditional glass-formers' according to Rawson (1).

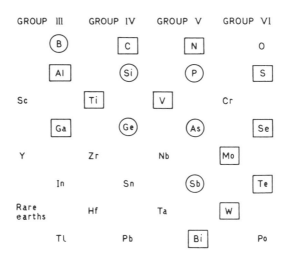

Fig. 1.6 Elements, the oxides of which are either glass-formers or conditional glass-formers.

Table 1.3
Ranges of glass formation in binary systems
(after Imaoka [13])

Metal oxide	Mol %			
	$B_2O_3{}^a$	$SiO_2{}^b$	$GeO_2{}^c$	$P_2O_5{}^d$
Li_2O	100–57.3	100–64.5	100–76.2	100–40
Na_2O	100–62.0	100–42.2	100–62	100–40
	33.5–28.5			
K_2O	100–62.3	100–45.5	100–40.5	100–53
Tl_2O	100–55.5	–	100–52.5	100–50
MgO	57.0–55.8	*100–57.5	–	100–40
CaO	72.9–58.9	*100–43.3	84.5–64.5	100–46
SrO	75.8–57.0	*100–60	86–61	100–46
BaO	83.0–60.2	*100.–60	100–90	100–42
			82.5–70.4	
ZnO	56.0–36.4		100–52	100–36
CdO	60.9–45.0			100–43
PbO	80.0–23.5		100–43	100–38
Bi_2O_3	78.0–37.0		100–66	

[a] 1–3 g material melted in Pt crucible and allowed to cool freely in air.
[b] 1–2 g material – as above.
[c] 1–3 g melt – as above.
[d] 1–3 g melt – as above.
* Involves extensive liquid–liquid phase separation.

The regions of glass formation in some simple binary systems are given in Table 1.3.

1.3 ATOMISTIC HYPOTHESES OF GLASS FORMATION

Glass formation is a kinetic phenomenon; any liquid, in principle, can be transformed into glass if cooled sufficiently quickly and brought below the transformation range. A good glass-forming material is then one for which the rate of crystallization is very slow in relation to the rate of cooling. As discussed in the earlier section, with conventional rates of cooling, some melts produce glass more easily than others. These facts lead many workers to postulate different atomistic hypotheses correlating the nature of the chemical bond, and the geometrical shape of the groups involved, with the ease of glass formation. It should be pointed out that, although these empirical hypotheses explain glass formation in some allied liquid systems, a unified hypothesis capable of explaining the phenomenon of glass formation in all the known systems has yet to be developed.

1.3.1 Goldschmidt's radius ratio criterion for glass formation

According to Goldschmidt [2] for a simple oxide of the general formula A_mO_n, there is a correlation between the ability to form glass and the relative sizes of the

9

Table 1.4
Limiting radius ratios for various coordination polyhedra

Polyhedron	Coordination number	Minimum radius ratio
Equilateral triangle	3	0.155
Tetrahedron	4	0.225
Trigonal bipyramid	5	0.414
Square pyramid	5	0.414
Octahedron	6	0.414
Cube	8	0.732

oxygen and A atoms. Glass-forming oxides are those for which the ratio of ionic radii R_A/R_O lies in the range 0.2 to 0.4. For ionic compounds the coordination number is often dictated by the radius ratio rule. From simple geometrical considerations of the maximum number of spherical anions packed around a cation maintaining anion–cation contact, the results set out in Table 1.4 can be calculated. Thus according to Goldschmidt, a tetrahedral configuration of the oxide is a prerequisite of glass formation. However, it should be pointed out that in glass-forming oxides the anion–cation bonding is far from purely ionic. Besides, as discussed earlier in the case of $9TeO_2$:PbO the coordination number of Te is six and not four. BeO with $R_{Be}/R_O \sim 0.221$ does not form glass.

1.3.2 Zachariasen's random network hypothesis

Since the mechanical properties and density of an oxide glass are similar to those of the corresponding crystal, the interatomic distances and interatomic forces must also be similar. Zachariasen [3] postulated that, as in crystals, the atoms in glass must form extended three-dimensional networks. But the diffuseness of the X-ray diffraction patterns show that the network in glass is not symmetrical and periodic as in crystals. For example, in the case of SiO_2 the only difference between the crystalline and glassy forms is that in vitreous silica the relative orientation of adjacent silicon–oxygen tetrahedra is variable whereas in the crystalline form it is constant throughout the structure. Such a difference is shown pictorially in Figure 1.7 for an imaginary two-dimensional oxide (A_2O_3) in both crystalline (a) and vitreous (b) forms.

Zachariasen proposed a set of empirical rules which an oxide must satisfy if it is to be a glass-former:

(1) No oxygen atom may be linked to more than two atoms of A.
(2) The number of oxygen atoms surrounding A must be small (probably 3 or 4).
(3) The oxygen polyhedra share corners with each other, not edges or faces.

If it is further required that the network be three-dimensional, a fourth rule must be added:

(4) At least three corners of each polyhedron must be shared.

Zachariasen's hypothesis has been more or less universally accepted; however, the

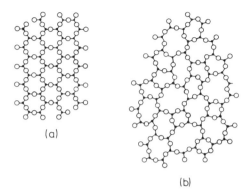

(a)

(b)

Fig. 1.7 Schematic two-dimensional representation of the structure of (a) a hypothetical crystalline compound A_2O_3 and (b) the glassy form of the same compound.

following limitations are pertinent and should be pointed out here.

(1) Although in most of the oxide glasses the coordination number of oxygen is two, Bray [4] has reported that in binary Tl_2O–B_2O_3 glasses with low Tl_2O content, the coordination number of oxygen may be three.

(2) The coordination numbers of silicon, phosphorus and boron in glass are 4, 4 and 3 or 4 respectively. However, as described earlier, the coordination number of tellurium in PbO–TeO_2 glasses is 6 with respect to oxygen. Alkali–phosphate glasses containing more than 50 mol % alkali oxide contain two-dimensional chains of various sizes, thus a three-dimensional network need not be a prerequisite for glass formation.

(3) Hagg [5] pointed out that an infinite three-dimensional network may not be a necessary condition for glass formation. He concluded: 'it seems as if a melt contains atomic groups which are kept together with strong forces, and if these groups are so large and irregular that their direct addition to the crystal lattice is difficult, such a melt will show a tendency to supercooling and glass formation'.

1.3.3 Smekal's mixed bonding hypothesis

According to Smekal [6] pure covalent bonds have sharply defined bond-lengths and bond-angles and these are incompatible with the random arrangement of the atoms in glass. On the other hand, purely ionic or metallic bonds completely lack any directional characteristics. Thus the presence of 'mixed' chemical bonding in a material is necessary for glass formation. According to Smekal, glass-forming substances with mixed bonding may be divided into three classes as follows:

(a) Inorganic compounds, e.g. SiO_2, B_2O_3, where the A–O bonds are partly covalent and partly ionic.
(b) Elements, e.g. S, Se having chain structures with covalent bonds within the chain and Van der Waals' forces between the chains.

11

(c) Organic compounds containing large molecules with covalent bonds within the molecule and Van der Waals' forces between them.

1.3.4 Sun's bond-strength criterion for glass formation

Since the process of atomic rearrangement which takes place during the crystallization of a material may involve the breaking and reforming of interatomic bonds, it may be reasonable to expect a correlation between the strength of these bonds and the ability of the material to form a glass. The stronger the bonds, the more sluggish will be the rearrangement process and hence the more readily will a glass be formed. This suggestion was first put forward by Sun [7], who showed that the bond strengths in glass-forming oxides are in fact particularly high. Some typical values from Sun's calculations are given in Table 1.5. It will be seen that the glass-forming oxides have single bond strengths greater than 90 kcal mol^{-1} and the modifiers have bond strengths less than 60 kcal mol^{-1}.

Table 1.5
Calculated single bond strengths of some oxides

Metal	Dissociation energy (E_d) (kcal mol^{-1})	Coordination number (N)	Single bond strength (E_d/N) (kcal mol^{-1})
B	356	3	119
		4	89
Si	424	4	106
Ge	431(?)	4	108
P	442	4	111
V	449	4	112
As	349	4	87
Sb	339	4	85
Zr	485	6	81
Zn	114	'2'	72
Pb	145	'2'	73
Al	317–402	6	53–67
Na	120	6	20
K	115	9	13
Ca	257	8	32

Sun himself pointed out that, although the bond strengths V–O, As–O and Sb–O are relatively high, the oxides are not good glass-formers. In fact V_2O_5 will not form a glass when melted alone. Sun suggested that 'small ring formation' may occur in the melts of these materials, which would result in easy crystallization.

A few other hypotheses, like Winter's p-electron criterion [8], Rawson's modification of Sun's hypothesis [9] etc., have been put forward from time to time. But none of these hypotheses are really capable of explaining glass formation to a more satisfactory extent than already described and will not be discussed further in this book.

1.4 KINETIC APPROACH TO GLASS FORMATION

Whether or not a given liquid will crystallize during cooling before T_g is reached is strictly a kinetic problem involving the rate of nucleation and crystal growth on the one hand and, on the other, the rate at which thermal energy can be extracted from the cooling liquid. In recent decades there have been several treatments of the conditions of glass formation, based on considerations of crystallization kinetics, and a good review is contained in the article *Under what conditions can a glass be formed?* by Turnbull [10]. This author pointed out that there are at least some glass-formers in every category of material, based on bond type (covalent, ionic, metallic, Van der Waals, and hydrogen). Cooling rate, density of nuclei and various material properties like crystal–liquid surface tension, and entropy of fusion etc. were suggested as significant factors which affect the tendency of different liquids to form glasses. This approach naturally raises the question *not* whether a liquid will form a glass on cooling, but rather *how fast* must a given liquid be cooled in order to avoid any detectable crystallization?

Uhlmann [11], developing Turnbull's idea, has provided some useful guidelines for glass formation by using theoretical time–temperature–% transformation (T–T–T) curves to specify critical cooling rates in terms of material constants. In the case of single-component materials or congruently melting compounds, if the nucleation frequency and rate of crystal growth are constant with time, then the volume fraction X crystallized in a time t may be expressed as:

$$X \sim \tfrac{1}{3}\pi I_v u^3 t^4 \text{ (for small values of } X)\qquad(1.1)$$

where I_v is the nucleation frequency per unit volume, and u is the rate of advance of the crystal–liquid interfaces per unit area of the interfaces (This is dealt with in more detail in Chapter 2). Both nucleation frequency I_v and rate of advance u are inversely proportional to the viscosity of the liquid.

The cooling rate required to avoid a given volume fraction crystallized may be estimated from equation (1.1) by the construction of T–T–T curves, an example of which is shown in Figure 1.8 for two different volume fractions crystallized. In constructing these curves, a particular fraction crystallized is selected, the time required for that volume fraction to form at a given temperature is calculated, and the calculations are repeated for other temperatures. The nose in a T–T–T curve, corresponding to the least time for the given volume fraction to crystallize, results from competition between the driving force for crystallization (which increases with decreasing temperature) and the atomic mobility (which decreases with decreasing temperature). The transformation times t_i are relatively long in the vicinity of the melting point as well as at low temperatures. The cooling rate required to avoid a given fraction becoming crystallized can be approximately represented by the relationship:

$$\left[\frac{dT}{dt}\right]_c \sim \frac{\Delta T_N}{\tau_N}\qquad(1.2)$$

13

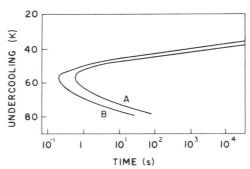

Fig. 1.8 Time–temperature transformation curves for salol corresponding to volume fractions crystallized of (A) 10^{-6} and (B) 10^{-8}.

where $\Delta T_N = T_E - T_N$; T_N is the temperature at the nose of the T–T–T curve; τ_N is time at the nose of the T–T–T curve; and T_E is the melting point. From Figure 1.8 it is apparent that the cooling rate required for glass formation is rather insensitive to the assumed volume fraction crystallized, since the time at any temperature on the T–T–T curve varies only as the one-fourth power of X.

Uhlmann's approach is useful in that it provides a clear basis for the common observation that the most obvious factor which may be correlated with success or failure in undercooling a liquid at some fixed rate is the magnitude of the viscosity at the temperature T_E (the true thermodynamic crystallization temperature). The importance of the value of T_E (for some specified viscosity–temperature relation) in deciding the possibility of obtaining a glass from the liquid is illustrated in Figure 1.9, using T–T–T curves for the molecular substance salol. From that figure it may be noted that, if T_E for Salol were raised by 40°C above the actual value, the critical cooling rate for glass formation would increase four orders of magnitude from $50\,\mathrm{K\,s^{-1}}$ to $10^5\,\mathrm{K\,s^{-1}}$. On the other hand, a lowering of T_E by 40°C would permit glasses to be formed even at cooling rates of $10^{-6}\,\mathrm{K\,s^{-1}}$.

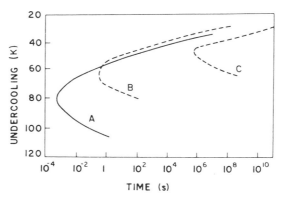

Fig. 1.9 Time–temperature transformation curves for salol-like materials having various melting points. The volume fractions crystallized are 10^{-6} in each case. (A) $T_E = 356.6\,\mathrm{K}$; (B) $T_E = 316.6\,\mathrm{K}$ and (C) $T_E = 276.6\,\mathrm{K}$.

A striking example of the increase in the number of glass-forming substances made possible by manipulating the cooling rate variable is provided by the recent experiments of Nelson et al. [12]. These workers have snap-melted powders of various refractory oxides using xenon flash and laser pulse heating techniques. Since the melting points of the substances investigated are very high, cooling by radiant energy loss from the tiny samples is extremely fast. Thus glasses were formed from such unlikely substances as La_2O_3.

REFERENCES

[1] Rawson, H. (1967), in *Inorganic Glass-forming Systems*, Academic Press, London, p. 9.
[2] Goldschmidt, V. M., Skrifter Norske Videnskaps Akad (Oslo), *I. Math-Naturwiss. Kl.* **8**, p. 7.
[3] Zachariasen, W. H. (1932), *J. Am. Chem. Soc.*, **54**, 3841.
[4] Baugher, J. F. and Bray, P. J. (1969), *Phys. Chem. Glasses*, **10**, 77.
[5] Hagg, G. (1935), *J. Chem. Phys.*, **3**, 42.
[6] Smekal, A. (1951), *J. Soc. Glass Technol.*, **35**, 411.
[7] Sun, K. H. (1947), *J. Am. Ceram. Soc.*, **30**, 277.
[8] Winter, A. (1955), *Verres Refract.*, **9**, 147.
[9] Rawson, H. (1956), in *Proc. IV Internat. Congr. on Glass*, Imprimerie Chaix, Paris.
[10] Turnbull, D. (1969), *Contemp. Phys.*, **10**, 473.
[11] Uhlmann, D. R. (1972), *J. Non-Cryst. Solids*, **7**, 337.
[12] Nelson, A. and Blander, M. (1974), *J. Non-Cryst. Solids*, **16**, 321.
[13] Imaoka, M. (1962), in *Advances in Glass Technology*, Part 1, Plenum Press, New York, p. 149.
[14] Staveley, L. A. K. (1955), *The Vitreous State*, Glass Delegacy of the University of Sheffield, p. 85.
[15] Mackenzie, J. D. (1960), in *Modern Aspects of the Vitreous State*, **1**, Butterworths, London, p. 188.
[16] Johari, G. P. (1974), *J. Chem. Educ.*, **51**, 27.

CHAPTER 2

Phase Transformations in Glass

The following types of phase transformation are usually observed in glass:

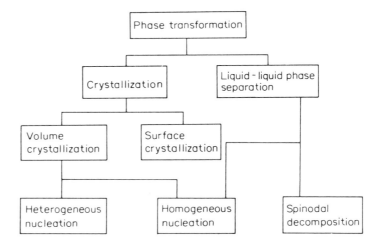

Crystallization: The growth of a crystalline phase(s) which may or may not have the same composition as the original liquid.

Surface crystallization: Crystal growth begins (i.e. nucleates) from the glass–atmosphere interface, and usually grows perpendicular to this interface.

Volume crystallization: Here crystal growth begins from 'nucleation sites' within the body of the material. The initiating site for crystallization may be a substance foreign to the bulk of the material, when it is called *heterogeneous nucleation*, but if the nucleus is the same as the bulk material it is called *homogeneous nucleation*.

Liquid–liquid phase separation: The growth of non-crystalline phases which will have a different composition from the original phase. N.B. A single-component system (e.g. SiO_2) cannot separate in this way.

Spinodal decomposition: Within a region which separates into two liquid phases

16

there will be a region where there is no energy barrier to nucleation and phase separation is, therefore, limited by diffusion only. (Pages 36–39 have a more detailed description.)

2.1 CRYSTALLIZATION

2.1.1 Homogeneous nucleation

When a liquid is cooled below its freezing point, crystallization occurs by the growth of crystals at a finite rate from a finite number of nuclei. Glass formation may be attributed to a low rate of crystal growth, a low rate of nuclei formation or a combination of both.

The stability of a particle of the new phase in homogeneous nucleation will depend on two contributions: one from a difference in free energy between the two phases and the other from the interfacial energy. At the melting point the free energy of a given quantity of a material is the same in the crystalline and in the liquid forms. At lower temperatures the crystalline form will have the lower free energy and the liquid will crystallize if nuclei are available.

If the increase in free energy per unit volume for the crystal–liquid transformation is $\overline{\Delta G}$, the volume of the nucleus (crystal) is $\frac{4}{3}\pi r^3$ (where r is the radius), the surface tension (or interfacial energy) is σ, the surface area $4\pi r^2$, then the free energy change for the nucleation ΔG is given by

$$\Delta G = -\tfrac{4}{3}\pi r^3 \cdot \overline{\Delta G} + 4\pi r^2\sigma. \tag{2.1}$$

At small values of r the surface term will dominate and ΔG will be positive. However, as r increases the volume term will dominate and ΔG will become negative. In Figure 2.1 the two terms in equation (2.1) are plotted as functions of r, with the summation shown as a solid line.

The critical radius, r^*, can be estimated by setting the derivative of ΔG with respect to r equal to zero and solving for r:

$$\frac{\mathrm{d}(\Delta G)}{\mathrm{d}r} = -\frac{12}{3}\pi r^2 \cdot \overline{\Delta G} + 8\pi r\sigma = 0$$

$$r^* = \frac{2\sigma}{\overline{\Delta G}}$$

Particles of radius smaller than r^* are called embryos and are unstable, owing to the decrease in free energy which accompanies their reduction in size. Particles of radius greater than r^* are called nuclei and are stable, since growth is accompanied by a decrease in free energy.

The free energy barrier, ΔG^*, associated with the formation of the critical-sized embryo is obtained by putting the value of r^* into equation (2.1)

$$\Delta G^* = \frac{16\pi\sigma^3}{3(\overline{\Delta G})^2} \tag{2.2}$$

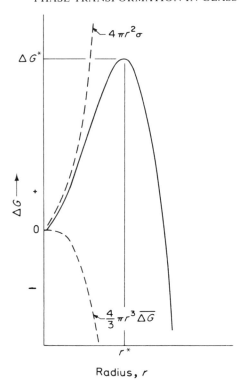

Fig. 2.1 Change in free energy of a spherical nucleus as a
function of radius.

Now, if v is the volume of one atom, the volume of a nucleus containing n atoms
is nv. So

$$nv = \tfrac{4}{3}\pi r^3$$

or

$$n = \tfrac{4}{3}\pi r^3/v$$

Substituting for $r*$ in the above equation, the number of atoms in the critical-
sized particle $n*$, is obtained as

$$n* = \frac{32\pi\sigma^3}{3v(\overline{\Delta G})^3}$$

Since the formation of an embryo involves a positive free energy change, the
probability of such an occurrence would be quite small. However, the entropy of
the solid–liquid system can be increased by the presence of a number of atom-
clusters in equilibrium with the atoms of the liquid. Assuming ideal mixing, the

18

entropy change equals the free energy change (since the heat of mixing in an ideal solution is zero). Thus,

$$\Delta Gn = N_r \Delta G_r + kT \left(\frac{N_r}{N_r + N} \ln \frac{N_r}{N_r + N} + \frac{N}{N_r + N} \ln \frac{N}{N_r + N} \right)$$

where N is the number of atoms per unit volume and N_r is the number of clusters of radius r per unit volume.

At equilibrium ΔGn is a minimum. Since $N_r \ll N$, we have

$$N_r = N \exp \left(-\frac{\Delta G^*}{kT} \right) \qquad (2.3)$$

By the addition of single atoms from the melt, embryos of critical size may grow to form nuclei. The rate of formation of nuclei is given by the product of the number of critical-sized nuclei present per unit volume and the rate at which atoms are attached to the embryo. The number of times per second that an atom attempts to cross the liquid–embryo interface is given by the vibrational frequency of the atom which is equal to kT/h, where h is Planck's Constant. The probability that an attempt will be successful is given by:

$$P = \exp \left(-\frac{\Delta G_a}{kT} \right)$$

Only those atoms which are adjacent to the interface are in a position to attach themselves to the embryo. If this number is N_s, then the number of atoms crossing the interface per second is

$$N_s \frac{kT}{h} \exp \left(-\frac{\Delta G_a}{kT} \right) \qquad (2.4)$$

The rate of nucleation, I, is given by the product of equations (2.3) and (2.4)

$$I = N_s \frac{kT}{h} \exp \left(-\frac{\Delta G_a}{kT} \right) \cdot N \exp \left(-\frac{\Delta G^*}{kT} \right) \qquad (2.5)$$

The second term of the equation, $\exp(-\Delta G^*/kT)$, gives the probability at the temperature T of forming a nucleus larger than the critical size. Turnbull and Cohen [1] call ΔG^* the *thermodynamic barrier to nucleation*. The first term, $\exp(-\Delta G_a/kT)$, is analogous to the diffusion of matter during the formation of the nucleus. ΔG_a is called (by Turnbull and Cohen) the *kinetic barrier to nucleation*.

The free energy of crystallization, ΔG, increases with falling temperature below the melting point at a rate given by the relation

$$\frac{d(\overline{\Delta G})}{dT} = -\Delta S_f$$

where ΔS_f is the entropy of fusion at the temperature T. If it is assumed that ΔS_f is independent of temperature then, at the melting temperature, T_m, we have

$$\overline{\Delta G}_m = \Delta H_f - T_m \Delta S_f = 0$$

19

where ΔH_f is the heat of fusion per unit volume. Thus at a temperature which is ΔT degrees below the melting point,

$$\overline{\Delta G} = \Delta T \cdot \Delta S_f = \Delta T \cdot \frac{\Delta H_f}{T_m} \tag{2.6}$$

Combining equations (2.2), (2.5) and (2.6)

$$I = N_s \frac{kT}{h} \exp\left(-\frac{\Delta G_a}{kT}\right) \cdot N \exp\left(-\frac{16\pi\sigma^3 T_m{}^2}{3\Delta T^2 \Delta H_f^2 \cdot kT}\right) \tag{2.7}$$

According to this equation the rate of nucleus formation is very sensitive to variations in temperature. Inserting data for water in equation (2.7), one finds that at $-40°C$, the nucleation rate increases by a factor of about 9 for 1 deg C fall of temperature.

The temperature dependency of the nucleation rate is shown in Figure 2.2. The shape of the curve can be understood qualitatively by examining the exponential terms of equation (2.5). The term $\exp(-\Delta G^*/kT)$ will be zero at the liquidus temperature and will increase as ΔT increases. At large undercoolings, the term $\exp(-\Delta G_a/kT)$ dominates and the nucleation rate decreases with further cooling after having passed through a maximum.

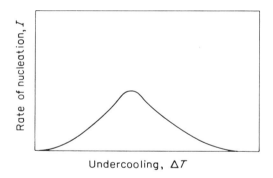

Fig. 2.2 Rate of homogeneous nucleation as a function of undercooling.

The nucleation theory discussed so far is often called the classical theory, because of several assumptions made during the derivation of the nucleation equation. First, it is assumed that embryos can be treated as bulk material, the free energy being independent of the size of the embryo. Secondly, it is assumed that each embryo has a sharp boundary with a well-defined surface energy. In fact there will be a gradient, both in degree of ordering and in chemical composition (in a system of more than one component) as one passes from the crystal phase to the liquid. For very small embryos the surface energy will also depend on the size of the embryo. Thirdly, the use of the Boltzmann function to describe the distribution of embryos is not strictly correct, although this requires only a slight adjustment to the classical theory.

2.1.2 Heterogeneous nucleation

In heterogeneous nucleation the nucleus develops on the surface of a foreign solid (substrate). The substrate may be the container wall or it may be a solid dispersed throughout the liquid. Figure 2.3 shows a cluster which has formed on a solid surface, the contact angle being θ. Equation (2.1) for the free energy change involved in forming a cluster now becomes

$$\Delta G = \frac{v_c}{v_m} \cdot \overline{\Delta G} + S_{lc} \cdot \sigma_{lc} + S_{cs}(\sigma_{cs} - \sigma_{ls}) \tag{2.8}$$

where v_c is the volume of the cluster,
  v_m is the molar volume of the crystalline phase,
  S_{lc}, S_{cs} are the surface areas of the liquid–crystal and the crystal–substrate interfaces respectively, and
  σ_{lc}, σ_{cs} and σ_{ls} are the interfacial energies between liquid–crystal, crystal–substrate, and liquid–substrate respectively.

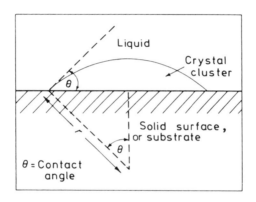

Fig. 2.3 Formation of a crystal cluster on a solid substrate.

The cluster is a spherical cap of radius r. Assuming a static equilibrium between the three phases, and also that the substrate is plane, the following equations can be derived:

$$\sigma_{ls} = \sigma_{cs} + \sigma_{lc} \cos \theta$$

$$v_c = \frac{4}{3}\pi r^3 \left(\frac{2 - 3 \cos \theta + \cos^3 \theta}{4} \right)$$

$$S_{lc} = 2\pi r^2 (1 - \cos \theta)$$

$$S_{cs} = \pi r^2 \sin^2 \theta.$$

Substituting these values in equation (2.8) and differentiating with respect to r, ΔG^* is found to be

$$\Delta G^*_{\text{het}} = \frac{16\pi\sigma^3_{\text{lc}} v^2_{\text{m}}}{3(\Delta G)^2} \cdot f(\theta)$$

where

$$f(\theta) = \frac{2 - 3\cos\theta + \cos^3\theta}{4}$$

and

$$r^* = -\frac{2\sigma_{\text{lc}} v_{\text{m}}}{\Delta G}$$

Hence the critical radius for the cap is the same as the critical radius for a sphere formed in homogeneous nucleation. However, the free energy ΔG^*_{het} involved in forming the cap is less, since $f(\theta) \leqslant 1$ for $0 \leqslant \theta \leqslant \pi$.

Thus the presence of the substrate causes a lowering in the thermodynamic barrier to nucleation. A low value of ΔG^*_{het} is obtained if there is a high degree of 'wetting' between the crystal phase and the substrate, i.e. θ is small.

Epitaxial growth
From the previous discussion it is clear that, if we introduce a suitable substrate into an undercooled liquid, we are able to reduce the energy barrier to nucleation. The question still remains as to what are the criteria for a suitable substrate? We have seen that the substrate needs to be thoroughly wetted by the liquid, but that is all we know. Crystal chemistry here comes to our aid in the form of 'epitaxial growth', sometimes called 'oriented overgrowth'. If we put into a undercooled liquid a 'seed' crystal which has a low-index plane in which the atomic spacing and arrangement are similar to those of one of the low-index planes in the crystal that 'wishes' to form, then the liquid will start to crystallize on that foreign nucleus. If there is no near match, then the liquid will not crystallize. The principle behind this is best seen in an example: it is known that platinum will nucleate lithium disilicate from the liquid. Lithium disilicate is an orthorhombic crystal having the unit cell dimensions $a = 5.80\,\text{Å}$, $b = 14.66\,\text{Å}$, and $c = 4.806\,\text{Å}$. These three axes are mutually perpendicular. Platinum is a cubic crystal with $a = b = c = 3.92\,\text{Å}$ What we should do now is to work out the likely planes in both platinum and lithium disilicate and compare to see if any two are very similar in size. In fact the (1 1 1) plane of platinum matches the (0 0 2) plane of lithium disilicate with about 6 per cent difference, called 'disregistry'.

The second question is: how much disregistry can be tolerated? The answer is based on practice rather than theory. It seems to be at least 15 per cent maximum for metal systems. In the case of glass-ceramics however, a disregistry of 8 per cent would seem to be the maximum which can be tolerated. Even so, after only six lattice spacings there will be complete mismatch.

The importance of heterogeneous nucleation in the glass field is seen in the production of glass-ceramics. It has been found that most glass-ceramics have crystals of about one micrometric size. This means that it is necessary to produce 10^{12} nuclei cm^{-3} of glass.

Two broad classes of nucleating agents can be distinguished in glass:
Group A: includes substances capable of forming minute crystals of low

solubility in the glass. They have a high tendency to reduce from the ionic form to the neutral state in the melt, e.g. Pt, Au, Ag, Cu.

Group B: comprises substances such as TiO_2, ZrO_2, P_2O_5 etc. which are soluble in silicate glasses, and large amounts (1–20 wt %) can be added before nucleation occurs. These substances rarely crystallize out as the independent oxides at the start of crystallization, but precipitate as a complex compound, e.g. $MgTiO_3$, Li_2TiO_3.

2.1.3 Rate of crystal growth

The rate at which a crystal nucleus grows will depend on the rate at which atoms arrive and remain at the surface of the nucleus. Initially, where the nucleus is microscopic (but greater than r^*), the growth rate will also be affected by the curvature of the nucleus–liquid interface, since as the nucleus grows there will be an increase in the interfacial surface energy. However, as the nucleus becomes macroscopic in size, i.e. when it can be considered a crystal, this increase in surface energy can be neglected in comparison with the decrease in free energy which occurs. Thus the crystal can be treated as having a planar interface.

(a) Normal growth
Consider the growth of a crystal where all atoms arriving at the crystal–liquid interface are able to join the crystal to become solid, this occurring uniformly at all points on the surface of the crystal. This process, referred to as normal growth, is shown qualitatively in Figure 2.4. In order that atoms may cross the interface between the crystal and surrounding liquid, they must acquire an activation energy, ΔG_a. The number of atoms crossing the interface, i, per unit time will be given by

$$\frac{di}{dt}(1 \rightarrow c) = sv_0 \exp\left(-\frac{\Delta G_a}{kT}\right)$$

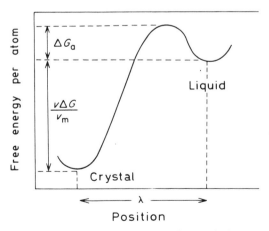

Fig. 2.4 Free energy per atom as a function of position relative to a crystal–liquid interface.

where s is the number of atoms in the liquid (l) facing the crystal (c) across the interface, and v_0 is the frequency at which each atom vibrates due to thermal energy. The exponential term in the above equation represents the probability of finding an atom with sufficient thermal energy to leave the liquid and join the crystal. Once the atom has crossed the interface to the crystal, its energy is reduced, relative to an atom in the liquid, by $\Delta G v / v_m$ where v is the volume occupied by an atom and v_m is the molar volume of the crystalline phase. Thus for an atom to jump from the solid to the liquid, the activation energy required is higher than for the reverse process and equal to $(\Delta G_a + \Delta G v / v_m)$. The number of such jumps per unit time is given by

$$\frac{di}{dt}(c \to l) = s v_0 \exp\left[-\left(\Delta G_a + \frac{v}{v_m}\Delta G\right)/kT\right]$$

Hence there is a net transfer of atoms from the liquid to the crystal given by

$$\frac{di}{dt}(l \to c) - \frac{di}{dt}(c \to l) = s v_0 \exp\left(-\frac{\Delta G_a}{kT}\right)\left[1 - \exp\left(-\frac{v\Delta G}{v_m kT}\right)\right]$$

If s atoms are transferred, then the crystal grows by λ, where λ is approximately one interatomic spacing. Thus the growth rate of the crystal, u, is given by

$$u = \lambda v_0 \exp\left(-\frac{\Delta G_a}{kT}\right)\left[1 - \exp\left(-\frac{v\Delta G}{v_m kT}\right)\right]$$

Since $v/v_m k = 1/R$, where R is the gas constant, the expression for u simplifies to

$$u = \lambda v_0 \exp\left(-\frac{\Delta G_a}{kT}\right)\left[1 - \exp\left(-\frac{\Delta G}{RT}\right)\right] \tag{2.9}$$

This is a modified form of the well known Hillig and Turnbull Equation [2]. This equation predicts that a maximum growth rate will be observed, because the thermodynamic term in square brackets increases with undercooling, whilst the kinetic term decreases. Furthermore, the growth rate should be experimentally observable in the vicinity of the melting point or liquidus temperature. For a particular temperature the growth rate will be constant; thus the growth will be linear with time.

Let us now consider two limiting cases: first, at small undercoolings

$$1 - \exp\left(-\frac{\Delta G}{-RT}\right) \approx \frac{\Delta G}{RT}$$

whence

$$u = \lambda v_0 \frac{\Delta G}{RT} \exp\left(-\frac{\Delta G_a}{kT}\right)$$

Thus u is directly proportional to the thermodynamic driving force, ΔG. For a one-component system $u \propto \Delta T$. Secondly, at large undercoolings

$$\exp\left(-\frac{\Delta G}{RT}\right) \ll 1$$

Thus $u = \lambda v_0 \exp(-\Delta G_a/kT)$. Hence $\log u$ is proportional to $1/T$.

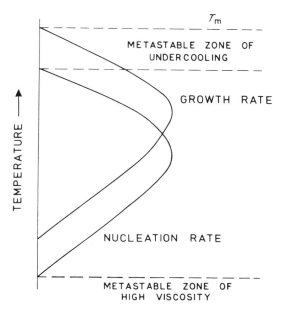

Fig. 2.5 Variation with temperature of homogeneous nucleation rate and growth rate of the second phase.

As in the nucleation process, the temperature corresponding to the maximum growth rate rises as the diffusion activation energy, ΔG_a, increases. However, the maximum growth rate occurs at much smaller undercoolings than is the case for the maximum nucleation rate: this is qualitatively shown in Figure 2.5. Finally, it should be mentioned that the diffusion activation energies for growth and nucleation may not necessarily be equal for a given system, since the atomic movements involved may be quite different for the two processes.

(b) Surface nucleation growth
In normal growth theory it has been assumed that all atoms which arrive at the crystal–liquid interface can be incorporated in the growing crystal. Thus, atoms can be added to (or leave from) any site on the crystal–liquid interface. In practice such a model may only apply when the interface is very rough on an atomic scale, i.e. when there are many imperfections present, providing a large number of equally favoured sites which does not change appreciably with temperature.

If the interface is atomically smooth, the first atom that arrives at this interface may join anywhere, since all sites are equivalent. The next atom, though, can either occupy an isolated site or it can take up a position adjacent to the first atom. The latter position is energetically more favourable because of the stronger bonding between the two atoms. Thus, growth will proceed by the spreading of a monolayer across the surface until the surface is covered.

For growth to continue further it is necessary for one or more atoms to take up initially unfavourable sites to start off another layer. Thus each layer has to be

25

nucleated separately and the process is referred to as surface nucleation growth. It can be shown [3] that the growth rate in this case is given by

$$u = A_1 \exp\left(-\frac{W^*}{kT}\right) \cdot \exp\left(-\frac{\Delta G_a}{kT}\right) \tag{2.10}$$

where A_1 is a constant, and the thermodynamic barrier, W^*, is given by

$$W^* = -\frac{K_1 \cdot \sigma_E^2 V_m}{\Delta G}$$

where K_1 is a shape factor, and σ_E is the specific edge surface energy. Equation (2.10) is derived in a similar way to the equations for crystal nucleation except that the nucleus is now only two-dimensional, and W^* is due to edge surface energy instead of area surface energy. The temperature-dependence of surface nucleation growth is more complex than for normal growth. The maximum growth rate does not occur as near to the melting point and, for small undercoolings, the growth rate is very low. This type of behaviour is not generally observed for silicate systems.

(c) Screw dislocation growth

Normal growth and surface nucleation growth models are based on the two extreme descriptions of the crystal–liquid interface, viz, atomically rough and atomically smooth. A third model is possible in which the number of interface imperfections (and therefore the number of sites available for growth) is limited. This occurs where screw dislocations intersect the interface. As shown in Figure 2.6 a ledge is provided which forms a set of sites. As atoms are added to the ledge, the latter 'spirals' round the dislocation line and continues to provide ledge sites as more atoms arrive at the interface. It can be shown [2] that for a one-component liquid the fractional number of such sites, f, on the interface is approximately given by

$$f \approx \frac{\Delta T}{2 \pi T_m}$$

The resulting growth rate is equal to the normal growth rate multiplied by f, i.e. f is the probability of an atom which arrives at the interface being able to find a

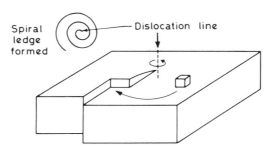

Fig. 2.6 Screw dislocation in a cubic solid showing the addition of an atom taking place during growth.

growth site. Thus for screw dislocation growth, u is given by

$$u \approx \frac{\Delta T}{2 \pi T_m} \cdot A \exp\left(-\frac{\Delta G_a}{kT}\right)\left[1 - \exp\left(-\frac{\Delta G}{RT}\right)\right]$$

Then for a one-component liquid at small undercoolings $u \propto \Delta T^2$, and at large undercoolings

$$u \approx \frac{\Delta T}{2 \pi T_m} \cdot A \exp\left(-\frac{\Delta G_a}{kT}\right)$$

In order to relate experimental data to these various growth models, the reduced growth rate is often utilized [4]. This criterion U_R, is defined as

$$U_R = \frac{U_\eta}{[1 - \exp\left(-\Delta G/RT\right)]}$$

where η is the viscosity at temperature T. It is assumed that the diffusion process involved in growth occurs by a mechanism similar to that of viscous flow. Thus the diffusion coefficient D can be related to η via the Stokes–Einstein equation given by

$$D = \frac{kT}{3 \pi \lambda \eta}$$

where λ is the molecular diameter, and

$$D = D_0 \exp\left(-\frac{\Delta G_a}{kT}\right)$$

The temperature-dependence of U_R gives information regarding the fraction of preferred growth sites at the interface. In terms of growth models, the U_R versus ΔT relation for normal growth would be a horizontal line for all ΔT values; for screw dislocation growth a straight line of positive slope passing through the origin (small ΔT), and for surface nucleation growth a curve passing through the origin which exhibits positive curvature (for small ΔT) will result.

It can be shown that the nature of the interface, i.e. whether it is rough or smooth, depends on the entropy of fusion, ΔS_f. A review of the theory is given by Uhlmann [4]. For materials of low $\Delta S_f (< 2R)$ the crystal–liquid interface should be quite rough and the growth kinetics are expected to be of the form predicted by the normal growth model. For materials of high $\Delta S_f (> 4–6R)$ the interface should be smooth and the growth kinetics are expected to be in agreement with the surface nucleation model or the screw dislocation model.

Finally, let us consider in brief the time-dependence of the crystal growth rate at a given temperature. A constant growth rate is assumed in one-component systems, since the interface is advancing into a region of constant composition. Thus the rate of the process controlling the growth is independent of the interface position and hence the time. However, in multi-component systems the crystallizing phases often differ in composition from the original liquid. Hence continual growth of the crystal phase would be expected to involve long-range diffusion over distances increasing with time. This would lead to a growth rate proportional to the square root of the time, since the effective diffusion is given by

$\sqrt{(Dt)}$, where D is the diffusion coefficient. It should be remembered, however, that as the remaining uncrystallized liquid becomes richer in the non-crystallizing components, the thermodynamic driving force, ΔG, will also change.

2.1.4 Growth and Nucleation

We have derived both the growth rate and the nucleation rate expressions as a function of temperature. Let us now consider a case in which nucleation and growth occur simultaneously. Let x be the fraction transformed, by which is meant the amount of material changing from liquid to crystal. Assuming the growth rate u to be constant, the linear dimension of the crystal at a time t is given by $u(t - \tau)$, where τ is the induction period for the formation of the crystal phase. If the growth rate is equal in all directions, a sphere will result, and volume of the particle will be

$$v_p = \tfrac{4}{3}\pi[u(t - \tau)]^3 \tag{2.11}$$

If we make the prior assumption that at the beginning of the transformation $v_{liquid} \gg v_{crystal}$, thus neglecting the effect of impingement and the decrease in volume of the liquid phase, the volume of the crystal phase at a given time t is given by

$$v_{crystal} = \int_{\tau=0}^{\tau=t} v_{liquid} \cdot v_p \cdot I \cdot d\tau$$

where I is the nucleation rate and is assumed to be constant for the present purpose. Then

$$\frac{v_{crystal}}{v_{liquid}} = x_t = \int_{\tau=0}^{\tau=t} v_p \cdot I \cdot d\tau$$

and

$$d\left(\frac{v_{crystal}}{v_{liquid}}\right) = dx_t = v_p dN \quad (\text{since } I = dN/dt)$$

To correct for the impingement and reduction in volume of the parent phase, v_{liquid} must be multiplied by the factor $(1 - \text{crystal/liquid})$. Thus

$$\frac{dv_{crystal}(t = t)}{v_{liquid}} = dx_t = v_p dN (1 - x_t)$$

$$\frac{dx_t}{1 - x_t} = v_p dN = -d\ln(1 - x_t)$$

and

$$\ln(1 - x_t) = -\int_{\tau=0}^{\tau=t} v_p dN$$

$$1 - x_t = \exp\left(-\int_{\tau=0}^{\tau=t} v_p dN\right) \tag{2.12}$$

Substituting equation (2.11) in equation (2.12) gives

$$1 - x_t = \exp\left\{ -\int_{\tau=0}^{\tau=t} \frac{4}{3}\pi\left[u(t-\tau)\right] I\, d\tau \right\} \qquad (2.13)$$

$$= \exp\left(-\frac{\pi}{3} u^3 I t^4 \right)$$

This is commonly known as the Johnson–Mehl equation [5]. This equation can be modified to a number of simplifying cases: during rapid nucleation, Avrami [6] assumed that I was a function of t. The N nucleation sites initially in unit volume of the phase disappeared due to growth in the following way

$$N_t = N \exp(-vt)$$

where v is the nucleation frequency of the sites. Then

$$I_t = N_t v = Nv \exp(-vt)$$

Substituting in equation (2.13) and assuming that vt is large, this integrates to

$$1 - x_t \exp(-K u^3 N t^3)$$

where K is a constant

Turnbull *et al.* have attempted to correlate different equations of nucleation and growth in various glass-forming systems: for details the reader is referred to the review article in *Modern Aspects of the Vitreous State* [1].

2.2 LIQUID–LIQUID PHASE SEPARATION

In many silicate and borate melts two-liquid phase formations can be observed. Immiscibility in these systems may occur above or below the liquidus; the latter is called sub-liquidus or metastable immiscibility. Common binary silicate systems which exhibit stable immiscibility above the liquidus usually contain divalent metal oxides like SrO, CaO, FeO, ZnO and MgO. Sub-liquidus immiscibility is often found in silicate melts that have an S-shaped liquidus, such as Na_2O, Li_2O and BaO. Some typical examples are shown in Figure 2.7.

An industrial application of liquid–liquid phase separation is the production of Vycor glass. This is made from a glass of approximate composition 8% Na_2O, 20% B_2O_3 and 72% SiO_2 which is heat-treated between 500° and 800°C (its liquidus is around 1100°C). Liquid separation is thereby induced, to form two interconnected phases, one rich in silica and the other rich in boric oxide. These two glasses differ in their solubility behaviour: when treated with dilute HCl, the silica-poor phase readily dissolves, leaving the silica-rich phase almost untouched. This results in a highly porous almost pure silica glass. The material has been used for biological sieves.

A more important application arises from further heating of the high-silica material at $\sim 1200°$C, whereby the material becomes more compact, the pores disappearing – eventually a dense type of silica glass (Vycor) is produced which only contains minor amounts of impurities (e.g. about 1 per cent boric oxide). By using this process silica glass articles can be made by conventional glass making methods, which are much easier than those starting from a pure silica melt.

29

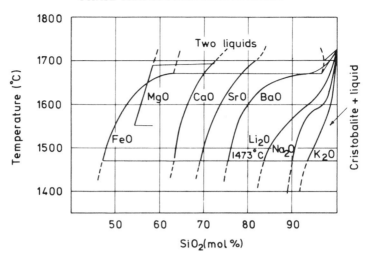

Fig. 2.7 Miscibility gaps and tendency towards immiscibility revealed by the phase diagrams for silica with various metal oxides.

2.2.1 Thermodynamics of liquid–liquid phase separation

The spontaneous occurrence of immiscibility in any system implies that the free energy of the system has been lowered by the separation into two phases. The free energy for mixing at a temperature T can be expressed as

$$\Delta G_m = \Delta H_m - T \Delta S_m$$

where ΔH_m and ΔS_m are both functions of the composition of the mixture. If ΔH_m has a negative or very small positive value, the general type of free energy curve shown in Figure 2.8 is obtained, and this is characteristic of two miscible liquids.

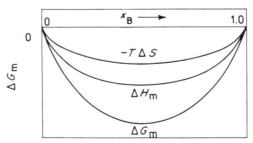

Fig. 2.8 A plot of free energy against composition for a solution when ΔH_m is negative.

If ΔH_m is large and positive, the curve shown in Figure 2.9(a) with deflection (Figure 2.9(b)) will result. With such a curve the free energy of the system will be

30

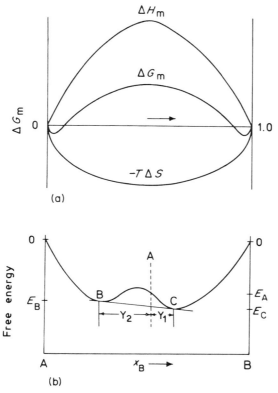

Fig. 2.9 A plot of free energy against composition for a solution (a) when ΔH_m is positive (b) for an immiscible system.

lowered by the separation of any liquid between B and C into two liquids of composition B and C, their relative concentrations being determined by the lever rule. As the temperature rises the $-T\Delta S$ term plays a larger role, and the magnitude of the deflection decreases. When the temperature is sufficiently high it disappears. The critical lowest temperature corresponding to disappearance of immiscibility is known as the consolute temperature, T_c. Now, we can plot the limiting phase composition as a function of temperature and arrive at a phase diagram with an 'immiscibility dome' as shown in Figure 2.10. This figure also contains an area within the immiscibility dome known as the spinodal. This inner dome (spinodal) is a result of the points of inflection of the free energy curve. A melt within this dome will spontaneously separate if the mobility of the ions is great enough. Whereas if the initial composition of the melt is outside the spinodal dome (shaded area in Figure 2.10) phase separation will not take place spontaneously, but will require the formation of nuclei.

Let us consider a quasi-chemical description of spinodal given by Swalin [7]. A homogeneous solution of n atoms with average composition x_B will deviate from

31

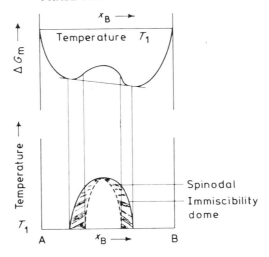

Fig. 2.10 Relationship between the spinodal and the free energy diagram.

the mean composition by an amount Δx_B due to thermal fluctuations. If $\overline{G}$ is the average and G the free energy of any one atom of this solution, then the probability of any atom having free energy G is given by Boltzmann statistics

$$P_{\Delta G} = A \exp\left(-\Delta G / kT\right)$$

where

$$\Delta G = G - \overline{G}$$

The probability that n atoms will have the same free energy simultaneously will be

$$P_{\Delta G}n = A \exp\left(-n\,\Delta G / kT\right) \tag{2.14}$$

If ΔG is small, G may be expanded by a Taylor series about x_B.

Thus

$$G = \overline{G} + \tfrac{1}{2}\frac{d^2 G}{dx_B^2}(x_B - \bar{x}_B)^2 + \ldots \tag{2.15}$$

If $(x_B - \bar{x}_B) \ll 1$, higher order terms may be neglected. By substituting (2.15) in (2.14) the probability of a small fluctuation involving n atoms is

$$P_{\Delta G}n = A \exp\left[-\frac{n(x_B - \bar{x}_B)^2 \left(\dfrac{d^2 G}{dx_B^2}\right)}{2kT}\right] \tag{2.16}$$

From equation (2.16) it is seen that a high probability of a composition fluctuation is associated with a small value of d^2G/dx_B^2, particularly if it is negative (Figure 2.11). Negative values of the second derivative of the free energy will give a positive exponent and will give a very large probability, which is to say, the separation will occur spontaneously. If, on the other hand, the second

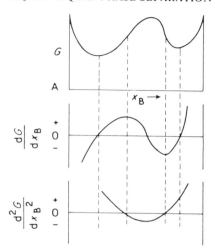

Fig. 2.11 Spinodal phase separation.

derivative is positive, the exponent will be negative and the probability will be less than unity. As the composition moves away from the spinodal the probability gets smaller and smaller. Consequently, phase separation is less likely to occur for such compositions.

To clarify the concept of spinodal further let us consider the free energy of mixing versus composition curve of a system at temperature below the consolute temperature [Figure 2.12(a) and (b)]. In this system any composition x_B between x' and x'' will separate into two phases. If x_B lies between the two points of inflection c and d corresponding to $d^2G/dx_B^2 \leqslant 0$, the system is unstable to infinitesimal fluctuation of composition; separation into two phases will occur with a continuous fall of free energy. This can be shown as follows: In Figure 2.12(a), at a very early stage of the process the separated phases will have compositions a and b: the average free energy of the system is G_2 which is lower than G_1, the initial value for the solution. Thus the phases will shift in composition and a continuous fall of free energy will take place until x' and x'' are reached, the average free energy of the system will then be G_3. This is a case of complete unmixing and is known as spinodal decomposition. However, if x_B lies outside the points of inflection c and d as shown in Figure 2.12(b), a small fluctuation of composition causes a rise in the average free energy. For example, the initial decomposition into compositions a and b results in an increase in free energy G_2. It is not until a large composition difference is obtained (over the hump, to say, e) that further decomposition will occur with a net drop in free energy. Thus here we have a nucleation barrier and separation will occur by classical nucleation and growth.

The general features of liquid–liquid immiscibility can be qualitatively described by the regular solution model, originally introduced by Hildebrand [8] where atoms A and B are randomly mixed on a regular lattice. In such a model, the

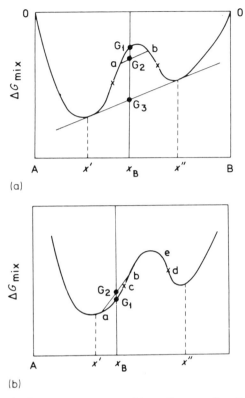

Fig. 2.12 Schematic free energy composition diagram for $T < T_c$ showing a graphical method of determining the thermodynamic driving force for liquid-phase separation.

free energy of mixing is given by

$$\Delta G_m = \alpha x(1-x) + RT[x \ln x + (1-x)\ln(1-x)]$$

where x is the mol fraction of B, and α is given by

$$\alpha = NZ[E_{AB} - \tfrac{1}{2}(E_{AA} + E_{BB})]$$

where E_{AA}, E_{BB} and E_{AB} are the energies of the various bonds between the atoms, N is Avogadro's number, and Z is the coordination number. It can be easily shown that α is related to the upper consolute temperature by $\alpha = 2RT_c$. For immiscibility to occur α (and hence ΔH_m) must be positive. According to this model the free energy of mixing curves are symmetrical about $x = 0.5$ and are similar to those shown in Figure 2.9. The binodal and spinodal curves are also symmetrical about $x = 0.5$.

This model does not describe the behaviour of common silicate and other glass forming oxide melts which are more complex than a simple random mixture of

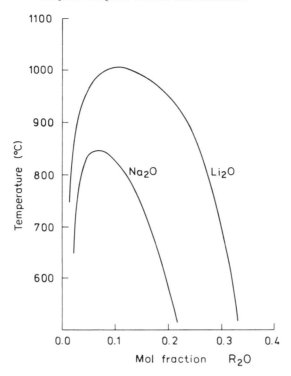

Fig. 2.13 Sub-liquidus immiscibility regions of soda–silica and lithia–silica systems.

atoms. Besides, the observed immiscibility domes in these systems are far from symmetrical; some typical examples are shown in Figure 2.13. In the binary Li_2O–SiO_2 and Na_2O–SiO_2 systems the critical composition occurs at $x \sim 0.1$ rather than 0.5. This occurs due to non-regular mixing of the components in these systems, and can be accounted for if the appropriate partial molar free energies of the components in these mixtures are known. Unfortunately such reliable data in glass-forming oxide systems are very rare. In absence of such data various sub-regular solution models with more than one adjustable parameter have been proposed by different workers [9] – which essentially are curve fitting exercises and will not be discussed here. For further information regarding different models the reader is referred to the review article by James [10].

In describing highly asymmetric miscibility gaps in many binary silicate systems Haller et al. [11] have proposed a modification of the regular solution model which considers the regular mixing of new end components – a complex molecule or 'multimer' $(SiO_2)_m$ and a stoichiometric compound $R_2O.nSiO_2$ at the limit of the miscibility gap on the alkali oxide side, where m and n are integers. By choosing appropriate values of m, n and a parameter ΔS (representing the additional entropy change due to changes in internal degrees of freedom of the liquids on mixing) Haller and his colleagues were able to produce a symmetric gap

35

when the experimental miscibility boundary data were plotted in terms of the mol fractions of the new end components. For Na_2O–SiO_2 system a good fit was obtained with $(SiO_2)_8$ and Na_2O, $3SiO_2$ as end members. Similarly, for Li_2O–SiO_2 the required values were $m = 6$, $n = 2$, and for BaO–SiO_2 system $m = 8$, $n = 2$ produced the best fit.

Similar analysis has been extended by Macedo et al. [12] in various binary borate systems. Good agreement with experimental miscibility boundary data could be shown by considering the regular mixing of an appropriate stoichiometric compound depending on the system (such as K_2O, $3B_2O_3$) and a complex boric oxide end member $(B_2O_3)_5$, the latter component being the same for all the systems.

The existence of complex structural groups such as $(SiO_2)_8$ and $(B_2O_3)_5$ in silicate and borate glasses is an interesting possibility. However, such big polymeric groups have not yet been experimentally confirmed. (For further details see Chapter 10).

2.2.2 Nucleation theory

The assumptions of uniform composition, and a chemical potential of the nucleus equal to that of the pure material, are the main objections to the classical nucleation model as discussed before. Many workers have tried to find ways of dealing with these objections but by far the most comprehensive are those of Cahn and Hilliard [13] who derived a general equation for the free energy of a system having a spatial variation in composition. This expression permits the properties of a critical nucleus to be calculated without any assumptions about its homogeneity, nor does it require the energy of the nucleus to be divided into a surface and volume term.

They assumed that the local free energy*, f, in a region of non-uniform composition will depend both on the local composition and on the composition derivatives of the immediate environment. Assuming that the composition gradient is small compared with the reciprocal of the intermolecular distance, it may be expanded in a Taylor series about $f(c)$, the free energy per molecule of solution of uniform composition 'c'. From such an expansion they obtained an expression for the total free energy of the form:

$$F = N_v \int_v [f(c) + k(\nabla c)^2 + \ldots] dV \tag{2.17}$$

The quantity in square brackets is the expansion for one molecule of the solution; this must be multiplied by the number of molecules in a given volume (mol) to give F. Here, v is the volume of one atom, V is the volume of the cluster and k is Boltzmann's constant.

There are numerous paths by which an initially unstable fluctuation (embryo) may grow into a stable one. Cahn and Hilliard considered only those paths passing over the lowest free energy barrier. The top of this barrier is a saddle point at which an embryo becomes a critical nucleus. The form of this nucleus is defined

---

* Note that Cahn and Hilliard used the Helmholtz free energy, F, rather than the Gibbs free energy used in preceding sections.

by the minimum value of equation (2.17), subject to the average composition remaining constant, i.e.

$$\int_V (c - c_0)\, dV = 0 \tag{2.18}$$

where c_0 is both the initial and average composition. Neglecting higher order gradient terms we thus have:

$$F = \int_V [f(c) + k(\nabla c)^2 + \lambda(c - c_0)]\, dV \tag{2.19}$$

where λ is a Lagrangian multiplier.

The minimum value (external) of equation (2.19) is obtained by use of the Euler equation which leads to the result:

$$2k\nabla^2 c + \left(\frac{\partial k}{\partial c}\right)(\nabla c)^2 = \frac{\delta f}{\delta c} + \lambda$$

Assuming that initial nucleation has little effect on the overall composition it can be shown that

$$\lambda = -\left(\frac{\partial f}{\partial c}\right)_{c = c_0}$$

Therefore

$$2k\nabla^2 c + \left(\frac{\partial k}{\partial c}\right)(\nabla c)^2 = \frac{\delta f}{\delta c} - \left(\frac{\partial f}{\partial c}\right)_{c = c_0}$$

The solution of this equation, subject to the necessary boundary conditions, describes the spatial composition variation in a critical nucleus.

It can be seen from Cahn's nucleation theory that as the spinodal is approached the work necessary to form a critical nucleus (W^*) tends to zero, and within the spinodal there is no barrier to nucleation except a diffusion barrier. The phase separation behaviour within the spinodal would therefore be expected to be different.

In a single-phase system of non-uniform composition the local free energy may be written as:

$$F = \int_V [f(c) + (\nabla c)^2]\, dV$$

By limiting our discussion to the initial stages of phase separation, higher order gradient energy terms will be negligible, except at very large gradients. Expanding $f(c)$ about the average composition (c_0):

$$f(c) = f(c_0) + (c - c_0)\left(\frac{\partial f}{\partial c}\right) + \frac{1}{2}(c - c_0)^2 \left(\frac{\partial^2 f}{\partial c^2}\right) + \cdots$$

since

$$\int_V (c - c_0)\, dV = 0$$

The difference between the initial homogeneous solution and the inhomogeneous solution will be:

$$\Delta F = \int \left[\frac{1}{2}(c - c_0)^2 \left(\frac{\partial^2 f}{\partial c^2} \right) + k(\nabla c)^2 \right] \mathrm{d}V \tag{2.20}$$

k must be positive, otherwise a negative surface tension will result, with no stability even outside the spinodal.

In order to simplify the development of the theory Cahn considered the Fourier components of the composition rather than the composition. Any arbitrary fluctuation can be described by its Fourier component, and because of the nature of Fourier components, ΔF is the sum of contributions from each Fourier component separately. For a Fourier component of the form

$$(c - \dot{c}_0) = A \cos \beta x$$

where β is the wavenumber, and x is a distribution parameter,

$$\frac{\mathrm{d}c}{\mathrm{d}x} = -A\beta \sin \beta x$$

From equation (2.20)

$$\Delta F = \left(\frac{\partial^2 f}{\partial c^2} \right) \int_{-\infty}^{+\infty} \left[\frac{1}{2}(A \cos \beta x)^2 \right] \mathrm{d}x + \int_{-\infty}^{+\infty} k(-A\beta \sin \beta x)^2 \mathrm{d}$$

$$\Delta F = \left(\frac{\partial^2 f}{\partial c^2} \right) \frac{VA^2}{4} + \frac{kA^2 \beta^2 V}{2}$$

Therefore

$$\Delta F = \frac{VA^2}{4} \left[\left(\frac{\partial^2 f}{\partial c^2} \right) + 2k\beta^2 \right]$$

It can thus be seen that when $(\delta^2 f/\delta c^2)$ is positive the mixture will be stable to all fluctuations, but when it is negative the solution will be unstable to fluctuations of a wavenumber less than some critical value, β_c.

The value of β_c is obtained when the value in square bracket is equal to zero, i.e.

$$\left(\frac{\partial^2 f}{\partial c^2} \right) = -2k\beta^2_c$$

or

$$\beta_c = \left[-\frac{(\partial^2 f/\partial c^2)}{2} \right]^{1/2}$$

As a spinodal is approached $(\partial^2 f/\partial c^2) \to 0$, and $\beta_c \to 0$ and so the critical wavelength tends to approach infinity.

Thus at a finite time after phase separation has begun, the composition of the solution will be described by a superposition of sine waves of fixed wavelength, but random in orientation, phase, and amplitude. In order to visualize the situation Cahn used a computer to generate random numbers with a flat distribution in the interval 0 to π for phase angles. The computed sections through the two-phase structure show interconnectivity of the two phases.

Finally, in summarizing, the following table gives a comparison between classical nucleation and spinodal decomposition.

Classical nucleation	Spinodal decomposition
New phase starts from small fluctuations (nuclei) which grow in extent. Composition remains constant with time.	Continuous variation of both extremes in composition with time until break down by spinodal decomposition. Growth is not in extent but in amplitude.
Interface between phases is always sharp during growth.	Initially the interface is very diffuse but gradually sharpens.
Tendency for random distribution of particle positions in matrix.	Regularity of second phase distribution both in size and position, characterized by a geometric spacing.
Tendency for the second phase to separate into spherical particles with little connectivity.	Tendency for the second phase to separate as non-spherical structure with high connectivity.

2.2.3 Coarsening kinetics

In the later stages of a phase separation process, the volume fraction of the separated phases has reached the equilibrium value, and the mode of growth at this stage is called *coarsening*. Here the larger particles grow at the expense of the smaller, the driving force of the process being a reduction in the surface energy.

Two approaches to the coarsening of spherical particles have been made. Both are basically dependent upon the Thomson–Freundlich formula for the solubility of spherical particles. Only one of these approaches, namely Lifshitz–Slyozov [14] will be discussed here. In this derivation the growth process is regarded as analogous to the evaporation of liquid droplets. Small droplets are found to evaporate rapidly because of their higher vapour pressures and subsequently the vapour condenses on the large particles. The solubility of a spherical particle of radius r_i is given by the Thomson–Freundlich formula:

$$C_i = C_\infty \exp\left[\frac{2\sigma M}{RT\rho}\right] \cdot \frac{1}{r_i}$$

This equation may be expanded in the form

$$C_i = C_\infty \left[1 + \frac{2\sigma M}{RT\rho} \cdot \frac{1}{r_i} + \ldots\right]$$

If at a given time the concentration at a point remote from the particle is C_t, the flux, J, at the surface of a particle of radius r_i will be:

$$J = D\frac{C_t - C_i}{r_i}$$

where D is the diffusion coefficient.

Now, defining a population of particles r^* which satisfy the condition

$$C_t = C_\infty \left[1 + \frac{2\sigma M}{RT\rho} \cdot \frac{1}{r^*} \right]$$

It is clear that particles of radius r smaller than r^* will dissolve, and those of radius greater than r^* will grow. The rate of change in the radius r_i of the ith particle may be shown to be:

$$\frac{dr_i}{dt} = DC_\infty \frac{2\sigma}{RT} \left(\frac{M}{\rho} \right)^2 \frac{1}{r_i} \left[\frac{1}{r^*} - \frac{1}{r_i} \right] \tag{2.21}$$

From this equation it can be seen that particles with twice r^* grow at the maximum rate, while those with radii less than r^* disappear.

It may also be shown that, irrespective of the initial population distribution, the distribution function tends asymptotically to a universal function, $f(Z)$ of the reduced variable, $Z = r/r^*$. This distribution function is characterized by one size of particle only, $r^* = \bar{r}$ and is invariant in the sense that the ratio of the largest radius of the distribution to the critical radius is always 3/2. Applying this condition to equation (2.21), i.e. $r_i = r_{max} = 3/2\bar{r}$, we obtain on integration

$$\bar{r}^3 - \bar{r}_0^3 = \frac{8\sigma}{9RT} \left(\frac{M}{\rho} \right)^2 DC_\infty t$$

i.e.

$$\left[\bar{r}^3 - \bar{r}_0^3 \right] \propto t$$

The precipitated volume fraction, V_f, which is constant for this process may be determined in terms of the average volume of the particles (v_p):

$$V_f = N v_p \text{ where } v_p = \frac{4\pi}{3} \bar{r}^3$$

Therefore

$$V_f = N \frac{4\pi}{3} \left[\frac{8\sigma}{9RT} \left(\frac{M}{\rho} \right)^2 DC_\infty \right] t$$

Thus

$$\frac{1}{N} - \frac{1}{N_c} = \frac{4\pi}{3V_f} \left[\frac{8\sigma}{9RT} \left(\frac{M}{\rho} \right)^2 DC_\infty \right] t$$

And

$$N \propto t^{-1}$$

$$V_f = \text{constant}$$

$$\bar{r} \propto t^{1/3}, \text{ provided } \bar{r}_0 \text{ is small compared to } \bar{r}.$$

Liquid–liquid phase separation in glass has been extensively studied by numerous workers over the last twenty years – and still is an ever expanding area, a coverage of which is outside the scope of this book. James [10] has summarized an up-to-date review in this field.

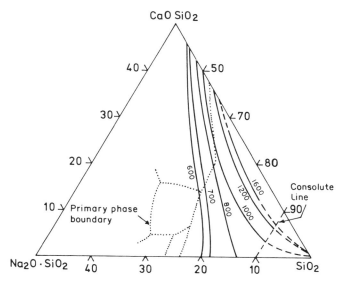

Fig. 2.14 Liquid–liquid immiscibility in the soda–lime–silica system. The isotherms join compositions with the same T_m (°C).

Finally, as a typical example Figure 2.14 shows the liquid–liquid immiscibility region in the most important commercial glass system, namely Na_2O–CaO–SiO_2.

2.3 GLASS-CERAMICS

Glass-ceramics are polycrystalline solids prepared by the controlled crystallization of glasses[15]. Crystallization is accomplished by subjecting suitable glasses to a carefully regulated heat-treatment schedule which results in the nucleation and growth of crystal phases within the glass. In many cases, the crystallization process can be taken almost to completion but a small proportion of residual glass phase is often present. This residual glass phase often exerts a marked influence on a number of important properties of glass-ceramics. For example, properties that involve diffusion of ions through the glass-ceramic structure and those involving viscous flow will be strongly dependent on the chemical composition and volume fraction of the glass phase. In addition, mechanical strength, mode of fracture propagation, chemical stability, and transparency of glass-ceramics to visible radiation are all influenced by the glass phase.

For some applications it is desirable to minimize the proportion of the glass phase whereas for others, deliberate steps are taken to retain a controlled amount of glass phase and to adjust its composition with the object of improving certain properties of the glass-ceramic.

Since the initial discovery by Stookey[16] that controlled nucleation and crystallization of glasses could be achieved for certain photosensitive glasses, there have been rapid strides in the technology and applications of these materials. The further discoveries that TiO_2 [17] and P_2O_5 [18] were effective nucleation catalysts in a wide range of glass compositions were followed by the investigation of other oxide catalysts such as ZrO_2, SnO_2, Cr_2O_3, and V_2O_5 amongst others.

Glass-ceramics can be produced from a wide range of glass types, some of the more important of which are given in Table 2.1. Materials having widely different and useful properties are possible. For example, glass-ceramics have been produced having thermal expansion coefficients ranging from negative values to high positive values close to those of metals such as steel or copper.

Table 2.1
Common glass ceramic systems*

System	Nucleating agents	Crystal phases
$Li_2O–Al_2O_3–SiO_2$ ($Al_2O_3 > 10\%$)	TiO_2; $TiO_2 + P_2O_5 + ZrO_2$	β-spodumene/β-eucryptite solid solutions
$Li_2O–Al_2O_3–SiO_2$ ($Al_2O_3 < 10\%$)	P_2O_5	Lithium disilicate, quartz
$Li_2O–ZnO–PbO–SiO_2$	P_2O_5	Lithium disilicate, lithium zinc silicate, quartz
$MgO–Al_2O_3–SiO_2$	TiO_2	Cordierite, cristobalite
$Na_2O–BaO–Al_2O_3–SiO_2$	TiO_2	Nepheline, hexacelsian
$ZnO–Al_2O_3–SiO_2$	P_2O_5; TiO_2	Willemite

* From McMillan [21].

The first essential structural requirement of a commercial glass-ceramic is that it can be melted and formed easily and without harm to the furnace refractories. This can be an acute problem with certain glass-ceramics because of the use of large quantities of lithia, in particular, which readily attack the refractories. Other compositions contain large quantities of alumina, and such glasses will need high temperatures and also will have a small working range. Another point which must be borne in mind is that the glass should be sufficiently stable so that it does not start to devitrify during the forming operation. If this happened, large crystals could be produced which would reduce the strength. Fortunately small additions of some components can slow the process down e.g. 0–3 % of alumina, zinc oxide, boric oxide, potash or soda are added to lithia glasses.

A second essential requirement is that the glass can be nucleated and crystallized quickly and economically to get the desired type of crystal. The heating schedule also need not be too accurately controlled. Crystallization must, however, not be too rapid as uncontrolled crystal growth induces strains in the resulting glass-ceramic, which reduces its mechanical strength.

A third, highly important requirement is that the resulting glass-ceramic has the desired properties. Today many compositions are known which crystallize to give a glass-ceramic, but for commercial success they must have: (1) high mechanical strength, (2) low thermal expansion, or some matched expansion, (3) good chemical durability, and (4) very high or low, electrical conductivity.

2.3.1 Microstructure

The most noticeable characteristic of glass-ceramics is their extremely fine grain size, and this feature is responsible for many of the valuable properties of the materials. The average crystal size in a glass-ceramic would usually be less than one micrometre and materials with crystals of 20 nanometres are known.

2.3.2 Mechanical properties

At room temperature glass-ceramics like ordinary glasses and ceramics are brittle materials, exhibiting no region of ductility or plasticity, and they show perfect elastic behaviour up to the load which causes failure.

The strength of glass is greatly dependent upon surface flaws or Griffith cracks which act as stress multipliers. These reduce the strength of glass from a theoretical value of about one million to $10\,000$ lb/in^2 for bulk glass.

Generally glass-ceramics are stronger than ordinary glasses and most conventional ceramics, as can be seen from Table 2.2:

Table 2.2
Strength of glasses, glass-ceramics and ceramics

Material	Modulus of rupture (lb/in)
Glasses	8 to 10 ($\times 10^{-3}$)
Glass-ceramics	10 to 50
Porcelain (unglazed)	10 to 12
Porcelain (glazed)	12 to 20

(a) *Influence of chemical composition*
McMillan has reported the mechanical properties of Li_2O–ZnO–SiO_2 glass-ceramic nucleated with P_2O_5, in which he replaced ZnO by Al_2O_3. Some of his results are shown in Figure 2.15. It appears that for this type of glass-ceramic the disappearance of β-spodumene and the appearance of quartz as the major crystalline phase is associated with an increase in mechanical strength.

(b) *Effect of heat treatment schedule*
Appreciable changes in physical properties of glass-ceramics may be achieved by varying the heat-treatment schedule. Watnabe heat-treated a lithium–alumino–silicate with silver as a nucleating agent, at various temperatures between 500° and 700°C and noticed a progressive increase in the modulus of rupture. Phillips, using a lithium–zinc-oxide–silica material nucleated with P_2O_5, showed a fairly sharp increase in strength to a maximum from which it then decreased (Figure 2.16).

43

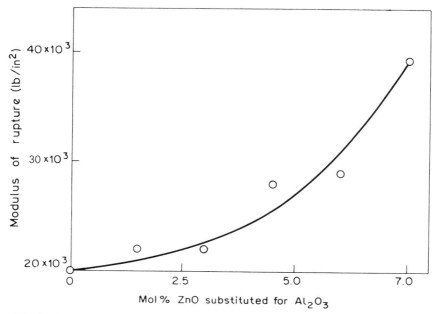

Fig. 2.15 Effects of ZnO and Al_2O_3 contents on the strength of glass-ceramics
Glass composition (mol %): $SiO_2 + Li_2O + K_2O = 91.5$
$P_2O_5 = 1.0$
$Al_2O_3 = (7.5 - x)$
$ZnO = x$
(after McMillan).

(c)   Effect of surface condition
A normal glass rod when tumbled with similar rods in a ball mill shows strengths only about 15 per cent of those of the unabraded specimens. Glass-ceramics subjected to a similar treatment show a decrease to only 80 per cent that of the unabraded – a result which suggests that glass-ceramics are much less susceptible to surface damage than are glasses and this may be one of the reasons for their greater strength.

(d)   Effect of temperature
Loss of strength with temperature is an important factor in the practical application of glass-ceramics because it limits the temperatures at which they can be used. Two general trends are observed as shown in Figure 2.17
 (i) a general decrease of strength with temperature (curve 1), and
(ii) a decrease of strength followed by a small increase at higher temperature
    (curve 2)
Figure 2.17 also shows the behaviour of a typical glass (curve 3).

44

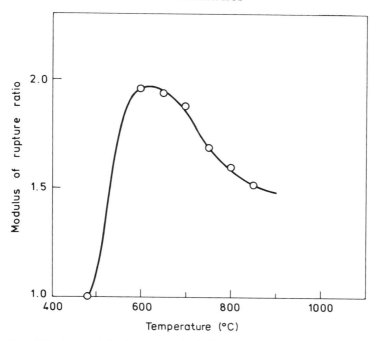

Fig. 2.16 Relative moduli of rupture of a lithium–zinc–silicate glass-ceramic heat-treated to different maximum temperatures (after McMillan).

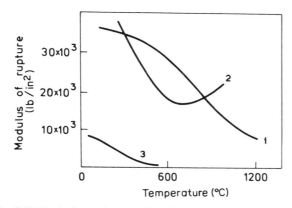

Fig. 2.17 Variation of strength with temperature for glass ceramics (1 and 2) and a typical glass (3).

The possible reasons for the high mechanical strengths of glass-ceramics are:
(1) Conventional ceramics contain pores which will weaken the material. Glass-ceramics, in general do not have such voids.

45

(2) Glass-ceramics are generally more resistant to abrasion than ordinary glasses, and so are less susceptible to surface damage which causes surface flaws.

(3) In glasses a propagated crack travels through a single homogeneous phase; in a glass-ceramic one may imagine the cracks being diverted, slowed down, or even stopped by phase boundaries. The crack may follow grain boundaries; the very fine grain size of glass-ceramics undoubtedly plays an important part in determining their mechanical properties.

2.3.3 Thermal properties

The dimensional changes which occur with change of temperature are of great importance from a number of points of view. First, the coefficient of thermal expansion should be as low as possible to minimize strain resulting from temperature gradients within the material. Second, a material joined to a different material must have a well-matched thermal expansion coefficient to prevent cracking of the joint on thermal changes. An amazing range of glass-ceramics are available with thermal expansion coefficients as low as zero, or even negative.

(a) Influence of crystal type
The thermal expansion coefficient of a glass-ceramic can be markedly different from that of the parent glass, and as a rule the expansion coefficient is dependent upon the crystal phase present, e.g. a glass-ceramic of composition: 57.2 per cent SiO_2, 27.8 per cent Al_2O_3, 11.1 per cent Li_2O nucleated with 3.0 per cent P_2O_5 has a thermal expansion of -38.7×10^{-7} per$°C$, the major crystal phase being β-eucryptite. A single crystal of β-eucryptite has an average thermal expansion of -64×10^{-7} per$°C$ between 20 to 1000°C, (-176×10^{-7} parallel to the C axis, $+82 \times 10^{-7}$ perpendicular to the C axis). It is the presence of β-eucryptite which gives the glass-ceramic its negative thermal expansion coefficient.

(b) Effect of heat-treatment schedule
The thermal expansion characteristics of a glass-ceramic can be markedly affected by the heat-treatment schedule since this determines the proportions and nature of the crystal phase present. This is strikingly illustrated by comparing curves B and D of Figure 2.18. Both of these glass-ceramics were prepared from the same parent glass composition. In one case, however, the heat-treatment schedule was adjusted to produce all of the crystalline silica in the form of cristobalite (curve B), and in the other case to give the silica entirely in the form of quartz. By variation of the heat-treatment process, glass-ceramics containing mixtures of quartz and cristobalite in various proportions could also be produced from the same glass with intermediate thermal expansions.

2.3.4 Some important applications of glass-ceramics

Glass-ceramics derived from $Li_2O–Al_2O_3–SiO_2$ and $CaO–Al_2O_3–SiO_2$ glasses, with thermal expansion coefficients approaching zero, have been used for a number of products where resistance to a sudden change in temperature is essential. These include oven dishes, cooker hot plates, laboratory bench tops, pipes, and valves. An important engineering application is in the manufacture of

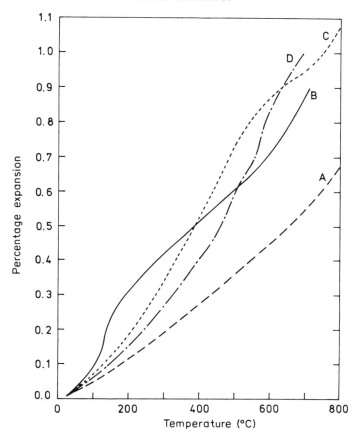

Fig. 2.18 Thermal expansion curves of glass-ceramics showing influence of phase inversions.

(A) Glass-ceramic containing neither quartz nor cristobalite.
(B) Glass-ceramic containing all of the crystalline silica in the form of cristobalite.
(C) Glass-ceramic containing a mixture of quartz and cristobalite.
(D) Glass-ceramic containing all of the crystalline silica in the form of quartz.

large rotary heat exchangers for gas turbines. In this case zero thermal expansion is necessary to ensure dimensional stability of the discs allowing satisfactory gas seals to be maintained.

The possibility of making glass-ceramics with similar thermal expansion coefficients as metals has led to their use in improved forms of vacuum tube envelopes [19].

An important application of glass-ceramics has been in the production of radomes for guided missiles. For this, the material should be transparent to microwaves and have a low temperature coefficient of permittivity. Furthermore,

47

high strength, good thermal shock resistance, and high rain erosion resistance are needed. Cordierite glass-ceramics fulfil these requirements, having loss tangents at 400°C and 10 000 MHz of less than 0.001. The change of permittivity between 20 and 400°C is also low, being less than 4 per cent. Glass-ceramics derived from $ZnO-Al_2O_3-SiO_2$ glasses with additions of divalent oxides such as CaO, BaO, or SrO also have a low dielectric loss. The glass-ceramics containing BaO have loss tangents as low as 0.0005 at 400°C and 10 000 MHz, and the change in permittivity between 20 and 400°C has been reported to be only 3 per cent.

Solder glass-ceramics have comparatively low sealing temperatures, but devitrification of the glass allows subsequent processing at higher temperatures. Glass-ceramics of the $PbO-ZnO-B_2O_3$ type can be used for sealing together glasses of thermal expansion coefficient 90×10^{-7} per °C and the manufacture of tubes for colour televisions depends on the use of sealing materials of this type.

Glass-ceramics having high capture cross-sections for thermal neutrons have also been developed [20] for use in reactor control rods. These include the oxides of cadmium, indium and boron.

Glass-ceramics containing copper oxide were induced to form a surface layer of adherent copper when heat-treated in a reducing atmosphere. This was used to produce printed circuit boards, resistors, and components for electron microscopes [21].

Glass-ceramic coatings for steel processing vessels using in the food and chemical industries have proved to have better acid resistance, improved thermal shock resistance, and higher strengths than the conventional vitreous enamel coatings. The chemical inertness of glass-ceramics has also generated interest in possible applications in the biomedical field. Production of artificial teeth from $Li_2O-ZnO-SiO_2$ glass-ceramics, and the use of similar glass-ceramic powder as a filler in polymeric composite restorative materials have already been reported [22]. Advantages are conferred by the high abrasion resistance of the glass-ceramic and also by the possibility of achieving close thermal expansion matching between the composite and the natural tooth material. The use of glass-ceramics with a high absorption for diagnostic X-rays is also a valuable advance [23]. Glass-ceramics have also been successfully used as bone substitutes in implants [24].

Although strong, refractory glass-ceramics can be made, their fracture toughness has to be improved for special engineering applications. This has been achieved by incorporating strong fibres in a glass-ceramic matrix. Aveston [25] has reported a cordierite type of glass-ceramic reinforced with carbon or silicon carbide fibres. For the carbon fibre reinforced material a modulus of rupture of 80 000 lb/m² was obtained for a composite containing 30 vol% fibre.

Glass-ceramics are not usually transparent to visible radiation owing to light scattering at crystal–glass interfaces within the material. The production of transparent glass-ceramics is achieved by ensuring that the crystals present are smaller than the wavelength of visible light or by minimizing the difference between the refractive indices of the crystalline and glass phases. Three systems from which transparent glass-ceramics can be derived have been reported [26]: these are $Li_2O-MgO-ZnO-Al_2O_3-SiO_2$ compositions in which the major phase has the β-eucryptite structure, $Al_2O_3-SiO_2$ compositions containing mullite as the major phase, and $ZnO-Al_2O_3-ZrO_2-SiO_2$ compositions in which the predominant phase is a spinel.

Transparent glass-ceramics of the eucryptite type have a thermal expansion coefficient of nearly zero and can be polished to an optical finish. One application of this special material has been in the manufacture of mirror blanks for large optical telescopes. The refractory mullite and spinel types of transparent glass-ceramics have potential as envelope materials for high-performance lamps.

Other transparent glass-ceramics containing ferro-electric crystal phases such as barium titanate or alkali niobates have been reported [27]. These materials exhibit electro-optical effects and therefore are of interest for possible applications in bistable optical switching devices.

Recently Corning Glass, USA, have reported a family of machinable glass-ceramics in which the major crystalline phase is fluorophlogopite mica, $KMg_3AlSi_3O_{10}F_2$ [28]. Glasses having wt% compositions of SiO_2 30–50, B_2O_3 3–20, Al_2O_3 10–20, K_2O 4–12, MgO 15–25, and F 4–10 are heat treated at 950–1050°C to develop a microstructure of interlocking mica crystals. Although they have fairly high thermal expansion coefficients ($50-115 \times 10^{-7}$ °C^{-1}) the glass-ceramics have good thermal shock resistance. They are also stable up to 700°C but at higher temperatures fluorine is lost, causing structural degradation. Because of their unique microstructure, the mica glass-ceramics are machinable to close tolerances; they can be drilled, sawn, or turned in a lathe. It is the combination of easy cleavage of the mica flakes and deflection blunting mechanisms, due to random orientation and interlocking of the mica crystals, that confers machinability to this material.

REFERENCES

[1] Turnbull, D. and Cohen, M. H. (1960), in *Modern Aspects of the Vitreous State*, 1, Butterworths, London, p. 38.

[2] Hillig, W. B. and Turnbull, T. (1956), *J. Chem. Phys.*, 24, 914.

[3] Hillig, W. B. (1966), *Acta Met.*, 14, 1868.

[4] Uhlmann, D. R. (1971), in *Advances in Nucleation and Crystallisation of Glasses*, American Ceram. Soc., Colombus, p. 91.

[5] Johnson, W. A. and Mehl, R. F. (1939), *Trans. Amer. Inst. Min. (metall) Engrs.*, 135, 416.

[6] Avrami, M. (1939; 1940; 1941) *J. Chem. Phys.*, 7, 1103; 8, 212; 9, 177.

[7] Swalin, R. A. (1972), in *Thermodynamics of Solids*, 2nd Edition, John Wiley and Sons, Inc., New York, p. 156.

[8] Hildebrand, J. H. (1929), *J. Amer. Chem. Soc.*, 51, 66.

[9] (a) Hammel, J. J. (1967), *J. Chem. Phys.*, 46, 2234.
  (b) Bardy, H. K. (1953), *Acta Met.*, 1, 202.
  (c) Lumsden, J. (1952), in *Thermodynamics of Alloys*, Inst. Metals, London.
  (d) Van der Toorn, L. J. and Tiedema, T. J. (1960), *Acta Met.*, 8, 711.
  (e) De Fontaine, D. and Hilliard, J. E. (1965), *Acta Met.*, 13, 1019.
  (f) Cook, H. E. and Hilliard, J. E. (1965), *Trans. Met. Soc., AIME*, 233, 142.

[10] James, P. F. (1975), *J. Mater., Sci.*, 10, 1802.

[11] Haller, W., Blackburn, D. H. and Simmons, J. H. (1974), *J. Amer. Ceram. Soc.*, 57, 120.

[12] Macedo, P. B. and Simmons, J. H. (1974), *J. Res. Nat. Bur. Stand.*, A78, 53.

[13] Cahn, J. W. and Hilliard, J. E. (1959), *J. Chem. Phys.*, 31, 688.

[14] Lifshitz, I. M. and Slyozov, V. V. (1961), *J. Phys. Chem. Solids*, **19**, 35.
[15] McMillan, P. W. (1964), in *Glass Ceramics*, American Press.
[16] Stookey, S. D. (1956), British Patent 752 243.
[17] Stookey, S. D. (1960), British Patent 829 447.
[18] McMillan, P. W., Hodgson, B. P. and Partridge, G. (1970), British Patent 943 599.
[19] McMillan, P. W., Hodgson, B. P. and Partridge, G. (1966), *Glass Technol.*, **7**, 121.
[20] McMillan, P. W., Partridge, G., Hodgson, B. P. and Heap, H. R. (1966), *Glass Technol.*, **7**, 128.
[21] McMillan, P. W. (1976), *Phys. Chem. Glasses*, **17**, 195.
[22] McMillan, P. W. and Hodgson, B. P. (1964), *Glass Technol.*, **5**, 142.
[23] Seki, S. (1974), *Proc. Tenth Int. Congr. Glass*, **14**, 88.
[24] MacCulloch, W. T. (1968), *Brit. Dent. J.*, **124**, 361.
[25] Muller, G. (1974), *J. Dent. Res.*, **53**, 1342.
[26] Hench, L. L. (1974), *Proc. Tenth Int. Congr. Glass*, **9**, 30.
[27] Aveston, J. (1972), in *The properties of fibre composites*, IPC Science and Technology Press, London, p. 63.
[28] Beall, G. H. and Duke, D. A. (1969), *J. Mater, Sci.*, **4**, 340.
[29] Borelli, N. F. (1967), *J. Appl. Phys.*, **38**, 4243.
Borelli, N. F. and Layton, M. M. (1969), *Trans. IEEE*, **ED-16**, 511.
[30] Beall, G. H. (1971), in *Advances in nucleation and crystallisation in glasses*, American Ceram. Soc., p. 251.
[31] Chyung, K., Beall, G. H. and Grossman, D. C. (1974), *Proc. Tenth Int. Congr. Glass*, **14**, 33.

CHAPTER 3

Glass Formation by the Sol–gel Process

I. Strawbridge

3.1 INTRODUCTION

As described in Chapter 1 glass, according to the ASTM definition, is 'an inorganic product of fusion, which has cooled to a solid state without crystallizing' [1]. However in recent years, several unconventional techniques for glass making, which avoid the use of the molten state, have received much attention. These unconventional routes to glass, which were described by Mackenzie [2], include solid state methods such as neutron bombardment and vapour phase techniques such as evaporation, RF sputtering, vapour phase hydrolysis and chemical vapour deposition. Solution techniques also exist, for example, electrolytic deposition of amorphous oxide films and most importantly from the point of view of the current discussion, hydrolysis and polymerization of compounds such as alkoxides by the sol–gel process.

Since 1969, with the publication of the comprehensive review of Schroeder [3], there has been extensive interest in the sol–gel process as a means of producing not only amorphous films, but also monoliths and glass fibres. However, the sol–gel process is not such a recent development, since much earlier in Germany it had been discovered that clear and stable films of silica could be obtained by evenly spreading colloidal solutions of silicic acid over a glass surface, preferably by spinning the substrate [4]. This process was used industrially in the early 1940s in Germany, but was abandoned when vacuum evaporation techniques proved more economical.

Due to the need to coat larger surfaces, interest was revived in the solution technique and Schroeder [5] demonstrated that the presence of colloidally precipitated hydroxide or oxide hydrates in a solution was a sufficient, although not necessary, requirement for the production of optically homogeneous inorganic films on solid surfaces. The only requirement was to start from solutions containing metallic compounds, which formed polymers or polysaturated groups in solution, yielding gels of low crystallization tendency during drying. Metal alkoxides which could undergo controlled hydrolysis were therefore particularly suitable for gel formation.

3.2   GENERAL CHEMISTRY

In terms of chemical mechanisms, sol–gel processing can be divided into two distinct areas, namely the alkoxide route and the aqueous route. The former generally requires the use of an organic solvent, normally an alcohol, in order to act as mutual solvent for the organometallic precursor and the water for hydrolysis, whereas the latter route employs the principles of colloid chemistry to generate colloid sized particles from ionic species in an aqueous medium.

The alkoxide route has attracted more attention in recent years and will therefore be covered in more detail. The aqueous systems do have some interesting applications making them worthy of some consideration.

3.2.1   The alkoxide route

The following equations describe the fundamental reactions which allow the conversion of monomeric organometallic precursors into gels and ultimately glasses or ceramics.

1) Hydrolysis

$$M (OR)_x + H_2O \underset{\text{catalyst}}{\overset{H^+/OH^-}{\rightleftharpoons}} M (OR)_{x-1}OH + ROH$$

$$\downarrow H_2O$$

$$M (OR)_{x-2}(OH)_2 + ROH$$

etc.

2) Condensation

$$\equiv M-OH + HO-M \equiv \; \rightleftharpoons \; \equiv M-O-M \equiv + H_2O$$
$$\equiv M-OH + RO-M \equiv \; \rightleftharpoons \; \equiv M-O-M \equiv + ROH$$
$$\equiv M-OR + RO-M \equiv \; \rightleftharpoons \; \equiv M-O-M \equiv + R-O-R$$

M is a metal chosen from Al, In, Si, Ti, Zr, Sn, Pb, Ta, Cr, Fe, Ni, Co and several others.
R represents an alkyl group.
The condensation reactions also facilitate the reaction of compounds of different elements leading to the synthesis of multicomponent gel systems [6].
All the reactions can be regarded as equilibria with the extent of the reaction being determined by the species involved and the conditions under which the reactions take place.

3.2.2   Reaction kinetics

A large proportion of the kinetic studies have been performed on silicon alkoxides, one of the earliest being by Aelion and co-workers [7] on the hydrolysis of tetraethoxysilane (TEOS). They studied the changes in the water and alcohol content of solutions as a function of time in order to determine the kinetics of hydrolysis in several solvents under both acidic and alkaline conditions.

Using hydrochloric acid as the catalyst and initial mole ratios $H_2O/TEOS$ greater than 4; the extent of the hydrolysis reaction:

$$Si\,(OEt)_4 + 4H_2O \rightleftharpoons Si(OH)_4 + 4EtOH$$

was found to be complete for acid concentrations greater than 0.005 molar in dioxane as solvent, and similar results were found with ethanol and methanol. In ethanol the extent of the hydrolysis reaction decreased at acid concentrations below 0.003 molar. Reactions performed in methanol, ethanol and dioxane showed similar values for the rate constant for hydrolysis suggesting that re-esterification was of little importance.

The hydrolysis reaction was shown to be second order with respect to water and TEOS, and the rate was directly proportional to the acid concentration. Mechanisms were proposed involving the formation of a quinquevalent transition state, followed by a charge transfer and dissociation:

A.　$\equiv Si - OR + H_3OB \rightarrow \equiv Si \quad - \quad \overset{..}{\underset{..}{O}}R \rightarrow \equiv Si - OH + HOR + HB$

$$H\,\overset{..}{O}H \qquad HB$$

or

$$HB$$

B.　$H_2O + \equiv Si - OR + HB \rightarrow H_2O \quad : \quad \overset{|}{\underset{/\backslash}{Si}} - \overset{..}{O}R \rightarrow \; \equiv Si - OH + HOR + HB$

where B is a base.

The insensitivity of the rate constant to the dielectric constant of the medium, but the reduction in the rate with the reduction in the strength of the acid, suggested the involvement of the undissociated acid, HB, in the transition state where the acid proton was donated in the formation of the alcohol. The rate-determining step was assumed to be the regrouping of bonds or the attachment to the hydrophobic silicate molecule. The effective rate in either case would depend on the product $[acid] \times [H_2O]$.

By comparison with the observation of an oxonium salt with sulphur dioxide [7], the existence of a similar preformed salt H_3OB was thought likely here, in which case mechanism A would outstrip B in view of its bimolecular nature.

A different hydrolysis mechanism was found to operate in alkaline catalysis since a large reduction in reaction rate was observed in solvents with decreasing dielectric constant, i.e., the participation of ions was indicated. At low concentrations of TEOS in methanol the reaction was shown to be first order with respect to TEOS, but at higher concentrations the reactions did not follow a simple order apart from in the early stages, but became complicated by secondary reactions. This resulted in the rapid precipitation of polymer and the development of heterogeneous reactions. A nucleophilic substitution mechanism of the form:

$$HO^- + \equiv Si - OR \quad \rightarrow \quad \overset{-1/2 \quad | \quad -1/2}{HO - \underset{/\backslash}{Si} - OR} \rightarrow \; HO - Si \equiv + {}^-OR$$

$$RO^- + HOH \rightarrow \quad HOR + {}^-OH$$

53

was suggested for dilute solutions, where the decomposition of the transition state was considered to be the most likely rate-determining step.

The extent of hydrolysis in ethanol and methanol was found to be incomplete for all but very low concentrations of TEOS, since precipitation of polymer occurred rapidly.

Studies by Schmidt and co-workers [8] suggested that in the formation of sodium borosilicate gels under conditions of alkaline catalysis, the ethanolic solvent participated in the hydrolysis equilibrium. Hence re-esterification did occur under alkaline conditions, contrasting with acid catalysis.

These observations can be summarized as follows.

Both acid and base act as a catalyst for the hydrolysis and condensation reactions.

But initial hydrolysis proceeds more quickly under acidic conditions, whereas condensation proceeds more rapidly under alkaline conditions.

Another factor which becomes important in the production of multicomponent gels is the relative rates of reaction of the alkoxides of different metals. Yoldas [9] has stated that the order of addition of the alkoxide components was important in producing a homogeneous gel, and the first component to be hydrolysed should be the most slowly reacting one – this generally being the silicon alkoxide. When this has been rendered an active polymerizing species, other alkoxides can be added which are then more likely to react with the already hydrolysed compounds rather than with themselves.

3.2.3 The aqueous route

It is also possible to produce amorphous materials using colloidal systems in an aqueous medium. These techniques rely on generating a stable dispersion of colloidal sized particles in a solvent (in this case water) and destabilizing this sol in a controlled manner to give a solid gel from which the solvent can be removed, and then sintering to give a dense amorphous mass.

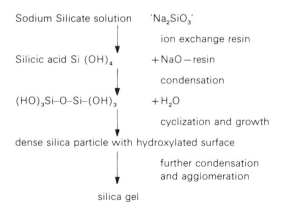

Fig. 3.1 Schematic of aqueous route to silica glass.

The system which has probably received the most attention is silica, and many workers have described methods of producing dense silica glass from silica sols [10, 11].

The methods for producing silica sols include growing particles from sodium silicate derived silicic acid, by means of condensation reactions (schematically illustrated in Figure 3.1) or by dispersing commercially available fumed silica in water [11].

3.3 SOL–GEL PROCESSING

For the preparation of amorphous materials from alkoxides the sol–gel process can be broken down into the following stages.

Precursors, dissolution and mixing.
Sol to gel transition.
Forming (e.g., casting, fibre drawing and film formation).
Drying.
Gel to glass transition.

Each of these stages will be dealt with in turn later, but first it would be useful to highlight the claimed advantages and disadvantages of glass formation by the sol – gel process [12] (Tables 3.1 and 3.2 respectively).

Table 3.1
Proposed advantages to sol–gel processing (after [12])

Advantages	Qualifications
1. Better homogeneity.	Mixing is performed in low viscosity solutions.
2. Better purity.	Raw materials can be purified by standard techniques, e.g., distillation.
3. Low temperature preparation: Savings in energy. Minimize evaporation losses. Minimize air pollution. No reaction with container. Bypass phase separation.	Glass or plastic (e.g. PTFE) can be used as required. Liquid–liquid immiscibility of certain melts can be avoided. High temperatures where rate of crystallization is at a maximum can be avoided.
4. New non–crystalline solids outside the range of normal glass formation.	This follows from bypass phase separation.
5. New crystalline phases from new non–crystalline solids.	
6. Better glass products from the special properties of gels.	
7. Special products, e.g., films and fibres.	

Table 3.2
Proposed disadvantages to sol–gel processing (after [12])

Disadvantages	Qualifications	
1. High cost of raw materials.	This may be remedied if the dem d increased and by choice of cost- ctive products.	
2. Large shrinkage during processing.		
3. Residual microporosity.		
4. Residual hydroxyl.	These may be overcome by refin heat treatment processes.	
5. Residual carbon.		
6. Health hazards of organic solutions.		
7. Long processing times.		
8. Difficulty in producing large pieces.		

3.3.1 Precursors, dissolution and mixing

The most commonly employed starting materials are metal alkoxides ᐧ   h are available for a large range of elements and with various alkoxy group: 13]. The solubility and reactivity of metal alkoxides varies widely, a   ᴐme experimentation is frequently necessary to obtain the required solui   ᴑn occasions it may be unavoidable to compromise on the purity advanta   ᴣn by alkoxide starting materials requiring the use of a metal salt, e.g.,   e, chloride or acetate, in order to get the metal precursor into solution.

In order to achieve the advantages of improved homogeneity the species must be dispersed uniformly into a single phase solution. Si alkoxides are sensitive to hydrolysis by water a non-aqueous solvent is required, but in order to convert the organometallic compounds int which will readily enter into condensation reactions, controlled hydrol be achieved. This necessitates a mutual solvent for the alkoxide, the po and any catalyst. The most widely-used solvents are alcohols – typica anol, ethanol and propanol.

It is important to realize that the choice of starting materials cai marked effect on the resulting gel and ultimately on the glass. To ach maintain good homogeneity in multicomponent systems care must be minimize self-condensation. This may be controlled by the processin; tions, e.g., the order of mixing, temperature and solvent type, as well as th of reactants; e.g., by using alkoxides of a similar reactivity [14]. Under ᐧ mixing could be included attempts at controlled prehydrolysis of tl reactive alkoxides prior to the addition of more reactive components.

Mixed alkoxides are also available for many systems, and they pr further addition to the armoury of starting materials. In many ca: solubility of the double alkoxide is higher than that of either single alkoxi example, the ethoxides of Al and Mg are only slightly soluble in cool e

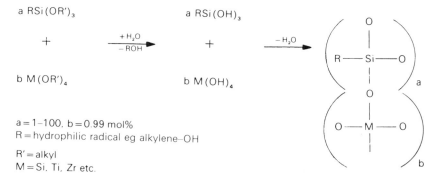

Fig. 3.2 Organically modified silicate (after [17]).

whereas the double alkoxide $Mg\{Al(OEt)_4\}_2$ is highly soluble [16]. This probably reflects the fact that double alkoxides are generally less associated in alcohols or other organic solvents than are the constituent alkoxides.

A further type of material which has received some attention is mixed alkyl alkoxy precursors [17]. These can enter into reactions schematically illustrated in Figure 3.2.

Such processes have enabled contact lens material to be produced where the organic modification has influenced the wetting and permeability properties.

To summarize, the careful choice of starting materials and mixing conditions enables a wide range of compositions to be prepared where the mixing of components should be uniform down to the molecular scale giving processing advantages over conventional melting techniques and even novel systems which would otherwise be unobtainable.

3.3.2 Sol to gel transition

Reactions in the sol
Having produced a uniformly dispersed solution the next stage is to transform this into a gel. Depending on the history of the sol this may occur naturally even if sealed because of the water that has already been added to the system, or more water may be necessary, possibly by absorption of water from the atmosphere if left open.

In the case of an aqueous colloidal system, a gel may be produced by destabilizing the sol in a controlled manner. This could be done by encouraging the particles to approach each other and overcome whatever stabilizing barrier is operating. Depending on the system, this might be achieved by heating the sol, freezing the sol, concentrating the sol, adjusting the pH or by adding electrolyte [10].

In recent years, there have been several publications which have attempted to understand the processes occurring during the sol–gel transition for alkoxide-derived systems, with silica being the most widely studied. The techniques employed include gas chromatography (GC), nuclear magnetic resonance (NMR), small angle X-ray scattering (SAXS) and viscosity measurements.

In a study by Brinker and co-workers [18], a two stage hydrolysis was performed. The first stage consisted of a partial hydrolysis using the mole ratio of components:

$$TEOS/EtOH/H_2O/HCl = 1/3/1/0.0007$$

followed by the reaction at either

$$pH\ 0.95\ with\ H_2O/TEOS = 5.1\ (A2)$$

or

$$pH\ 7.8\ with\ H_2O/TEOS = 3.7\ (B2)$$

In the first stage it was observed that hydrolysis was rapid but incomplete due to insufficient water. Measurements by gas chromatography suggested species of the type $SiO(OR)_2$, but the more reliable NMR studies indicated a distribution of species of the kind $Si\,(OR)_{4-n}\,(OH)_n$ where $n = 1$, 2 and possibly 3. Dimers were detected in 90 minutes, whereas dimers, chains and cyclic polymers were present after 24 hours. It was therefore proposed that for acid catalysis and low water levels condensation is promoted before hydrolysis is complete leading to lightly cross-linked polymers.

The second stage led to hydrolysis being completed under acidic conditions, whereas under alkaline conditions unhydrolysed TEOS remained after condensation reactions were complete. It was suggested that phase separation may have caused this unhydrolysed residue.

The solutions from the second stage reactions were examined using SAXS by Schaefer and Keefer [18, 19]. Dense colloidal silicas were not detected in either solution but instead fractal clusters were observed. A fractal species has a structure which becomes progressively indistinct as its radius r increases, hence its mass increases proportionally with r^d where d is less than 3. The structures detected are illustrated in Figure 3.3 [18]. The polymers from the acid solution were found to be lightly cross-linked compared with the clusters from the alkaline solution.

The correlation length ξ in a concentrated solution represents the mean separation between polymer clusters, but in a dilute solution ξ approaches the Guinier radius of gyration R_g as the polymer species untangle. In the acid system, dilution leads to a difference to appear between ξ and R_g. This is not the case in the alkaline system because of the presence of relatively compact clusters which do not interpenetrate even when concentrated.

Intrinsic viscosity measurements were performed by Sakka et al. [20], who found linear polymers were produced in acid catalysed solutions when the $H_2O/TEOS$ ratio was 1 or 2, but spherical or 3 dimensional polymers resulted if the ratio was 5 and 20. Sakka and Kamiya [21] reported particle-like species as a result of acid and base catalysis for $H_2O/TEOS = 20$.

On the other hand, there has been a body of opinion with the view that either dense particles [10, 22] or structures such as triple chains [23] are formed in such systems.

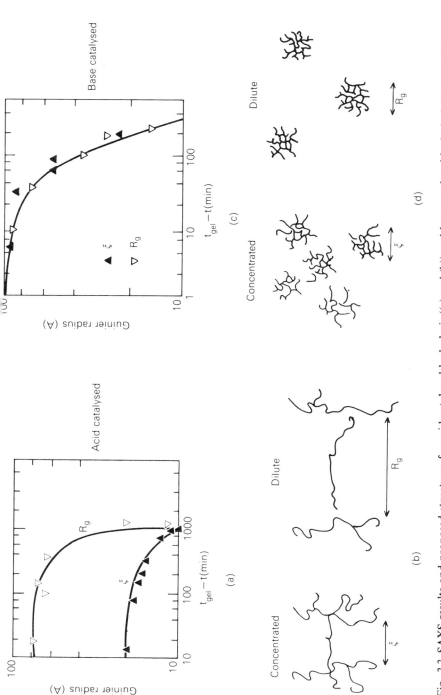

Fig. 3.3 SAXS results and proposed structures for acid-catalysed hydrolysis ((a) and (b)) and base catalysed hydrolysis ((c) and (d)).

An interpretation has been proposed by Keefer [24] and supported by Scherer [25] for the effect of reaction mechanism on polymer structures.

In base catalysis nucleophilic attack occurs on the silicon of alkoxide (Figure 3.4a) leading to a quinquevalent transition state with maximum negative charge separation. The alkoxy groups around the silicon hinder attack by OH by virtue of both their size and negative charge, but once one alkoxy group has been substituted attack becomes easier and therefore once started hydrolysis goes to completion. Above pH 2 (the isoelectric point of silica), the condensation reaction proceeds as in Figure 3.4b [10] by the nucleophilic attack by an ionized silanol group on a un-ionized one. Therefore it was suggested that more densely cross-linked polymers would result from fully hydrolysed monomers.

Under acid conditions attack is by an electrophilic mechanism with the positive attacking group attracted to OR (Figure 3.4c). The first attack would therefore be expected to be easier than subsequent substitutions. This means that condensation reactions are likely to begin before hydrolysis is complete giving more lightly cross-linked species. For these reasons Keefer preferred base catalysis, since this led to densely cross-linked polymers, whereas acid catalysis resulted in linear lightly cross-linked polymers.

Other factors as well as pH are also important. The water concentration plays a crucial role because of its participation in both hydrolysis and condensation. High water concentrations will favour hydrolysis but inhibit condensation. Therefore even at low pH, high water concentration will lead to high cross-link

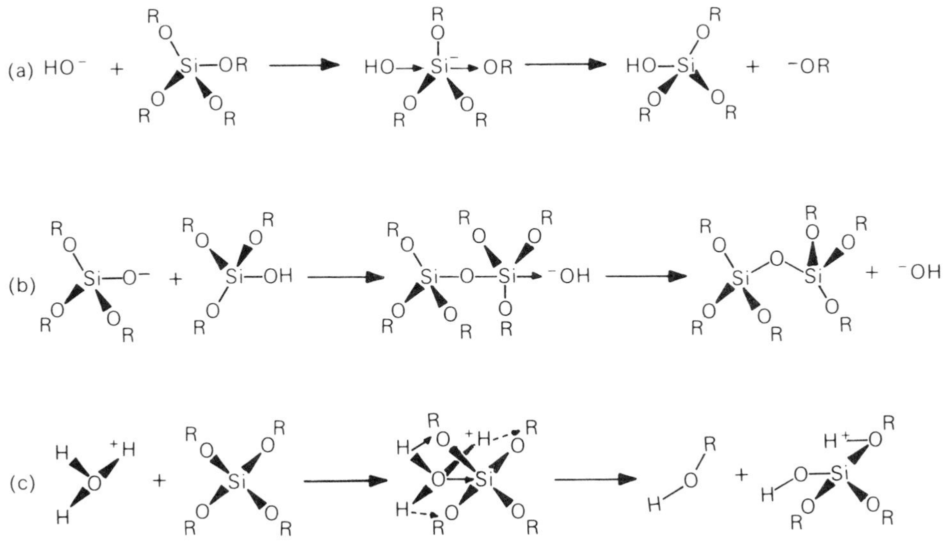

Fig. 3.4 Reaction mechanisms: (a) base catalysed hydrolysis of silicon alkoxide by nucleophilic reaction mechanism; (b) nucleophilic condensation reaction mechanism; (c) acid catalysed hydrolysis of silicon alkoxide by electrophilic mechanism. R = H, Et or $Si(OR)_3$ (after [24]).

60

densities and at pH 2, for example, where condensation is at a minimum, hydrolysis may be complete before condensation begins. Alternatively, at low water concentrations even at high pH, condensation reactions may commence before hydrolysis is complete.

Scherer [25] proposed that in the basic system polymer growth is by addition of monomers to clusters giving dense particles, but under acid conditions growth occurs by aggregation of clusters leading to highly branched structures.

It was further suggested that in acid, re-esterification of the gel may occur even though hydrolysis may have gone to completion in the sol reactions. This process tends to be driven by the alcohol used as the solvent. This leads to SiOR retained in the gel which may cause problems during subsequent drying. This difficulty suggests the use of excess water during hydrolysis.

If these comments are compared with the results of Aelion et al. [7] discussed in section 3.2.1, there may appear to be some contradictions. These may best be explained by the fact that the majority of earlier work was based on reactions in their initial stages. On the other hand, the more modern studies have investigated much further into the reaction process and have made allowances for the increased acidity of silanol groups with the extent of hydrolysis and condensation.

The alcohol employed as solvent can also be a factor affecting reactions. As well as an alcohol being a reaction product thus possibly promoting a reverse reaction, alcohol groups can exchange rapidly from the solvent to the alkoxide [18, 26]. The size of alkoxy groups will also influence reactions. Aelion et al. [7] showed that the hydrolysis rate decreased with alkoxy group size, hence if TMOS is dissolved in EtOH this would lead to some exchange and then the ethoxy group would react more slowly than the original methoxy groups. Schmidt and co-workers (27) reported that hydrolysis of components of the type $R'Si(OR)_3$ when R' is an alkyl group, was faster than $Si(OR)_4$ when acid catalysed, but slower when under basic conditions.

The $H_2O/TEOS$ ratio (r) has been proved to be important in determining the structure of the resulting gel. In a closed system increasing this ratio reduced the time to gelation [28]. In an open system however, where the sol is free to lose solvent by evaporation and absorb water from the atmosphere, the gel time was observed to increase for r greater than 10 [29]. This was accounted to the increased dilution with respect to the alkoxide, as were the observations by Yoldas [30] where ^{29}Si NMR showed a decrease in the condensation rate when diluted by alcohol.

It is clear that increasing r reduces the retained organics in dried gels [22, 31, 32] and r has been demonstrated to affect the surface area and density of the dried gels [32, 33, 34].

Gelation

As species grow, the viscosity of the sol will increase until a point when the network extends throughout the liquid volume and a sudden increase in viscosity is noted (Figure 3.5). The factors which have been shown to reduce the gelation time (t_g) include increasing temperature [35, 36, 37]; increasing $H_2O/TEOS$ ratio [28, 36]; increasing alkoxide concentration [32, 36] and increasing pH to give a minimum at pH 8 [37].

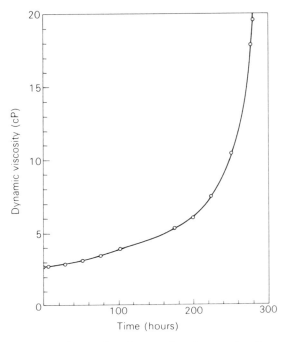

Fig. 3.5 Dynamic viscosity, η (in cP or mPa. s) as a function of time from preparation for solution with H_2O/TEOS mole ratio 1.74 (after [29]).

3.3.3 Forming

Now we have discussed the processes occurring as the mobile sol transforms into an elastic gel, it would be worthwhile at this stage to indicate the methods available to form potentially useful shapes.

Monoliths
The most obvious forming technique is casting, so that a monolithic gel can be produced. Glass or PTFE vessels are frequently employed for this purpose. Because of the large shrinkages associated with the sol–gel process for making glass it is normally only possible to cast simple shapes such as discs and cylinders. The art in monolith production generally revolves around the drying stage and this will be discussed later.

Thin films
Another area where sol–gel finds applications is in the formation of thin films – either as coatings on a substrate or as free-standing films. The techniques available for producing uniform thin films described by Schroder [3] include the following.

Dipping processes, in which the parts to be coated are withdrawn from the solution and simultaneously covered with a liquid film.

'Lowering' processes, in which the substrate remains stationary while the liquid level is lowered.

Spinning techniques, where the liquid spreads from the centre of a rotating substrate. This is only applicable to smaller discs.

Spraying processes.

The dipping processes are the most economical and the most universally applicable.

It is also possible to produce thin free-standing films, for example by withdrawing a suitably shaped wire from a solution [20]. In order to produce such films the composition of the solution and the timing of the drawing is critical. The solution must be one which will form linear polymeric species, i.e., acid catalysed with low water levels (H_2O/TEOS $= 1.5 - 4$) and the viscosity normally needs to be in the range 10–100 P.

Fibres

There are two main methods of producing fibres by the sol–gel route. Either a monolith may be formed, dried and sintered, then a fibre may be drawn at elevated temperatures by what would be conventional means. Alternatively it may be done by pulling fibres from the solution during the appropriate viscosity period. Again the latter method requires careful control of the solution chemistry and viscosity. The conditions for fibre drawing from solution are broadly similar to those for drawing thin films described in the previous section [38].

Powders

Particularly in the early days of sol–gel processing, the alkoxide route led to gels which uncontrollably cracked, therefore the fragments were frequently ground into a powder which could then be hot pressed into a larger amorphous body [6, 39]. However, more recently the sol – gel process has been recognized as a technique for producing high purity ceramic powders with controlled particle sizes [40].

3.3.4 Drying

Conventional drying to give xerogels

The most critical stage for producing sol–gel glasses of any significant size is during drying. It is during this process, illustrated in Figure 3.6 for an acid catalysed gel, a base catalysed gel, a colloidal gel under conditions of high silica solubility and a colloidal gel with weak interparticular bonding [41], that cracking is most likely to occur. These difficulties can be minimized to some extent by using very slow drying rates over periods up to many months and a controlled atmosphere of the solvent to be removed [42] or under hypercritical conditions [43]. Iler [10] has discussed in detail the problems pertaining to the drying of colloidally derived gels. According to Iler shrinkage can be avoided as the liquid phase is removed by the following five methods.

Strengthening the gel by reinforcement, increasing the strength of the inter-particle bonds, thus opposing the shrinkage forces.

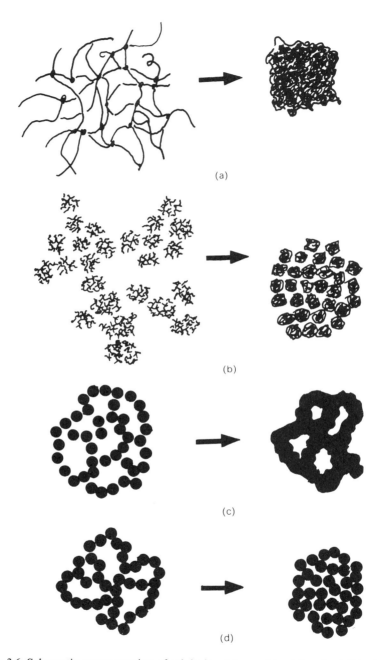

Fig. 3.6 Schematic representation of gel drying: (a) an acid catalysed gel; (b) a base catalysed gel; (c) a colloidal gel aged under conditions of high silica solubility; (d) a colloidal gel with weak interparticular bonding. After [41].

Reducing the surface tension forces by enlarging the pore diameter by an ageing or hydrothermal process.

Replacing the water with a polar liquid of lower surface tension, for example water by alcohol.

Heating the liquid-filled gel under pressure to above the critical point where no liquid–vapour boundary exists and releasing the vapour.

Making the surface of the silica hydrophobic.

During the initial stages of drying the evaporation of the liquid from the micropores in the gel causes large capillary stresses to develop which can initiate cracking. These capillary forces depend on the rate of evaporation which is related to the solvent vapour pressure, and also depend inversely on the pore size. Therefore larger pore sizes and a stronger gel network tend to reduce the degree of cracking. During the final stages of drying, cracking is a result of non-uniform contraction of the body [43]. This can be due to temperature gradients, compositional inhomogeneities and different local rates of reaction. A distribution of pore sizes is also harmful since small pores will be subject to larger stresses and given a distribution of pore radii, cracks will develop because of differential stresses.

Drying control chemical additives
Another method for reducing the tendency towards cracking on drying employs organic additives, termed drying control chemical additives (DCCA), which are capable of controlling the vapour pressure of the solvents in the gel pores. The additive studied by Wallace and Hench [44] was formamide (NH_2CHO) which has a vapour pressure of 0.1 Torr at 40°C compared with 100 Torr for methanol at 21°C.

Although the mechanism for the action of DCCA is not fully understood, the reduced evaporation rate of such liquids during initial drying was believed to reduce the differential contraction which caused cracking. Hench and co-workers have produced several papers describing the use of these additives [44–47]. Formamide has also been shown to influence reactions during the sol to gel transition by decreasing the hydrolysis rate and increasing the condensation rate for tetramethoxysilane [48–50]. These gels have larger pores with a narrower distribution and improved strength (45). Figure 3.7 schematically illustrates how the presence of a narrow pore size distribution can lead to reduced tendency to cracking [44]. Uncertainties about how the changes in the reaction rates influenced the structures were highlighted by Scherer [25]. He pointed out that coarser gels are produced in systems with higher pH, but formamide decreases the pH [50]. Larger pore volumes were reported for gels prepared using low vapour pressure solvents, e.g., formamide and glycerol [45, 51], which may have resulted to some extent from the rapid ageing (i.e., consolidation or strengthening) caused by the increased rate of condensation reactions [50].

Hypercritical drying to give aerogels
One of the above methods referred to by Iler [10] for drying colloidally-derived gels has found applications with materials prepared from alkoxides to give dried

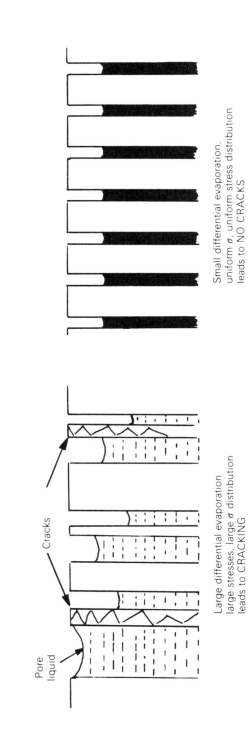

Cracks

Pore liquid

Large differential evaporation. large stresses. large σ distribution leads to CRACKING

Small differential evaporation. uniform σ. uniform stress distribution leads to NO CRACKS

Fig. 3.7 Proposed mechanism for the cracking of gels on drying, as influenced by pore size distribution (after [44]).

bodies with lengths of several tens of centimeters [52]. In the hypercritical methods the gel is heated above the critical temperature T_c and critical pressure of the liquid phase under which conditions no interface exists between liquid and gas. There is therefore no difference between solid–liquid and solid–vapour interfacial energies, thus the capillary pressure in the liquid is zero removing the driving force for shrinkage.

The method involves placing the gel in an autoclave along with excess solvent in which it is then heated to above the critical temperature. The liquid is intended to produce a high partial pressure of the solvent to prevent premature drying of the gel during heat-up. When the system has exceeded the critical conditions, the pressure is released while keeping the temperature above T_c. The drying process is thus greatly accelerated from days or weeks to hours.

A modification to this process was described by Tewari and co-workers [53], who rinsed the wet gel in liquid CO_2 under pressure until all the solvent was replaced. The CO_2 was then removed above its critical point. The benefit here is that the equipment can be more simple since $T_c = 31\,°C$ and $P_c = 73$ atm for CO_2 compared with 243°C and 63 atm for ethanol, and 240°C and 78.5 atm for methanol.

Since the tendency to shrink is removed, very low density silica gels (aerogels) have been prepared by this technique [10].

Drying models
For a more in-depth discussion of the mathematics involved in drying gels the interested reader is referred to a series of papers by Scherer [54–60]. The assumptions for the model describing the drying of a wet inorganic gel are outlined below.

Fluid is transported by flow down a pressure gradient according to Darcy's Law:

$$J = (D/\eta_L)\Delta P$$

where J is the flux (volume/area × time), D is the permeability (with units of area), η_L is the viscosity of the liquid and ΔP is the pressure gradient in the liquid. For a pure liquid the driving force for transport is proposed to be the pressure gradient rather than a diffusion mechanism [25].

The contraction is driven by the interfacial energy.

The solid phase of the gel is viscous.

The model produces the following general results for a viscous gel. The stresses which develop in the gel body as it is dried are proportional to the evaporation rate and the size of the body, and inversely proportional to the permeability and the bulk modulus. This explains why slower drying prevents cracking. Additionally, it may explain the effect of DCCA, which increase the pore size and hence giving higher permeability and the hardness which is related to the bulk modulus of the gel. The stress is shape-dependent, decreasing in the order: plate > cylinder > sphere.

3.3.5 Gel to glass transition

Processes during the gel to glass transition
We have now discussed how to produce a gel with all the free solvents and reaction products substantially removed and hopefully without any cracking. However, the body will still retain some 'non-glassy' components, especially residual alkoxy and silanol groups. There will also be a significant amount of porosity, both on a molecular scale determined by how densely cross-linked the primary particles or clusters are, and on a larger scale related to how the particles aggregated to form a gel and how this packing changed during subsequent ageing and drying processes. Our purpose now is to convert this dried porous gel into a fully densified and homogenous glass.

This conversion requires a treatment at elevated temperatures up to somewhere in the region of the glass transformation temperature T_g. Under certain conditions higher temperatures may be necessary but these should be below the temperature employed for conventional melting.

The low temperature zone of the heat treatment up to ca. 500°C is predominantly concerned with the removal of residual alkoxy and silanol groups leading to some densification on a molecular scale by virtue of condensation reactions. However, depending on the amount of residual volatiles and the rate of heating, some unreacted groups may remain trapped as larger porosity begins to close at higher temperatures. These trapped impurities may lead to bloating as heating progresses. Sintering of porosity will proceed as the transformation temperature is approached, the process being governed by viscous flow and the driving force being supplied by the decrease in the surface energy of the porous gel. Finally, the sintering process must be complete before significant crystallization of the glass has chance to occur. The rates of both sintering and crystallization increase with temperature and may be related to the viscosity of the glass.

Decomposition of volatiles
A wide range of techniques have been employed to monitor the loss of volatiles during heat treatment. The most common methods have been thermogravimetric analysis (TGA) differential thermal analysis (DTA) and infrared spectroscopy. Results presented by Gonzalez-Oliver *et al.* [61] illustrate clearly the changes observed by TGA and DTA during heat treatment of two silica gels prepared from TEOS (Figure 3.8). Here it can be seen that the weight loss from these gels corresponds to endothermic DTA peaks where physically adsorbed water and organics are being removed. Around 260–280°C sharp exothermic DTA peaks are visible on the O_2 traces caused by the oxidation of organics, not surprisingly these are absent when a nitrogen atmosphere is used. Some weight loss continues above 400°C resulting from the slow elimination of water from condensation reactions.

In the same study, Gonzalez-Oliver *et al.* [61] also presented infrared spectra for TEOS-derived silica gel after various heat treatments (Figure 3.9). The intensity of absorptions caused by the SiOH stretching vibration (3660 cm^{-1}) and the presence of absorbed water (3400 cm^{-1} and 1640 cm^{-1}) decrease as treatment temperature increases. The bands also shift to higher wavenumbers

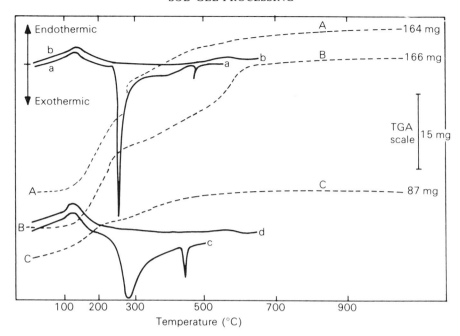

Fig. 3.8 DTA and TGA traces for two silica gels (derived from TEOS). The runs were carried out in both O_2 and N_2 gas flows. Traces (a), A (100% O_2) and (b), B (100% N_2) correspond to gel 1; traces (c), C (100% O_2) and (d), (100% N_2) correspond to gel 2. After [61].

and the 950 cm^{-1} shoulder diminishes suggesting a gradual strengthening in the silica network.

Densification behaviour
The main techniques used to study densification behaviour have been bulk density measurements, dilatometry, gas adsorption, SAXS and transmission electronmicroscopy (TEM).

A study by Nogami and Moriya [62] presented results for the densification behaviour for silica gels prepared by both HCl and NH_4OH catalysed hydrolysis and condensation of TEOS. In both cases they employed a H_2O/TEOS mole ratio of 10. Figure 3.10 compares the density of the gels at low temperature and their different sintering ranges. The acid-catalysed system produced a high density gel (ca. 1.8 g/ml) which sintered to 2.2 g/ml at 700 °C, whereas the alkaline system gave a much lower density (0.8 g/ml) and sintering did not occur until around 1000°C. Studies by nitrogen adsorption suggested a bimodal pore size distribution from the acidic gel with peaks around 2 nm and 7 nm. The smaller pores gradually collapsed with increasing temperature upto 600°C, but the larger pores required a temperature around 700°C leading to the abrupt shrinkage.

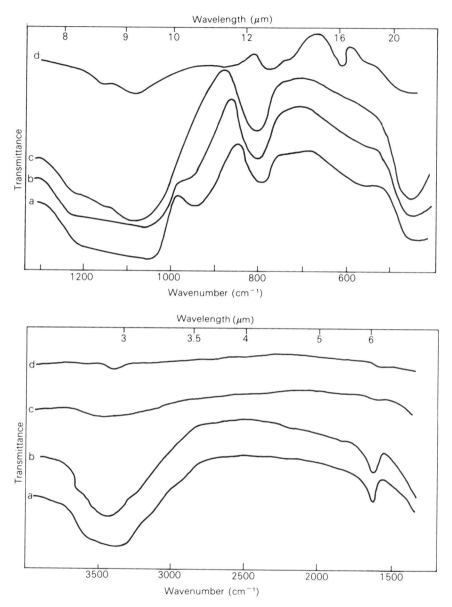

Fig. 3.9 Infrared transmission spectra for silica gel derived from TEOS (1% powdered gel in KBr). After drying specimens heated at 49°C followed by (a) no further treatment; (b), (c) and (d) heated (at 2°C/min) to 600°C, 800°C and 1500°C respectively. After [61].

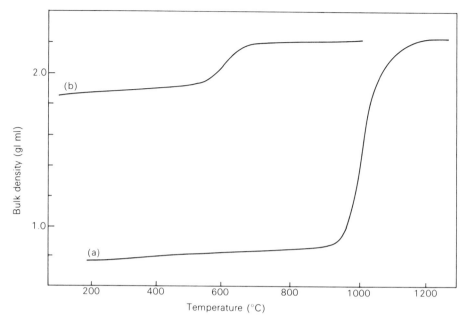

Fig. 3.10 Effect of increasing heat treatment temperature on bulk density of gels prepared by hydrolysing TEOS with (a) ammonia and (b) HCl. After [62].

In the alkaline gel, TEM indicated the presence of spherical particles with diameters around 20 nm. This contrasts with the acidic gel which was essentially featureless. The gel to glass conversion occurred at 1050°C and above following a sintering process controlled by viscous flow. The activation energy for viscous flow obtained using the Frenkel Equation was 170 kcal/mole which is similar to that of fused silica. However, the viscosities calculated from this analysis were about ten times lower than fused silica, probably caused by the relatively high water contents in the gel.

Similar densification results were reported by Yamane [63]. Dilatometry shows clearly the temperature at which the contraction occurs (Figure 3.11) and again the alkaline system requires higher temperatures to promote sintering.

For a more extensive collection of references on the gel to glass transition the reader should consult a review by James [64].

Early interpretations of the processes occurring during densification relied heavily on viscous sintering as the main mechanism. However, reviews by Scherer [25] and Brinker and Scherer [41] of more recent studies suggest a more complex mechanism involving at least four phenomena contributing to shrinkage as follows.

Capillary contraction.
Condensation-polymerization.

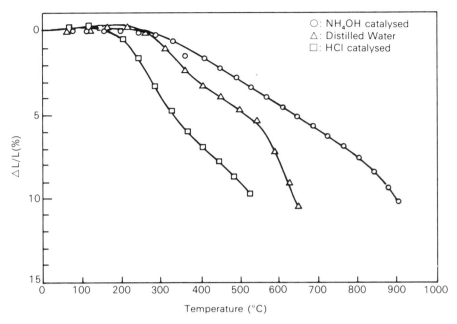

Fig. 3.11 Linear shrinkage of gels prepared under different catalysis conditions (after [63]).

Structural relaxation.
Viscous sintering.

The structural chemistry of the initial gel formation is also of great importance.

The starting point for densification will be a gel formed by the agglomeration of strong or weakly cross-linked clusters (depending on the catalysis route) which have been further consolidated during the drying process (Sections 3.3.2 and 3.3.4). The structures will still contain porosity on a molecular scale, since the skeletal density of the gel is lower than for a melted glass, and between the agglomerated clusters as larger more conventional pores. There is widespread experimental evidence for a reduced skeletal density compared to conventionally melted glasses in both silica gels and multicomponent systems [33, 41, 65, 66]. This reduced skeletal density would lead to a higher free energy compared with a colloidal gel and conventional glass as represented in Figure 3.12 [41]. This increase in free energy will influence the subsequent sintering behaviour of alkoxide derived gels.

Brinker et al. [65] proposed to divide heat treatment into three temperature ranges.

In Region I (< 200°C) there are substantial weight losses with little shrinkage.
In Region II (200 – 600°C) both weight loss and shrinkage are considerable.
In Region III (> 600°C) there is large shrinkage principally due to sintering, but little weight loss.

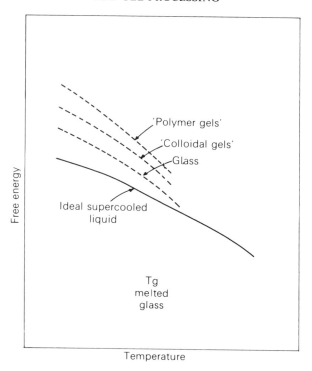

Fig. 3.12 Schematic free energy against temperature relationships for 'polymer' and 'colloidal' dried gels, glass and an ideal supercooled liquid of the same oxide composition (after [41]).

In Region I where weight loss occurs from desorption of physically-absorbed water and alcohol, the minor shrinkage is caused by capillary contraction. In Region II, removal of water from condensation reactions and oxidation of residual organics leads to the large weight loss. Shrinkage in this range results from skeletal densification rather than particle rearrangement or viscous sintering. Skeletal densification was proposed to occur by two mechanisms: by condensation reactions giving additional Si–O–Si linkages and by structural relaxation caused by atomic diffusion in the polymeric network, but without the release of water associated with the first mechanism. Finally in Region III, viscous sintering leads to rapid shrinkage and the collapse of the larger porosity, although a contribution from the mechanisms initiated in Region II would still be expected.

An example of some results from Brinker *et al.* [67] are shown in Figure 3.13 where weight losses and shrinkages are compared for an acid catalysed, a base catalysed and a colloidal silica gel. The behaviour corresponds to the predictions. The colloidal gel consists of fully densified particles with only surface hydroxyl groups. Hence there is little weight loss, and shrinkage occurs at high temperatures as coarse porosity is eliminated by viscous sintering. The acid

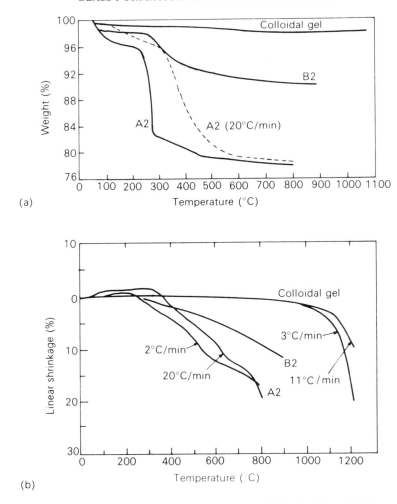

(a)

(b)

Fig. 3.13 (a) Percentage of original weight for colloidal gels B2 and A2 heated at 2°C/min. Dotted line is A2 heated at 20°C/min. (b) Linear thermal shrinkage at constant heating rates for gels as in (a). After [67].

catalysed gel shows more weight loss than the basic one, since it was prepared with a relatively low $H_2O/TEOS$ ratio of 5 and has been subject to re-esterification during drying. The basic gel was more densely crosslinked, therefore it expanded less at low temperatures than the acid gel and also contracted less in Region II, due to its higher skeletal density. The expansion at low temperatures is usually a sign of non-bridging oxygens and increases with their number. Faster heating for the acid gel delays the shrinkage in Region II, but leaves it with a higher hydroxyl content and a less relaxed structure: this accelerates the rapid shrinkage during sintering at around 800°C.

Sintering xerogels
The driving force to contraction of a porous gel is the reduction of its interfacial area, thus porosity is eliminated by viscous flow of the solid phase. The rate of viscous sintering is determined by relating the energy reduction on reducing surface area to the energy dissipated by viscous flow during contraction. In a gel the high surface area means that sintering can commence at relatively low temperature – near T_g where the viscosity is about 10^{12} Pa. s. It is generally agreed [25] that the contraction rate $\dot{\varepsilon}$ is given by:

$$\dot{\varepsilon} \propto \gamma_{sv} n^{1/3}/\eta$$

where:

γ_{sv} is the solid-vapour interfacial energy.
η is the viscosity.
n is the number of pores per unit volume.

This means that for a given relative density n increases as the pore size decreases, therefore the shrinkage is rapid for small pores or low viscosity. Additionally, the smaller the pores the higher the viscosity at which sintering will occur and if a distribution of sizes is present the small pores will sinter first.

Using a model which represents the microstructure as a network of solid cylinders enclosing a continuous pore space, a theoretical relationship can be derived such that a plot of reduced time $[K(t-t_0)]$ against actual sintering time t should give a straight line with gradient K.

Here

$$K = \gamma_{sv} \cdot (\rho_s/\rho_0)^{1/3}/\eta l_0$$

where

l_0 is the initial edge length of a unit cell of the structure.
ρ_s/ρ_0 is the initial relative density.
t_0 is the fictitious time when $\alpha/l = 0$.
α and l being the cylinder radius and length respectively.

Brinker and Scherer [68] showed that the model is not obeyed at the beginning of sintering of their multicomponent system (Figure 3.14). The viscosity increased by up to two orders of magnitude over about eight hours during the isothermal sintering experiment. This was attributed to condensation reactions eliminating non-bridging oxygens and to structural relaxations. At longer times agreement with the theory was good.

This model was applied to other studies by Brinker *et al.* [67], where isothermal sintering was conducted on an acid-catalysed, base-catalysed and colloidal silica gel. Under isothermal conditions the acidic gel showed a sharp increase in viscosity as the gel densified, the viscosity of the basic gel increased and then reached a plateau, whereas the colloidal gel had viscosities which were essentially constant throughout the densification.

This behaviour was explained in terms of the degree of cross-linking in the gel. The colloidal gel is constituted of dense silica particles which do not undergo chemical or structural alterations during sintering. By comparison, the acid gel experiences condensation reactions and structural relaxations during sintering, and the base gel is composed of clusters close to fully densified – a process which

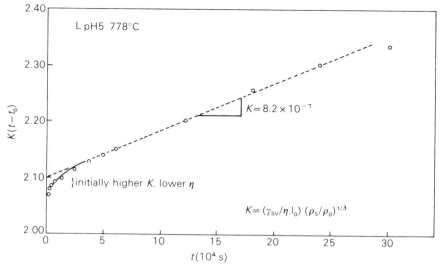

Fig. 3.14 Experimentally determined $K(t-t_0)$ versus time plots at 778°C (after [68]).

is then completed before sintering commences. It was also noted that the viscosity of the different gels at the beginning of the isothermal treatment was almost independent of the temperature. This was ascribed to the increased quantity of hydroxyl remaining in gel at lower temperatures. During heating the removal of OH increases the viscosity, but at the same time increasing the temperature serves to reduce the viscosity – the two effects almost balancing.

These effects can be used to produce a heat treatment which is optimum for sintering. Gallo and co-workers [66] have presented constant viscosity plots that can be used to select a temprature-time schedule keeping the viscosity constant during sintering. If an inappropriate isothermal treatment were to be employed the viscosity could rise to such an extent that the contraction would stop before the completion of densification. On the other hand if too great a heating rate were chosen, volatiles might be trapped leading to bloating and cracking.

Sintering aerogels
Silica aerogels have been found to sinter rapidly at around 1100 °C, but due to their high hydroxyl content foaming occurred at 1200 °C [69]. A chlorine treatment was therefore employed to remove the OH, thereby preventing bloating. Even though the viscosity rises because of the OH removal by the chlorine treatment, all the gels could be sintered at a viscosity of ca. 10^{10} Pa. s.

Gels in the systems $SiO_2-B_2O_3$, $SiO_2-P_2O_5$ and $SiO_2-B_2O_3-P_2O_5$ were sintered by Woignier and co-workers [70] in the same viscosity range as the silica aerogel. However, crystallization tended to occur in the phosphorus-containing systems.

3.4 APPLICATIONS

In this section it is intended to present a survey of some of the areas where sol–gel technology has found application or where applications have been proposed.

3.4.1 Variety of compositions

By way of introduction to the applications of sol–gel processing it would be useful to list the number of systems which have received attention (Table 3.3). This extensive, although not necessarily exhaustive, collection was assembled by James [64]. The majority of these compositions were used to produce bulk samples by slow drying and sintering, but some were processed by other methods; for example hot pressing or hypercritical drying. Not all these systems are necessarily amorphous.

Table 3.3
Systems prepared by sol–gel (after [64])

Single component systems
SiO_2
Al_2O_3

Binary systems

$SiO_2-Al_2O_3$	SiO_2-RmOn (R is Cr, Mn, Fe, Co, Ni, Cu or V)
$SiO_2-B_2O_3$	SiO_2-SrO
SiO_2-CaO	SiO_2-TiO_2
$SiO_2-Fe_2O_3$	$SiO_2-Y_2O_3$
SiO_2-GeO_2	SiO_2-ZrO_2
SiO_2-Li_2O	$B_2O_3-Li_2O$
SiO_2-Na_2O	GeO_2-PbO
SiO_2-PbO	$P_2O_5-Na_2O$
$SiO_2-P_2O_5$	

Ternary systems

$SiO_2-Al_2O_3-B_2O_3$	$SiO_2-B_2O_3-P_2O_5$
$SiO_2-Al_2O_3-CaO$	$SiO_2-B_2O_3-TiO_2$
$SiO_2-Al_2O_3-Li_2O$	$SiO_2-B_2O_3-ZnO$
$SiO_2-Al_2O_3-MgO$	$SiO_2-CaO-Na_2O$
$SiO_2-Al_2O_3-Na_2O$	$SiO_2-TiO_2-ZrO_2$
$SiO_2-B_2O_3-Na_2O$	$SiO_2-ZnO-K_2O$
$SiO_2-B_2O_3-PbO$	$SiO_2-ZrO_2-Na_2O$

Multicomponent systems
$SiO_2-Al_2O_3-B_2O_3-K_2O-Na_2O$
$SiO_2-Al_2O_3-TiO_2-Li_2O$
$SiO_2-Al_2O_3-ZrO_2-P_2O_5$
$SiO_2-Al_2O_3-Li_2O-Na_2O$
$SiO_2-Al_2O_3-(Tl_2O, Cs_2O, Ag_2O)$
$SiO_2-B_2O_3-Al_2O_3-Na_2O-BaO$
$SiO_2-B_2O_3-Na_2O-Al_2O_3$
$SiO_2-B_2O_3-Na_2O-V_2O_5$
$SiO_2-La_2O_3-Al_2O_3-ZrO_2$
$SiO_2-TiO_2-BaO-ZrO_2$

3.4.2 Monoliths

The difficulties involved in producing densified monoliths of any significant size means that the sol–gel process using alkoxides as starting materials is not an obvious choice. Some success was achieved by Susa et al. [71] who manufactured large rods of silica glass used for drawing low loss optical fibres. These had to be drawn from the rod at very high temperatures which may reduce the value of the technique.

Most success in the production of monolithics has been realized using dispersions of colloidal particles. Early studies of this kind were performed by Shoup [72] who prepared gels from mixtures of sodium or potassium silicate and colloidal silica sol (Ludox). The dispersion was gelled by controlling the pH with the addition of formamide. The alkali content was reduced to less than 0.02 wt % by leaching in weak acid. The gel was formed from spherical particles with the average pore diameter being in the range 10 to 350 nm, and with a narrow pore size distribution. On drying and sintering at temperatures above 1400°C large optically-clear fused silica monoliths were produced. Linear shrinkage was up to 50%.

An alternative technique to control shrinkage on drying has been described by Rabinovich et al. [73]. First, a gel was formed from a colloidal sol of fumed silica (Cab–O–Sil). After drying and firing to 900°C, fragments of the gel were redispersed in water and cast between concentric silica glass tubes acting as a mould. When gelled the gel tube was removed, dried in an ambient environment, before further dehydration using a helium and chlorine treatment and subsequent sintering to transparent glass at temperatures between 1260 and 1300°C. The success of the method relies on the formation of aggregates of larger size than the original fumed silica during the second dispersion. This leads to a gel of mixed aggregate sizes which undergoes minimal linear shrinkage (2–4%) on drying. The distribution of aggregate sizes produces a network of pores in the second gel (diameters 1–8 μm), which are larger than the primary interparticle pores (diameters 10–20 nm). This allows rapid removal of water from the body during drying without the development of sufficient stress to cause cracking.

Similarly, Scherer and Luong [74] dispersed a colloidal silica (Aerosil OX–50) in a non-polar solvent (chloroform) using n-decanol as a dispersion stabilizer. The alcohol adsorbed onto the hydroxylated silica surface, such that the long alkyl chain rendered the particles lyophilic, hence dispersible in the non-polar solvent. A dispersion of this kind is very sensitive to ionic concentration, therefore the addition of inert electrolytes such as ammonia or amines led to gelation by forming electrostatic bridges between the silica particles. Drying was achieved readily in about three days, primarily because of the relatively large pore sizes (60–100 nm). A treatment to 1000°C removed any residual organics and the sample was finally sintered in an He/Cl_2 atmosphere at 1450°C.

The production of pure silica glass plates of dimensions 20 cm × 20 cm × 1 cm has been reported by Seiko-Epson Co [75] for photo mask applications. Again the technique employs the addition of colloidal silica (e.g., aerosil OX–50) to an acid hydrolysed sol of TEOS, and adjusting the pH to the range 4 to 6 to promote gelation without subsequent cracking [76]. The incorporation of the colloidal silica resulted in larger pore sizes (ca. 30 nm when a 65% addition of colloidal silica was added) which led to drying being achieved in ten days. For the addition of silica greater than 80%, the gels cracked during drying and it was proposed that the optimum powder content would be around 50–60%. The monoliths were sintered at 1200°C *in vacuo* or in helium.

3.4.3 Fibres

Currently a great deal of interest is being shown in the sol–gel technique for the production of optical fibres. The advantage of this is that a variety of glass compositions can be formed as fibres with good homogeneity and tightly controlled composition. Fibres can be derived by two basic methods using sol–gel processing. They can either be drawn from the alkoxide sol under the correct conditions or they can be drawn at elevated temperatures from a preformed monolith.

Methods for producing fibres from sols have been extensively studied by Sakka and co-workers. A wide range of compositions have been covered including SiO_2 [38, 77]; SiO_2–TiO_2 (TiO_2 10–15 mol%); SiO_2–Al_2O_3 (Al_2O_3 10–30 mol%); SiO_2–ZrO_2 (ZrO_2 10–33 mol%) and SiO_2–Na_2O–ZrO (ZrO_2 25 mol%) [21].

In order to use this technique the sol must exhibit spinnability. That is to say if a rod is dipped into the reacting solution, it possesses the correct rheology for a thin strand to be pulled from the liquid as the rod is withdrawn. It has been shown that in order for the solution to be spinnable essentially linear polymers must be present. These are only formed when the hydrolysis of the alkoxide is performed under acidic conditions using a low H_2O/TEOS ratio, i.e., 1.5 to 4. The viscosity value is important, with the range 10–100 P being typical. The cross section of the fibre has been demonstrated to be affected by the H_2O/TEOS ratio and the viscosity at drawing. The former influences the cross sectional shape, with the composition which shrinks least on drying (i.e., H_2O/TEOS = 4 and high alkoxide concentration, e.g., TEOS/EtOH = 1) giving the most circular cross section. However the latter determines the overall diameter–the greater viscosity giving the greater diameter.

Alternatively, optical fibres can be manufactured by drawing a monolith at elevated temperatures. The monolith can either be produced by techniques described in Section 3.4.2 or by collapsing a glass tube whose inside has been coated by another glass composition by means of a sol–gel coating technique. Puyane and co-workers [78, 79] reported a technique of using silicon and germanium alkoxide sols to deposit layers on the inside of silica tubes. Each coating was gelled, dried and cured by a fast heat treatment (ca. 30 mins) prior to the deposition of the next layer. The preforms were sintered, collapsed and pulled to give fibres with losses of 22 dB/km at 0.85 μm wavelength. In another method, Sudo *et al.* [80] employed silicon and germanium alkoxides to produce sintered feed particles for fusing in an oxyhydrogen flame. The fused glass was fabricated by a 'rod in tube' method to a preform which produced a fibre having a loss of 7 dB/km at 0.8 μm.

While optical fibres have received considerable attention, there is increasing interest in the development of ceramic fibres by sol–gel for mechanical reinforcement and for insulation. Sowman [81] has described the preparation and properties of fibres in the SiO_2–Al_2O_3–B_2O_3 system. Also referred to were application areas such as ceramic fibre-metal composites [82], ceramic fibre-polymer composites [83], ceramic-ceramic composites [84] and high temperature fibres [85, 86, 87].

3.4.4 Coatings

The production of thin silica films is perhaps the area of sol–gel processing which has the most potential for commercial applications. The likely applications for sol–gel derived films have been summarized by Sakka [88] (Table 3.4).

Already there are a wide range of applications predominantly for optical purposes. For some time, glass panes with a coating to act as a neutral density filter which helps to control solar radiation entering buildings have been available from Schott Glaswerke [89]. This system employs a double sided single layer coating of titania containing dispersed colloidal palladium which serves to determine the desired absorption.

Rear view mirrors for motor vehicles are produced by Deutsche Spezialglas AG [90] where a multilayer coating of TiO_2–SiO_2–TiO_2 is employed to improve contrast and reduce glare. A summary of the areas of application of the fifty or so products already manufactured by Schott Glaswerke was given by Dislich [91].

Anti-reflective films on silicon solar cells have been described by Sandia National Laboratories and Westinghouse [92, 93]. Heat transport through window glass can be controlled by using thin indium-tin oxide (ITO) films [94]. These films are highly transparent to solar radiation, but highly reflective in the infrared. Therefore windows composed of ITO coated glass make good passive solar collectors. Good environmental stability has been reported.

Table 3.4
Possible applications of sol–gel coatings (after [88])

Substrate	Effects
Glass	Chemical durability
	Alkali resistance
	Mechanical strength
	Reflectivity control
	Colouring
	Electrical conduction
Metal	Corrosion resistance
	Oxidation resistance
	Insulation
Plastic	Surface protection
	Reflectivity control

Optoelectric PLTZ films (Pb, La, Ti, Zr oxides) have been proposed [95] and alkali barriers for electric bulbs can be produced by treating with a solution containing Sb $(OBu)_3$, B $(OBu)_3$, Ti$(OBu)_4$ and Al$(OsecBu)_3$ [96].

The sol–gel process also readily lends itself to the production of coloured films by the incorporation of transition metals into the coating compositions. Such studies were performed by Sakka and co-workers [20, 97] where the elements copper, iron and nickel were added to silica or silica–titania coating solutions. The transition metals were added as halides or formates.

It has generally been accepted that when using alkoxide sols, it is difficult to produce defect-free densified films with greater thicknesses than ca. 0.5 μm [3, 29, 97, 98]. Multiple dipping procedures may yield greater thicknesses. However, attempts have been made to overcome this problem by the incorporation of low melting glass powder into the oxide solution. Here the alkoxide serves as a binder and thus a thick homogeneous film is formed by subsequent heat treatment to above the softening point of the glass powder [99].

3.4.5 Ceramics

As indicated earlier, the sol–gel technique offers several processing advantages over many material production methods, namely high purity, good homogeneity and control over particle size distribution.

These are ideal properties for the production of ceramic precursor materials. The aqueous colloid route is frequently chosen for this purpose, or, in the case of alkoxides, they tend to be alkali catalysed. The sol–gel processing of colloidal dispersions of hydrous oxides and alkoxides to oxide ceramics has been reviewed by Segal [100], who outlined the manufacture of a range of ceramic materials including spherical particles of oxide nuclear fuels. Other ceramic materials formed from gels include transparent porous Al_2O_3 [101], barium titanate [102, 103], infrared transparent mullite [104] and pure homogeneous lead zirconate-titanate (PZT) ceramics [105]. A survey of electronic ceramics prepared by sol–gel processing was presented by Blum [106].

3.4.6 Refractories

By suitable choice of starting materials sol–gel techniques have found some applications in specialized refractory products. Emblem and co-workers [107, 108] have employed technical grade ethyl silicate [SiO_2 content = 40%, cf in TEOS SiO_2 content = 28%] to bond various refractory aggregates such as alumina, mullite, sillimanite and zircon. Normally base catalysis (e.g., piperidine) is employed to produce refractory shapes which gives gel times as low as several minutes to allow rapid mould stripping. After air drying and baking to remove volatiles, the product is fired to generate the ceramic bond. Shapes including burner qualls, well blocks and sliding gate plates have been made to close dimensional tolerances.

Prehydrolysed ethyl silicates and silica sols have also found applications in the foundry industry where they are employed to bond fine aggregates, e.g., zircon and molochite to form ceramic shells into which molten metal can be cast [109].

3.4.7 Other applications

There are still many areas where sol–gel technology may have application, but some which have been discussed recently include superionic conductors [110], hollow glass microspheres [111] and filters and membranes [112].

3.5 CONCLUSIONS

Sol–gel technology has become a vast area of study which has employed a wide range of characterization techniques in order to attempt to understand the processes and polymer structures involved. In such a review it would have been impossible to do justice to all the work that has been conducted in the last fifteen years or so, therefore the interested reader is referred to the proceedings of the four International Workshops [113–116] and several other conferences on the subject [117–119]. However, a survey has been presented of some of the more important literature which has given an informed insight into the physics and chemistry of the technique, and the wide range of applications (both current and proposed) have been highlighted. It is clear that sol–gel processing has huge potential to produce a range of specialized products in established compositions as well as novel compositions not readily accessible by conventional manufacturing techniques.

REFERENCES

[1] ASTM definitions appear in ASTM designation C162–166.
[2] D. R. Secrist and J. D. Mackenzie, *Modern Aspects of the Vitreous State* 3 Ed. J. D. Mackenzie, Butterworths, London 1964.
[3] H. Schroeder, *Physics of Thin Films*, **5** (1969) 87–141.
[4] W. Geffcken and E. Berger, *Deutsche Reichspatent* 736, 411 (1939). (Jenaer Glaswerk Schott u. Gen., Jena GDR).
[5] H. Schroeder, *Opt. Acta.*, **9** (1962) 249–54.
[6] H. Dislich, *Angen. Chem. Ed. Eng.*, **10** (6) (1971), 363–70.
[7] R. Aelion, A. Loebel and F. Eirich, *J. Am. Chem. Soc.*, **72** (1950) 5705–5712.
[8] H. Schmidt, H. Scholze and A. Kaiser, *J. Non-Cryst Solids*, **48** (1982) 65.
[9] B. E. Yoldas, *J. Mat. Sci.*, **14** (1979) 1843–9.
[10] R. K. Iler, *The Chemistry of Silica*, J. Wiley, New York (1979).
[11] E. M. Rabinovich, J. B. MacChesney, D. W. Johnson Jnr, J. R. Simpson, B. W. Meager, F. V. DiMarcello, D. L. Wood and E. A. Sigety, *J. Non-Cryst. Solids.* 63 (1984) 155–61.
[12] J. D. Mackenzie, *J. Non-Cryst. Solids.*, **48**, (1982) 1–10.
[13] D. C. Bradley, R. C. Mehrotta and D. P. Gaur, *Metal Alkoxides*, Academic Press (1978).
[14] M. Yamane, S. Inoue and K. Nakazawa, *J. Non-Cryst. Solids*, **48** (1982) 153–9.
[15] L. Levene and I. M. Thomas, U. S. Patent 3640 093 (1972).
[16] S. Govil and R. C. Mehrotta, *Syn. React. Inorg. Metal-Org. Chem.*, **5** (1975) 267.
[17] G. Philipp and H. Schmidt, *J. Non-Cryst. Solids*, **63** (1984) 282–292.
[18] C. J. Brinker, K. D. Keefer, D. W. Schaefer, R. A. Assink, B. D. Kay and C. S. Ashley, *J. Non-Cryst. Solids*, **63** (1984) 45–69.
[19] D. W. Schaefer and K. D. Keefer in *Better Ceramics Through Chemistry* Eds C. J. Brinker, D. E. Clark and D. R. Ulrich, North Holland New York (1984).

REFERENCES

[20] S. Sakka, K. Kamiya, K. Makita & Y. Yamamoto, *J. Non-Cryst. Solids*, **63** (1984) 223–35.

[21] S. Sakka and K. Kamiya, *J. Non-Cryst. Solids*, **48** (1982) 31–46.

[22] B. E. Yoldas, *J. Non-Cryst. Solids*, **51** (1982) 105–21.

[23] M. F. Bechtold, R. D. Vest and L. Plambeck Jnr, *J. Am. Chem. Soc.* **90** (1968) 4590–8.

[24] K. D. Keefer, in ref 19, 15–24.

[25] G. W. Scherer, *Yogyo-Kyokai-Shi*, **95** [1] (1987) 21–44.

[26] L. W. Kelts, N. J. Effinger and S. M. Melpolder, *J. Non-Cryst. Solids*, **83** (1986) 353–74.

[27] H. Schmidt, H. Scholze and A. Kaiser, *J. Non-Cryst. Solids*, **63** (1984) 1–11.

[28] V. Gottardi, M. Guglielmi, A. Bertoluzza, C. Fagnano and M. A. Morelli, *J. Non-Cryst. Solids*, **63** (1984) 71–80.

[29] I. Strawbridge, PhD Thesis, Sheffield (1984).

[30] B. E. Yoldas, *J. Non-Cryst. Solids*, **83** (1986) 375–90.

[31] C. J. Brinker and S. P. Mukherjee, *J. Mat Sci* **16** (1981) 1980–8.

[32] J. C. Debsikdar, *Advanced Ceramic Materials* **1** [1] (1986) 93–8.

[33] I. Strawbridge, A. F. Craievich and P. F. James, *J. Non-Cryst. Solids*, **72** (1985) 139–57.

[34] L. C. Klein & G. J. Garvey in ref 19, 33–39.

[35] E. M. Rabinovich, *J. Mat. Sci.*, **20** (1985) 4259–97.

[36] M. F. Bechtold, W. Mahler & R. A. Schunn, *J. Polym. Sci., Chem. Ed.* **18** (1980) 2823–55.

[37] C. B. Hurd, *Chem. Revs.*, **22** (1938) 403–22.

[38] S. Sakka, K. Kamiya and T. Kato, *Yogyo-Kyokai-Shi* **90** [9] (1982) 555–6.

[39] R. Jabra, J. Phalippou and J. Zarzycki, *Rev. Chem. Miner.* **16** (1979) 245–64.

[40] A. Ayral and J. Phalippou, *Advanced Ceramic Materials* **3** [6] (1988) 575–79.

[41] C. J. Brinker and G. W. Scherer, *J. Non-Cryst. Solids*, **70** (1985) 301–22.

[42] L. C. Klein and E. J. Garvey, *Ultrastructure Processing of Ceramics, Glasses and Composites* Eds. L. L. Hench and D. R. Ulrich, J. Wiley (1984).

[43] J. Zarzycki, in ref 42, 27–42.

[44] S. Wallace and L. L. Hench in ref 19, 47–52.

[45] L. L. Hench in *Science of Ceramic Chemical Processing* Eds L. L. Hench and D. R. Ulrich, J. Wiley (1986) 52–64.

[46] S. H. Wang and L. L. Hench in ref 19, 71–77.

[47] L. L. Hench and G. Orcel, *J. Non-Cryst. Solids*, **82** (1986) 1–10.

[48] I. Artaki, T. W. Zerda and J. Jonas, *Mater. Lett.*, **3** (1985) 493–6.

[49] I. Artaki, M. Bradley, T. W. Zedra, J. Jones, G. Orcel. & L. L. Hench in ref 45, 73–80.

[50] G. Orcel and L. L. Hench, *J. Non-Cryst. Solids*, **79** (1986) 177–94.

[51] J. D. Mackenzie, in ref 45, 113–22.

[52] J. Zarzycki, M. Prassas and J. Phalippou, *J. Mater Sci.*, **17** (1982) 3371–9.

[53] P. H. Tewari, A. J. Hunt and K. D. Lofftus, *Mater. Lett.* **3**, [9, 10] (1985) 363–7.

[54] G. W. Scherer, *J. Non-Cryst. Solids*, **87** (1986) 199–225.

[55] G. W. Scherer, *J. Non-Cryst. Solids*, **89** (1987) 217–238.

[56] G. W. Scherer, *J. Non-Cryst. Solids*, **91** (1987) 83–100.

[57] G. W. Scherer, *J. Non-Cryst. Solids*, **91** (1987) 101–21.

[58] G. W. Scherer, *J. Non-Cryst. Solids*, **92** (1987) 375–82.

[59] G. W. Scherer, *J. Non-Cryst. Solids*, **99** (1988) 122–144.

[60] G. W. Scherer, *J. Non-Cryst. Solids*, **99** (1988) 324–58.

[61] C. J. R. Gonzales-Oliver, P. F. James and H. Rawson, *J. Non-Cryst. Solids*, **48** (1982) 129–52.

83

[62] N. Nogami and Y. Moriya, *J. Non-Cryst. Solids*, **37** (1980) 191–201.
[63] M. Yamane in *Sol–gel Technology for Thin Films, Fibres, Preforms, Electronics and Speciality Shapes* Ed. L. C. Klein, Noyes Publications, New Jersey (1988) 200–25.
[64] P. F. James, *J. Non-Cryst. Solids*, **100** (1988) 93–114.
[65] C. J. Brinker, G. W. Scherer and E. P. Roth, *J. Non-Cryst. Solids*, **72** (1985) 345–68.
[66] T. A. Gallo, C. J. Brinker, L. C. Klein and G. W. Scherer in ref 19, 85–90.
[67] C. J. Brinker, W. D. Drotning and G. W. Scherer in ref 19, 25–32.
[68] C. J. Brinker and G. W. Scherer in ref 42, 43–59.
[69] J. Phalippou, T. Woignier and J. Zarzvcki, in ref 42, 70–87.
[70] T. Woignier, J. Phalippou and J. Zarzycki, *J. Non-Cryst. Solids*, **63** (1984) 117–30.
[71] K. Susa, I. Matsuyama, S. Satoh and T. Suganuma, *Electron. Lett* **18** [12] (1982) 499–500.
[72] R. D. Shoup, *Colloid and Interface Science* Ed. M. Kerker, **3** Academic Press, New York (1976) 63.
[73] E. M. Rabinovich, J. B. MacChesney, D. W. Johnson Jnr, J. R. Simpson, B. W. Meagher, F. V. DiMarcello, D. L. Wood & E. A. Sigety, *J. Non-Cryst. Solids*, **63** (1984) 163–72.
[74] G. W. Scherer and J. C. Luong, *J. Non-Cryst. Solids*, **63** (1984) 163–72.
[75] J. D. Mackenzie, *J. Non-Cryst. Solids*, **100** (1988) 162–8.
[76] M. Toki, S. M. Miyashita, T. Takeuchi, S. Kambe and A. Kochi, *J. Non-Cryst. Solids*, **100** (1988) 479–82.
[77] S. Sakka and H. Kozuka, *J. Non-Cryst. Solids*, **100** (1988) 142–53.
[78] R. Puyane, A. L. Harmer and C. J. R. Gonzalez-Oliver, *European Conf. on Optical Comm.*, C-24 (1982).
[79] A. L. Harmer, R. Puyane, C. J. R. Gonzalez-Oliver, *Proc. Int. Conf. Optical Fibre Comm.*, (1982) 40–44.
[80] S. Sudo, M. Nakahara and N. Inagaki, *4th Int. Conf. on Integrated Optics and Optical Fibre Comm. Tech. Digest* 27.A3-4 (1983) 14–15
[81] H. G. Sowman in ref 63, 162–83.
[82] Y. Abe, S. Horikiri, K. Fujimura and E. Ichiki in *Progress in Science and Engineering of Composites* Eds T. Hayashi, K. Kawata and S. Umekawa, ICCM-IV, Tokyo 1982.
[83] R. B. Pipes, D. D. Johnson and K. A. Karst in *Proceedings of the 31st Annual Reinforced Plastics Technical Conference*, Washington DC (1976).
[84] D. B. Leiser and H. E. Goldstein US Patent 4148962 (1979) assigned to NASA.
[85] L. J. Korb, C. A. Morant, R. M. Calland and C. S. Thatcher, *Am. Ceram. Soc. Bull.* **60** (11) (1981) 1188–93.
[86] L. R. White, D. L. O'Brien and G. A. Schmidt, in *Proc. 3rd Conf. on Fabric Filter Technology for Coal-fired Power Plants* (EPRI) Scottsdale Arizona.
[87] SAFFIL FIBRES, brochure published by ICI Mond Division, New Ventures Group, The Heath, Runcorn, Cheshire WA7 4QF.
[88] S. Sakka, 'Gel Method for Making Glass' in *Treatises on Material Science* **22** Academic Press (1982).
[89] Prospect 'Calorex' and Prospect 'IROX' (trademarks) Schott Glaswerke, Mainz, FRG (1983).
[90] Prospect *Glass Ohne Reflexe*, Deutsche Spezialglas KG, Grunenplan, FRG.
[91] H. Dislech, in ref 63, 50–79.
[92] B. E. Yoldas, European Patent Application 0.008215 (1978). Westinghouse Electric Corp., USA.
[93] C. J. Brinker and M. S. Hartington, *Solar Energy Mat.*, **5** (1981) 159–72.
[94] N. J. Arfsten, *J. Non-Cryst. Solids*, **63** (1984) 243–9.
[95] A. V. Novoselova, N. Ya. Turova, E. P. Turevskaya, M. J. Yanovskaya, V. A.

REFERENCES

Kozunova and N. J. Kozlova, Izv. Acad. Nauk. SSSR, *Neorganicheskie Materialy*, **15** (1979) 1055.

[96] E. E. Hammer and W. C. Martyny, DOS 24 02 244 (1974), General Electric Co., USA.

[97] Y. Yamamoto, K. Kamiya and S. Sakka, *Yogyo-Kyokai-Shi* **90** (1982) 328–33.

[98] I. Strawbridge and P. F. James, *J. Non-Cryst. Solids*, **86** (1986) 381–393.

[99] C. J. Brinker and T. R. Scott, US Patent 4476156 (1983) assigned to US Department of Energy, Washington DC.

[100] D. L. Segal, *J. Non-Cryst. Solids*, **63** (1984) 183–91.

[101] B. E. Yoldas, *Am. Ceram. Soc. Bull.* **54** [3] (1975) 286–8.

[102] K. S. Mazdiyasni, R. T. Dolloff & J. S. Smith, *J. Am. Ceram. Soc.*, **52** [10] (1969) 523–6.

[103] A. Mosset, I. Gautier-Luneau, J. Galey, P. Strehlow & H. Schmidt, *J. Non-Cryst. Solids*, **100** (1988) 339–44.

[104] B. Sonuparlak, *Adv. Ceram. Materials* **3** [3] (1988) 263–7.

[105] Z. Q. Zhuang, M. J. Haun, S-J. Jang and L. E. Cross, *Adv. Ceram. Materials.* **3** [5] (1988) 485–90.

[106] J. B. Blum, in ref 63, 296–302.

[107] H. G. Emblem, *Trans. J. Brit. Ceram. Soc.* **74** (1975) 223–8.

[108] H. G. Emblem, *Trans. J. Brit. Ceram. Soc.*, **79** No 4 Ivi-Ivii.

[109] *Monsanto Technical Bulletin* ref 53–40 (E) ME-2 (1986).

[110] J. P. Boilot and P. Colomban, in ref 63, 303–29.

[111] R. L. Downs, M. A. Ebner and W. J. Miller in ref 62, 330–81.

[112] L. C. Klein, in ref 62, 382–99.

[113] V. Gottardi, Ed. Proc. First Int. Workshop on Glasses and Glass Ceramics from Gels, Padua, *J. Non-Cryst. Solids*, **48** (1982).

[114] H. Scholze, Ed. Proc. Second Int. Workshop on Glasses and Glass Ceramics from Gels, Wurzburg, *J. Non-Cryst. Solids*, **63** (1, 2) (1984).

[115] J. Zarzycki, Ed. Proc. Third Int. Workshop on Glasses and Glass Ceramics from Gels, Montpellier, *J. Non-Cryst. Solids*, **82** (1–3) (1986).

[116] S. Sakka, Ed. Proc. Fourth Int. Workshop on Glasses and Glass Ceramics from Gels, Kyoto, *J. Non-Cryst. Solids*, **100** (1988).

[117] C. J. Brinker, D. E. Clark and D. R. Ulrich, eds. *Better Ceramics through Chemistry* Vol 32, Proc. Materials Research Soc. Symp. Albequerque, New Mexico (North Holland, New York 1984).

[118] L. L. Hench and D. R. Ulrich, Eds *Ultrastructure Processing of Ceramics, Glasses and Composites* (J. Wiley, New York 1984).

[119] L. L. Hench and D. R. Ulrich, Eds *Science of Ceramic Chemical Processing* (J. Wiley, New York 1986).

CHAPTER 4

Physical Properties

4.1 DENSITY OF GLASSES

Density is the most frequently measured property of glass. It can be me.
quickly and accurately (up to three places of decimals even in a mod
equipped laboratory), and is quite often used in the quality control of comm
glass manufacture. Systematic studies of the variation of the properties of si
glasses with composition have produced large amounts of data, which
hoped would provide information pertinent to the problem of the constitut
glass. This hope has not always been justified.

The fact that the specific volume of glass is approximately an additive pr
of its constituent oxides has long been recognized within a narrow compo
range. Calculation of density by means of additive factors often gave
approximation to the correct value. A detailed survey of the older literatur
the different factors used may be found in Morey's *Properties of Glass* [

4.1.1 Boric oxide and binary borates

The density of boric oxide glass has been measured by several workers a
known to be critically dependent on the amount of dissolved water tl
contains. A value of $1.812 \, \text{g/cm}^3$ has been reported by Napolitano [2] an
density changes non-linearly with temperature. The non-linear variation o
molar volume of B_2O_3 with temperature has been attributed to different t
models [3] such as:

(a) the sharing of two oxygens by adjacent borons,

$$2 - O - B \overset{O}{\underset{O}{\diagdown}} \; \rightleftharpoons \; - O - B \overset{O}{\underset{O}{\diamondsetminus}} B - O -$$

(b) the change of the coordination number of boron from three to four
    increasing temperature,
(c) the formation of $-B = O$ groups etc.,

The density of alkali–borate glasses at room temperature ($25°$ C) is show
Figure 4.1 (a) as a function of alkali content. The density of these glasses incre
with increasing concentration of metal oxides as does the thermal expansiv
The density of alkali–borate melts at $900°$ C, however, increases with increas
concentration of metal oxides up to a maximum and then decreases (Fig

86

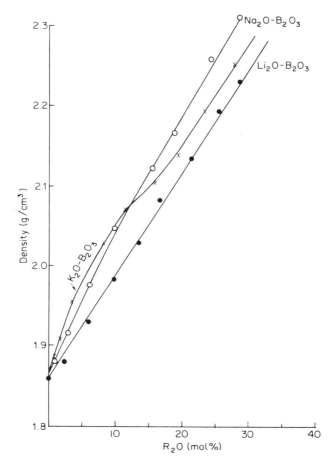

Fig. 4.1 (a) Variation of density of binary alkali–borate glasses at 25°C for different compositions.

4.1b). From the volume of liquid containing one gram-atom of boron, Shartsis *et al.* [10] have estimated the average B–B distance in these borate melts as a function of the composition. They have shown that the average B–B distance decreases at first, followed by an increase with increasing alkali oxide concentration (see Figure 4.2).

4.1.2 Alkali–silicate glasses

The partial molar volume of silica in binary alkali–silicates is nearly equal to that of pure fused silica and is independent of the cation over large ranges of composition. Figure 4.3 shows the density of binary alkali–silicates at 20°C after Peddle [1], and at 1400°C, after Bockris *et al.* [7]. The thermal expansion of

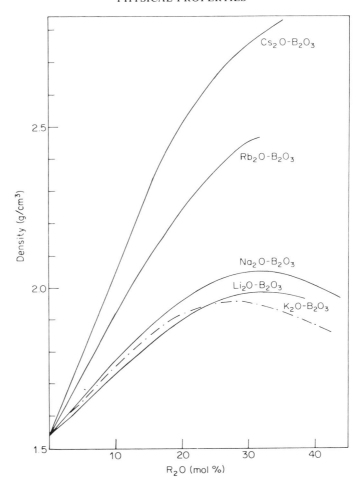

Fig. 4.1 (b) Variation of density of binary alkali–borate melts at 900°C for different compositions.

binary alkali–silicates increases suddenly when the amount of metal oxide exceeds 10–12 mol %. Similar behaviour has also been observed for alkaline-earth metals; typical results are shown in Figure 4.4. This sudden change in the slope of the graph indicates, according to some authors, a major modification of the structure of the liquid silicates at about this composition.

4.1.3 Alkali–germanate glasses

The density of binary alkali–germanate glasses at room temperature has been determined by Murthy et al. [13], and is shown in Figure 4.5 as a function of

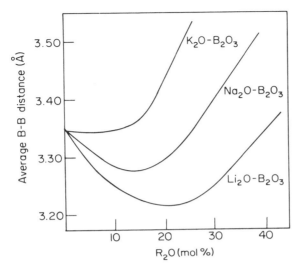

Fig. 4.2 Average B–B distance in binary alkali–borate melts at 1000°C.

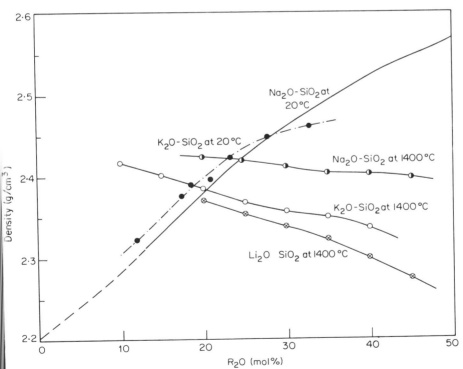

Fig. 4.3 Density of binary alkali–silicate glasses and melts.
(The points for 1400°C have been shifted upwards by 0.20 unit)

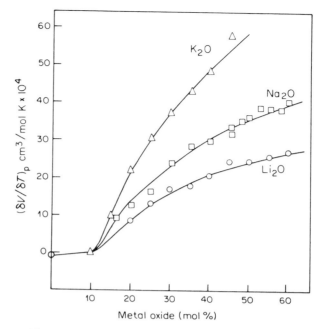

Fig. 4.4 Thermal expansion of alkali–silicates.

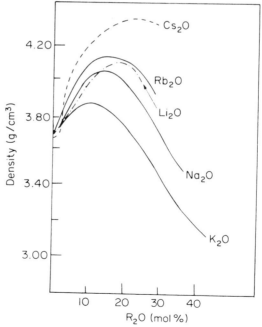

Fig. 4.5 Density of binary alkali–germanate glasses at room temperature.

composition. The maxima in these curves are related to the change of coordination number of germanium from six to four.

4.1.4 Alkali–phosphate glasses

Binary alkali–phosphate glasses are extremely hygroscopic and thus the accurate determination of density for these glasses is very difficult. The approximate densities of lithium- sodium- and potassium-phosphate glasses are shown in Figure 4.6. Less than 1 wt % water was found to be present in these glasses, by chemical estimation.

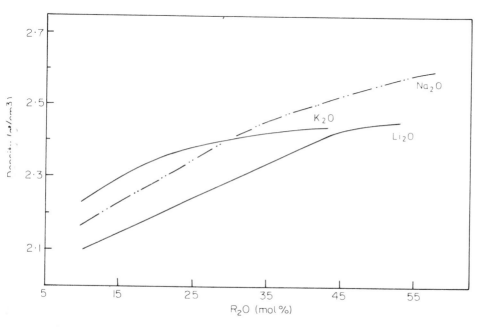

Fig. 4.6 Density of binary alkali–phosphate glasses at room temperature.

4.2 PARTIAL MOLAR VOLUME OF CONSTITUENT OXIDES IN GLASSES AND MELTS

Extensive thermodynamic properties, such as molar volume, enthalpy, or free energy of pure substances, are functions of temperature and pressure; corresponding properties of their components in solutions also depend on concentration. Thus the total volume of n mols of a pure substance of molar volume V is simply nV at a fixed temperature and pressure, while that of a solution containing n_1 mols of the first component, n_2 mols of the second, and so on, in general is *not* $n_1 V_1 + n_2 V_2 + \ldots$. Two functions are useful in describing the behaviour of such a solution: the *apparent molar property* of the solute, and the *partial molar properties* of solute and solvent.

91

If a solution of volume V contains n_1 mols of solvent and n_2 mols of solute, and if v_1 is the molar volume of the pure solvent, then the apparent molar volume of the solute, Q, is defined as:

$$Q = \frac{V - n_1 v_1}{n_2} \quad (T \text{ and } P \text{ constant}) \tag{4.1}$$

Q may be calculated from the densities of solvent and solution. In the same solution the partial molar volumes of the components are defined as:

$$\bar{v}_1 = \left(\frac{\partial V}{\partial n_1}\right)_{n_2, T, P} \text{ and } \bar{v}_2 = \left(\frac{\partial V}{\partial n_2}\right)_{n_1, T, P}$$

Each equals the change in total volume of solution, at constant temperature and pressure, on addition of one mol of the specified component to a volume of solution so large that the concentration is not appreciably changed. If the solution is made by simultaneous addition of n_1 mols of solvent and n_2 mols of solute – the concentration being kept fixed – the total volume will be:

$$V = n_1 \bar{v}_1 + n_2 \bar{v}_2$$

The partial molar volumes of components in solution are additive.

The relationship between apparent and partial molar volume can be deduced as follows:

From equation (4.1): $\quad V = n_1 v_1 + n_2 v_2$

Differentiating with respect to n_2 at constant T, P, and n_1 yields

$$\bar{v}_2 = \left(\frac{\partial V}{\partial n_2}\right)_{n_1, T, P} = Q + n_2 \left(\frac{\partial Q}{\partial n_2}\right)_{n_1, T, P}$$

Thus $\bar{v}_2$ and Q approach identity as n_2 approaches zero; that is, as the solution approaches infinite dilution.

The apparent molar volume has little thermodynamic utility, but it is often used as a step towards the determination of the partial molar volume. There are different methods for determining partial molar volume in solution; the most commonly used is the graphical method of intercept described by Lewis and Randall [4]: first calculate the apparent molar volume of the solution as a whole; plot this apparent molar volume against the mol fraction as shown in Figure 4.7. Draw a tangent to the curve at any point; the intercept of this tangent upon the ordinate of $N_1 = 1$ is equal to $\bar{v}_1$ and the intercept corresponding to $N_2 = 1$ equals $\bar{v}_2$ at this composition of mixture.

In practice, however, instead of the molar volume of the solution, the deviation from ideality in molar volume, $\Delta \bar{v}$, is plotted as a function of composition (in mol %); and partial molar volumes are calculated from the equations:

$$\bar{v}_1 = v_1 + \Delta \bar{v}_1 \text{ and } \bar{v}_2 = v_2 + \Delta \bar{v}_2$$

Numerous density studies on glasses and melts of known chemical composition have been made by various workers, but only in a limited number of cases have attempts been made to isolate the partial molar volume factor which is thermodynamically meaningful. However, it should be pointed out that satisfac-

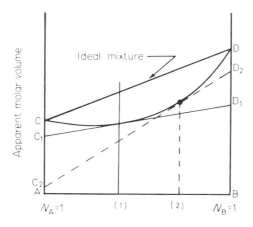

Fig. 4.7 Estimation of partial molar volume from apparent molar volume of a mixture. The apparent molar volume of A is AC and that of B is BD. For composition (1) the partial molar volumes of A and B are AC_1 and BD_1 respectively. Compare AC_1 and BD_1 with AC_2 and BD_2, the partial molar volumes for another composition (2).

tory separation of this parameter is possible only in the case of binary solutions; this can be extended to ternary solutions where the third component is an 'inert solute' or makes ideal solutions with the other two components separately. Unfortunately no such 'ideal' oxide is used in glass manufacture. Further complications arise in molar volume calculations in 'solid glasses', for the density of a glass depends on its thermal history.

4.2.1 Binary alkali–silicates

Partial molar volume (in cm^3) of component oxides in $Na_2O–SiO_2$ and $K_2O–SiO_2$ glasses were first calculated by Callow [5] using the density data for these glasses at $25°$ C reported by Morey and Merwin [6]. His calculated results are shown in Figure 4.8. The partial molar volumes of alkali oxides and silica in binary alkali–silicate melts were also determined by Bockris et al. [7] at $1400°$ C; these follow almost the same trend to those shown in Figure 4.8. The partial molar volume of silica, $\bar{v}_{SiO_2}$, remains practically unaltered up to 10–12 mol% R_2O, and thereafter it decreases slightly with increasing alkali oxide. The partial molar volume of alkali oxide, $\bar{v}_{R_2O}$, changes sharply with composition. Up to 12 mol% R_2O, $\bar{v}_{R_2O}$ decreases by as much as 5–6 cm^3 (which corresponds to about 15% volume contraction), and thereafter it increases rapidly with alkali oxide content. According to Bockris, the reduction of molar volume in alkali–silicates is to be anticipated because Si–O bonds have a strong direction-ality producing the infinite network structure of pure SiO_2 containing a large free space, whereas alkali oxides are ionic. Upon mixing, cations can be accommodated in the free space in the silica lattice. The sudden change in partial

93

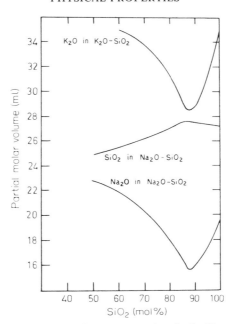

Fig. 4.8 Partial molar volumes of components in alkali–silicate glasses at 25°C.

molar volume at 10–12 mol% R_2O corresponds to the termination of the 'enclosed' situation of R^+ ions with respect to Si–O–Si cages. Discrete anionic groupings are formed at this composition range. These anionic groupings, according to Bockris, must contain Si–O–Si bond angles near in magnitude to those of vitreous SiO_2. The melt is a damaged random network up to 12 mol% R_2O containing discrete anions of formula $[Si_nO_{2n+3}]^{6-}$ up to 50 mol% and chains of formula $[Si_nO_{3n+1}]^{(2n+2)-}$ up to 66 mol% R_2O. However, for the composition range 12–33 mol% R_2O, an alternative model, in which discrete SiO_2 islets having diameters of about 10–40 Å are separated by thin films of melt of approximate composition: R_2O, $2SiO_2$, is also consistent with the partial molar volume results.

4.2.2 Binary alkali–borates

The partial molar volumes of Na_2O in sodium–borate melts at 1100°C and 1300°C were determined by Riebling [8] from density data; the results at 1300°C are shown in Figure 4.9. The partial molar volume of B_2O_3, $\bar{v}_{B_2O_3}$ at 1300°C decreased from 46.6 to 32.3 cm^3 at the 45 mol% Na_2O. It has been suggested that this volume decrease of 14.3 cm^3 mol^{-1} is associated with a coordination change of boron from 3 to 4. It is known that the $\alpha \rightarrow \beta$ transformation of B_2O_3, which corresponds to a coordination change of boron from 3.2 to 3.4, causes a partial molar volume change of 4.7 cm^3. Therefore the complete coordination change from planar BO_3 triangles to BO_4 tetrahedra would involve about 24 cm^3

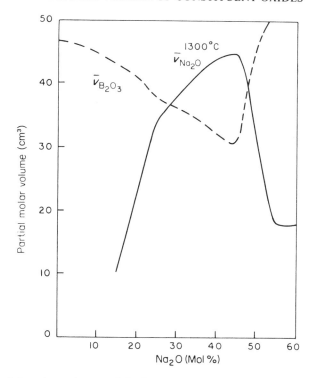

Fig. 4.9 Partial molar volumes of B_2O_3 and Na_2O in sodium–borate melts at 1300°C.

volume decrease; the partial molar volume decrease of 14.3 cm³ suggests that about 50% of total boron has been changed to 4-coordination in this composition range. From n.m.r. investigations on $Na_2O-B_2O_3$ glasses, Bray et al. [9] came to the same conclusion, that about 50% of the boron was in tetrahedral coordination. However, it should be pointed out that n.m.r investigations were made on 'solid' glasses at room temperature and the effect of temperature on the change of boron coordination is not yet definitely known.

The partial molar volume of B_2O_3 in $Na_2O-B_2O_3$ glasses at room temperature when calculated from the density data of Shartsis et al. [10] shows a similar trend to that of corresponding melts, the only difference being in their numerical values. As pointed out earlier, due to the uncertain thermal history and water content of these glasses, excessive weight should not be placed on the quantitative aspect of these results at room temperature.

4.2.3 Binary alkali–germanates

The partial molar volumes of GeO_2 and R_2O (R refers to Li, Na, K and Rb) at 1300°C in binary alkali–germanate melts have been calculated by Riebling [11] and the results are shown in Figures 4.10 and 4.11. One mol of molten GeO_2

95

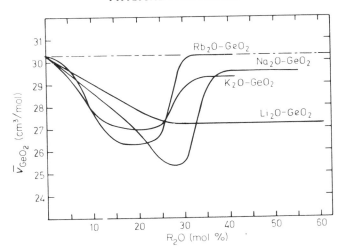

Fig. 4.10 Partial molar volume of GeO_2 in alkali–germanate melts at $1300°C$.

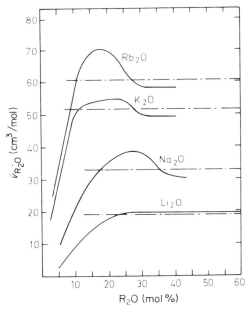

Fig. 4.11 Partial molar volumes of alkali oxides in alkali–germanate melts at $1300°C$.

occupies about 11 per cent more volume than one mol of SiO_2 at $1300°C$. This larger volume for GeO_2 could arise partially from the larger Ge–O distance and larger Ge–O–Ge angle compared to that of SiO_2. The addition of R_2O to molten GeO_2 produces significant contractions in the effective volume ($\bar{v}_{GeO_2}$) occupied

96

one mol of GeO_2 in these mixtures. This shrinkage amounts to as much as cm^3, out of 30.3 cm^3 molar volume, for GeO_2 in a melt containing 28 mol% Na_2O. These volume contractions are of the order of magnitude expected if the oxygens in these molten mixtures adopt an octahedral instead of a tetrahedral configuration around significant numbers of germanium atoms. The molar volume of the rutile form of GeO_2 (coordination number 6) is 5.47 cm^3 less than the molar volume of the quartz form in crystals at room temperature. A radius ratio of about 0.41 allows either a tetrahedral or octahedral configuration for GeO_2. The octahedral configuration occupies a smaller volume because it involves a more efficient packing of oxygens around each germanium atom. Ivanov and Yevstropov[12] have reported a refractive index maximum for germanate glasses containing about 15mol% Na_2O. They suggested that the addition of Na_2O to GeO_2 produced a change of coordination number from 4 to 6 for germanium. The refractive index maximum has also been reported at about mol% Na_2O by Murthy and Ip[13].

According to Riebling the gradual decrease of $\bar{v}_{GeO_2}$ with increasing R_2O content suggests that equilibria exist between octahedrally and tetrahedrally coordinated germanium atoms at each composition and temperature. It appears from Figure 4.10 that octahedrally coordinated germanium is present in the largest amount for melts containing between 10 and 25 mol% Rb_2O and K_2O, 15 and 32 mol% Na_2O, and more than 20 mol% Li_2O. Germanium seems to retain significant amounts of octahedral configuration for all melts between 20 and 65 mol% Li_2O.

Additional support for these postulates can be found in the positive inflections of $\bar{v}_{R_2O}$ of the three larger cations (Figure 4.11). These positive deviations show the same sequence of composition regions observed for the $\bar{v}_{GeO_2}$ contractions in Figure 4.10. The more efficient packing of oxygens around each germanium is apparently accompanied by a slightly more voluminous packing of oxygens around each cation. The coordination increase for germanium is not accompanied by a measurable decrease of $\bar{v}_{Li_2O}$. This could be a result of the rather high cation–oxygen attraction exhibited by Li^+.

The general trends of $\bar{v}_{GeO_2}$ and $\bar{v}_{R_2O}$ for germanate melts containing more than mol% R_2O are similar to those reported by Bockris et al.[7] for the corresponding silicate melts. This suggests that discrete anions are also capable of formation in germanate melts that are rich in alkali oxides. The formation of discrete germanate ring anions may be due to large oxygen–oxygen repulsions arising from (a) their octahedral configuration around the germanium atoms, and (b) the packing of these octahedra in relatively small rings which could make the octahedral arrangement unstable with respect to the tetrahedral configuration for germanium in these postulated ring anions, for melts between 30–50 mol% alkali oxide.

4.4 Ternary soda–alumino–silicate melts

The density of different Na_2O–Al_2O_3–SiO_2 melts has been determined by Riebling[14] in the temperature range 1500°–1700°C. The molar volume at 1500°C of different compositions having constant SiO_2 content is shown in Figure 4.12. The molar volume increases with increasing Al_2O_3 content.

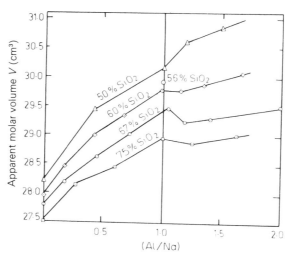

Fig. 4.12 Apparent molar volumes of Na_2O–Al_2O_3–SiO_2 melts having constant SiO_2 and different Al/Na ratios at 1500°C.

However, this volume increase tends to become less pronounced for higher SiO_2 content melts with Al/Na > 1.0. If it is assumed that silica retains a tetrahedral configuration ($\bar{v}_{SiO_2} = 27.24$ cm³ at 1500°C) in all the ternary melts, then the number of unknowns in the following equation is two ($\bar{v}_{Na_2O}$ and $\bar{v}_{Al_2O_3}$):

$$\text{Molar volume} = X_{Na_2O} \cdot \bar{v}_{Na_2O} + X_{Al_2O_3} \cdot \bar{v}_{Al_2O_3} + X_{SiO_2} \cdot \bar{v}_{SiO_2}$$

where X is the mole fraction.

At 1500°C, $\bar{v}_{Al_2O_3}$ in tetrahedral form has been estimated to be 31.44 cm³ and $\bar{v}_{Al_2O_3}$ in octahedral form is 26.74 cm³ at that temperature. A remaining unknown is $\bar{v}_{Na_2O}$ as a function of composition. But, as will be discussed in Chapter 8, Na_2O may react with Al_2O_3 as well as with SiO_2 in these melts; in fact simultaneous chemical equilibria are involved and, unless the relative magnitude of the equilibrium constants are known, there is no feasible method to isolate $\bar{v}_{Na_2O}$ and $\bar{v}_{Al_2O_3}$ in these melts.

4.3 REFRACTIVE INDEX OF GLASSES

Extensive measurements of the refractive index (by which is usually meant the refractive index for sodium light = 589.3 nm) have been made on glasses of simple composition, as part of the studies to find the relation between composition and refractivity. Attempts were made, mainly by Russian workers[15] to calculate the molar refraction of simple glasses, and to deduce from that the type of chemical bonding in glass.

From the vast amount of data that has been accumulated, mainly in organic chemistry, it has been found that the refraction increases with increasing molecular weight in an homologous series, and that the difference between the values for cessive homologous compounds diminishes on rising through such a series. hough the conclusion was that a relationship exists between refractivity and nposition, it was not possible to establish its form as long as one was limited to specific refraction. In 1856 Berthelot multiplied the specific refraction, $\bar{r}$, by molecular weight to obtain a characteristic which he designated as the *molar action*, R_m:

$$R_m = M \cdot \frac{n^2 - 1}{n^2 + 2} \cdot \frac{1}{d} \qquad (4.2)^*$$

:re d is the density (gcm^{-3}) and n is the refractive index.
Berthelot's molar refraction* proved to be directly related to chemical nposition. Thus the value of R_m for homologous compounds is proportional he number of methylene groups and increases by approximately 18 units for a. It has also been found that changes in temperature or state of aggregation of ter (solid, liquid or gas) do not alter the molar refraction as long as there is no mical transformation. Polymorphic transformations are always accompanied some sort of structural change. These changes markedly alter the packing of atoms and the volumes which these atoms occupy. Molar refractions of a aber of oxides are given in Table 4.1. In this table, the volumes of the cations practically constant, so that the increase in the polarizability of the oxides is portional to the rise in volume of the oxygen.

Table 4.1
Molar refraction of some polymorphic compounds

Formula	Structure	Molar refraction
SiO_2	Quartz	7.18
	Cristobalite	7.40
	Tridimite	7.50
TiO_2	Rutile	12.82
	Brookite	12.95
	Anatase	13.17
GeO_2	Tetragonal	8.5
	Hexagonal	9.6
Al_2O_3	α-form	10.7
	γ-form	11.3
$Al_2O_3 \cdot H_2O$	Diaspore	14.00
	Boehmite	14.50
$Al_2O_3 \cdot SiO_2$	Kyanite	17.61
	Sillimanite	18.55
$LiAlSi_2O_6$	α-form	21.80
	β-form	23.38

: molar refraction actually derived by Berthelot was

$$R = M \cdot \frac{n^2 - 1}{d}$$

the formula of Newton and Laplace. Equation (4.2) has been used by Fajan with the ız–Lorentz equation.

It has been found that the molar refractions of most complex silicates are not exactly equal to the sum of the constituent oxide refractions. The deviation from additivity is at once an indication of chemical union between the constituent oxides and a measure of the extent of this union. The deviation of the true refraction from that calculated will be the greater, the greater the difference in the polarizing action of the cations on the constituent oxides. Table 4.2 gives a comparison of the experimentally determined refractions for a number of minerals with refractions calculated for assumed 'silicate of aluminium' (aluminium coordination number six) and 'alumino silicate' (aluminium co-ordination number four) structures. It is interesting to note that the actual refractions of sillimanite, celsian etc. fall between the values calculated for the assumed structures. In fact sillimanite is known to have exactly this kind of structure where one half of the aluminium ions have a coordination number of four, and the other half have a coordination number of six.

Table 4.2
Determination of the coordination number of aluminium with molar refraction data

Compound	Molar refraction (calculated)		Molar refraction (experimental)	ΔR	
	(4)*	(6)*		(4)*	(6)*
Albite $NaAlSi_3O_8$	31.22	30.32	31.04	0.18	0.72
Orthoclase $KAlSi_3O_8$	33.23	32.33	33.08	0.15	0.75
Anorthite $CaAl_2Si_2O_8$	33.61	31.81	33.66	0.05	1.85
Celsian $BaAl_2Si_2O_8$	38.26	36.46	37.36	0.90	0.90
Sillimanite Al_2SiO_5	19.48	17.68	18.85	0.93	0.87
Kyanite	19.48	17.68	17.61	1.87	0.07
Spodumene $LiAlSi_2O_6$	23.26	22.36	21.95	1.31	0.41
Spodumene (high temperature)	23.26	22.36	23.38	0.12	1.02
Eucryptite $LiAlSiO_4$	16.08	15.08	14.92	1.16	0.26
Eucryptite (high temperature)	16.08	15.08	16.40	0.32	1.22
Pyrophyllite $Al_2(OH)_2Si_4O_{10}$	43.58	41.78	42.22	1.36	0.44

* Coordination number.

The molar refraction of a series of alkali and alkaline-earth silicates is shown in Table 4.3. It may be noticed that as the polarizing capacity of the cations like M^{2+}

100

Table 4.3
Molar refraction of oxygen in some silicate ｇ

Composition	Density (g/cm^3)	n_D	
$2\ Na_2O \cdot 4SiO_2$	2.472	1.5070	˒
$2\ K_2O \cdot 4SiO_2$	2.453	1.5055	5.
$Na_2O \cdot Li_2O \cdot 4SiO_2$	2.483	1.5230	40
$K_2O \cdot Li_2O \cdot 4SiO_2$	2.426	1.5145	45..
$Na_2O \cdot K_2O \cdot 4SiO_2$	2.469	1.5085	47.8
$Li_2O \cdot MgO \cdot 4SiO_2$	2.421	1.5310	39.62
$Li_2O \cdot CaO \cdot 4SiO_2$	2.492	1.5520	41.80
$Li_2O \cdot SrO \cdot 4SiO_2$	2.815	1.5655	43.32
$Li_2O \cdot BaO \cdot 4SiO_2$	3.150	1.5990	45.87
$Na_2O \cdot MgO \cdot 4SiO_2$	2.425	1.4995	41.45
$Na_2O \cdot CaO \cdot 4SiO_2$	2.601	1.5340	42.79
$Na_2O \cdot SrO \cdot 4SiO_2$	2.832	1.5400	44.99
$Na_2O \cdot BaO \cdot 4SiO_2$	3.035	1.5735	49.42
$K_2O \cdot MgO \cdot 4SiO_2$	2.401	1.4990	45.73
$K_2O \cdot CaO \cdot 4SiO_2$	2.466	1.5385	49.50
$K_2O \cdot SrO \cdot 4SiO_2$	2.740	1.5355	49.81
$K_2O \cdot BaO \cdot 4SiO_2$	3.011	1.5528	51.71

(After Malkin et al. [81]).

R^+ increases, the molar refractivity decreases. This will be further discussed in apter 8 in relation to acid–base concepts in glass.

.1 Boric oxide and alkali–borate glasses

hough the refractive index of B_2O_3 glass has been studied by several workers, re has not yet been a study of the complete removal of water from these glasses. e best value of the refractive index of B_2O_3 is probably 1.458 reported by rey and Merwin.

Refractive indices of $Na_2O–B_2O_3$ glasses, a typical example of the ali–borate series, are shown in Figure 4.13. The refractive index increases with reasing Na_2O. Although the plots of $n_D–R_2O$ give a non-linear curve, no tinct curvature could be seen around the point commonly known as 'boric de anomaly'.

.2 Alkali–silicate glasses

fractive indices of $R_2O–SiO_2$ glasses at 25°C are shown in Figure 4.14. Similar borate glasses, refractive index increases with increasing R_2O, and for a istant molar concentration of alkali oxide, $Li_2O–SiO_2$ has the highest, and ₂$O–SiO_2$ the lowest, refractive index.

101

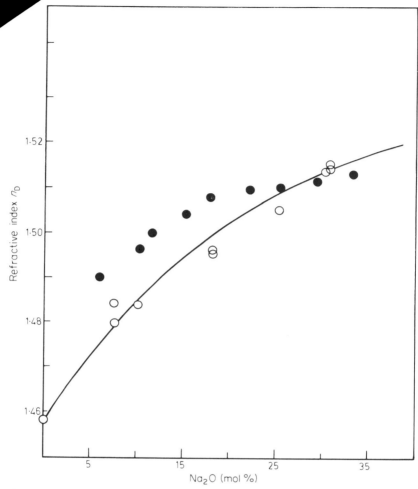

Fig. 4.13 Refractive indices of some binary sodium–borate glasses. (Open circle: Jenekel, *Z. Electrochem*, (1935), **41**, 211 and solid circle: Wulff and Majumdar, *Z. Physik. Chem.*, (1936), **31B**, 319).

4.3.3 Alkali–germanate glasses

Refractive indices of R_2O–GeO_2 glasses are shown in Figure 4.15 after Murthy *et al*. The refractive indices first increase with R_2O up to 15–20 mol % and then decrease with further increase of R_2O. This has been explained by a coordination change of germanium from four to six.

4.4   THERMAL EXPANSION OF GLASSES

Thermal expansion originates from the anharmonic vibrations of atoms in a solid. The repulsive and attractive forces in a simple crystalline solid are

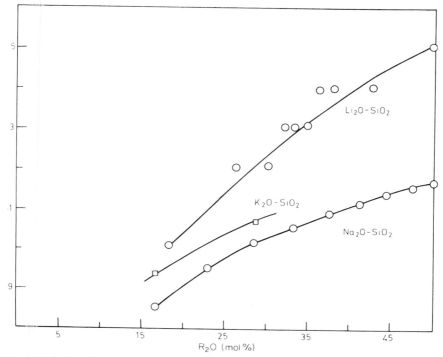

Fig. 4.14 Refractive indices of binary alkali–silicate glasses at room temperature.

resented in Figure 4.16. The point at which the solid curve intersects the rizontal axis corresponds to the normal interatomic distance, designated a. minimum in the curve at distance a' indicates the maximum distance to which atoms can be separated in tension. If this distance is exceeded, the cohesive ce decreases rapidly and the material breaks.

It is clear from the figure (see insert in Figure 4.16) that a greater force is uired to move two atoms a given distance, x, toward each other than to rease the distance between them by the same amount. The atoms of a stance at any temperature above 0 K possess some thermal motion; that is y vibrate about their equilibrium position. If the temperature is raised, the plitude of this vibration increases, but because the repulsive and attractive ces are not symmetrical (anharmonic vibration), the atoms are drawn back h greater force from the position $(a - x)$ than they are from $(a + x)$. In nsequence, the centre of vibration is displaced toward $+x$; that is, the eratomic distance increases and the substance expands.

In addition to this expansion, an expansion of the lattice resulting from a ange of bond angle with temperature without significant change in bond length lso possible. For example Megaw [16] has pointed out that, for diamond to nain cubic on heating, the tetrahedral angle must also remain. On the other nd, in silicon carbide the angle varies without violating the symmetry, so that

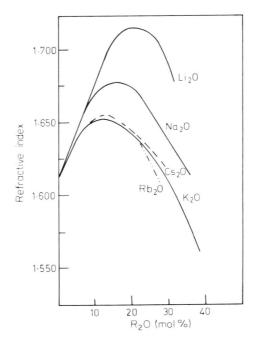

Fig. 4.15 Refractive indices of binary alkali–germanate glasses at 20°C.

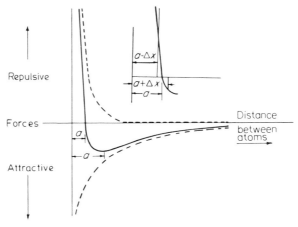

Fig. 4.16 Diagram illustrating schematically the action of interatomic forces which give rise to thermal expansion.

expansion can be considerable without any great increase of interatomic
ance.

he thermodynamic coefficient of thermal expansion, β, is defined as:

$$\beta = \frac{1}{V}\left(\frac{\partial V}{\partial T}\right)_P$$

re the volume, V, is a function of temperature, T, at constant pressure, P. The
fficient of linear thermal expansion, α, is defined as:

$$\alpha = \frac{1}{L}\left(\frac{\partial L}{\partial T}\right)_P$$

ere the length, L, is a function of temperature, at constant pressure. For
ropic materials like glass:

$$\beta = 3\alpha$$

The thermodynamic coefficient of thermal expansion can be calculated from an
iation that represents the expansion of the material, or by graphically
ermining the slope of the expansion data at a particular temperature and
iding by the volume at that temperature. The difficulty with the first procedure
that while equations (usually polynomials) can be obtained in limited
nperature ranges that appear to fit the data, a closed equation will not in fact
ictly represent the thermal expansion. This is especially true in the lower
nperature range, as can be seen in Figure 4.17. Equations derived from
une sen's theory of the solid state are probably the best to use for representing
ermal expansion.

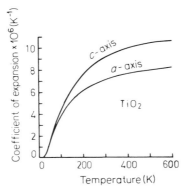

Fig. 4.17 Expansivity of single-crystal rutile as a function of
temperature (c-axis and a-axis refer to the TiO_2
crystal).

From thermodynamics we have the following set of definitions and
elationships:

$$k_T = -\frac{1}{V}\left(\frac{\partial V}{\partial P}\right)_T, \quad \beta = \frac{1}{V}\left(\frac{\partial V}{\partial T}\right)_P$$

105

and

$$\frac{\beta}{k_T} = \left(\frac{\partial P}{\partial T}\right)_V = \left(\frac{\partial S}{\partial V}\right)_T = -\frac{\partial^2 F}{\partial V \cdot \partial T} \tag{4.3}$$

where k_T is the isothermal compressibility, S is the entropy, and F is the free energy. Applying statistical mechanics to the quasi-harmonic approximation (the interaction constants of the harmonic theory are allowed to be volume-dependent) for a set of harmonic oscillators it can be shown that

$$\frac{\partial^2 F}{\partial V \cdot \partial T} = -\frac{1}{V} \sum_{i=1}^{3N} \gamma_i C_i$$

where

$$\gamma_i = -\frac{V \, d\omega_i}{\omega_i \, d V}$$

and

$$C_i = k(\hbar \, \omega_i/kT)^2 \exp(\hbar \, \omega_i/kT)/[\exp(\hbar \, \omega_i/kT) - 1]^2$$

C_i is the contribution of each vibrational mode to the heat capacity, k is Boltzmann's constant, $\hbar = h/2\pi$ where h is Planck's constant, and ω_i is the frequency of the ith mode. If γ is the weighted average of the individual γ_i terms,

$$\gamma = \sum_{i=1}^{3N} \gamma_i C_i \bigg/ \sum_{i=1}^{3N} C_i,$$

where N is Avogadro's number, then

$$\frac{\partial^2 F}{\partial V \cdot \partial T} = -\frac{\gamma_c}{V} v \tag{4.4}$$

Gruneisen's relation is obtained from equations (4.3) and (4.4)

$$\beta = \gamma \frac{k_T C_v}{V} \tag{4.5}$$

where γ, the Gruneisen parameter, is a measure of the average anharmonic interaction. At temperatures higher than the Debye characteristic temperature, θ_D, the value of γ is nearly constant, since the whole spectrum of vibrational frequencies is excited and γ is merely the arithmetic mean of the γ_i terms,

$$\gamma_\infty = \frac{1}{3N} \sum_{i=1}^{3N} \gamma_i$$

In the low-temperature limit as $T \to 0\mathrm{K}$, the Debye continuum is a valid model and again γ will approach a constant value:

$$\gamma_0 = -\frac{V}{\theta_0} \cdot \frac{d\theta_0}{d V}$$

where θ_0 is the limiting value of θ_D at low temperature and is directly related to the elastic constants. At intermediate temperatures the behaviour of γ may be stigated through Gruneisen's relation:

$$\gamma(T) = \frac{V.\beta}{k_T C_v} = \frac{V\beta}{k_s C_p}$$

ince the free energy is an additive function of state, equation (4.5) can be ressed as a sum of the contributions from the lattice vibrations (both the ustical and optical), the conduction electrons in metals, magnetic interaction,

$$\beta = \frac{k_T}{V}(\gamma_{aco}C_{aco} + \gamma_{opt}C_{opt} + \gamma_e C_e + \gamma_m C_m + \ldots) \qquad (4.6)$$

the case of non-magnetic metals, the heat capacity of the conduction electrons d the lattice can be approximated by:

$$C_e \simeq aT \text{ and } C_1 \simeq bT^3$$

very low temperatures.
quation (4.6) can therefore be written as:

$$\beta = \frac{k_T}{V}(\gamma_e aT + \gamma_i bT^3)$$

the terms are rearranged,

$$\frac{\beta V}{k_T T} = \gamma_e a + \gamma_i bT^2$$

nd the values of $(\beta V)/(k_T T)$ are plotted against T^2, a straight line should result.
The separation of the free energy terms for the acoustical and optical vibrations n a polyatomic crystal can also be obtained. The heat capacity, C, of the optical nodes is determined from Einstein's function using the value of the infrared eso ance frequency at the absorption maximum:

$$C_{opt} = f\left(\frac{\theta_E}{T}\right) \text{ where } \theta_E = \frac{h}{k}$$

The heat capacity of the acoustical modes, C_{aco}, is taken as the difference between obs rved values of C_v and calculated values of C_{opt}. If values of $\beta V/C_{aco} k_T$ are plot ed against the value of C_{opt}/C_{aco}, a straight line should result. Such a linear plot for TiO_2 has been found experimentally.

From the standpoint of structural chemistry, atoms which are held together by a strong chemical bond should not separate as much as two atoms joined by a weak bond. A convenient way to test this hypothesis is to compare the expansion coefficient with enthalpy, ΔH, which is a criterion of bond energy. Henglein [17] cal ulated the results for alkali halides and this is shown in Table 4.4 which de onstrates that in this case $\alpha \Delta H$ is essentially constant. A comparison of this ki d on more complex molecules is difficult to make, partly because of the lack of rel able data and partly because of the presence of several different kinds of bond. Sc ttered data on oxides of relatively simple structure suggest, however, that this co relation has a limited validity.

Table 4.4
The expansion coefficients and heats of formation of the
alkali halides

Compound	$\alpha \times 10^6$ (0 to 79°C)	$\alpha \times \Delta H$ (arbitrary units)
LiF	92	11.0
LiCl	122	11.8
LiBr	140	12.2
LiI	167	11.9
NaF	98	10.9
NaCl	110	10.7
NaBr	119	10.8
NaI	135	10.3
KF	100	10.9
KCl	101	10.5
KBr	110	10.7
KI	125	10.6

In a compound of complex structure the forces associated with different bonds may differ greatly, resulting in a marked anisotropy in expansion. As an example, Austin and Pierce [18] measured the thermal expansion of a single crystal of $NaNO_3$ and found that the thermal expansion along the a-axis is much smaller than that along the c-axis.

Experimentally it has been found that a simple close-packed lattice has a greater expansion than a less symmetrical and more complicated structure. Thus oxides such as MgO and CaO have coefficients of expansion which are relatively high; the slightly more complex oxide, ZrO_2, when it has the cubic fluorite structure, has a slightly lower coefficient of expansion. The group comprising BeO, Al_2O_3, $MgAl_2O_4$, and $BeAl_2O_4$ all have structural similarities and have about the same expansion coefficient, which is again lower.

According to Megaw [16], for crystals having the same structure, the expansion coefficient is inversely proportional to the square of the valency (z); and when substances of different structures are considered, for the same valency the mean expansion is directly proportional to the square of the coordination number. This relationship has achieved a considerable degree of success, as shown in Table 4.5. Attempts to apply this rule to more complex compounds have yielded unsatisfactory results.

Although some of the widest exceptions can be explained on a structural basis, there is, as yet, no set of broad principles which enables us to predict quantitatively, or even to explain, the thermal expansion of complex molecules.

Certain properties are known to be additive in solution (ideal!), for instance parachor. In view of the analogy between glass and a solution, efforts have been made from time to time to develop additive relationships between the chemical composition of glass and thermal expansion. Several sets of empirical factors have been proposed for the additive equations (see *Properties of Glass* by Morey) [1]. Unfortunately these arbitrary empirical factors change their numeri-

Table 4.5
Test of relation between expansion coefficient (α) and
electrostatic share* (Q)

Compound		Q	$\alpha \times 10^6$	$\alpha Q^2 \times 10^6$
CsCl		1/8	53	0.87
NaCl		1/6	40	1.11
CaF$_2$	2/8 =	1/4	19	1.19
CuBr		1/4	19	1.19
MgO	2/6 =	1/3	10	1.11
ZrO$_2$	4/8 =	1/2	4.5	1.12
ZnS	2/4 =	1/2	6.7	1.67

* The electrostatic share, according to Pauling, is the valency
divided by the coordination number.

va es in different systems, and even in a single system these hold constant
ly (er a limited concentration range – thus making these factors unsuitable for
y r l scientific use.

Th thermal expansion of glass is important in many ways: (a) when high
rr l endurance is necessary, the coefficient of expansion must be small;
w n high internal stresses are desired, as in tempered glass, the coefficient of
pa ion should be large; (c) when glass is joined to glass, as in laboratory
pa tus or bifocal spectacle lenses, the coefficient of the several glasses must
at ' or nearly so; (d) the same is true for glass to metal seals; and (e) volumetric
ar s in laboratory apparatus.

T coefficient of thermal expansion of glass is very sensitive to changes in
er cal composition, particularly in alkali content. In the literature, the thermal
p sion is often quoted as a mean value over a temperature interval, say from 0°
) 0°C. Table 4.6 shows the thermal expansion along with other physical
ro rties of a few commercial glasses. The coefficient of expansion of fused
li 5.5 × 10^{-7} per°C is very small. Other glasses invariably expand more
ap ly. The lime glasses, which have high properties of alkali, have coefficients of
xp sion ranging from 80 to 90 × 10^{-7} per°C, as do many of the lead alkali
ili tes. B$_2$O$_3$, in quantities up to 10–15 wt %, reduces the thermal expansion.
Co equently low-expansion glasses are almost always high in silica and low in
lk i content, with boric oxide usually present.

e rate of linear expansion of glass with temperature is *almost* constant up to
h nnealing range of the glass. More accurate measurements, however, require
eq tions of three or more constants to represent them; and when such accuracy
is question, the thermal history of the glass becomes of importance. According
to lorey, 'at higher temperatures the expansion of glass becomes a less simple
m ter, and its dimensional changes over the entire range up to the temperature at
w ch it becomes a liquid of low viscosity cannot be represented in any simple
m nner.'

typical thermal expansion curve for a barium flint glass is shown in Figure
4 3 after Peters and Cragoe [19]. The expansion is nearly linear up to 510°C and
r roducible for the different determinations. Between 510° and 530°C

109

Table 4.6

Physical properties of some commercial glasses

Type of glass	Viscosity data			Coefficient of linear thermal expansion ×10⁷ (0–300°C)	Refractive index (n_D)	Log (volume resistance) (ohm–cm at 250°C)	Power factor	Dielectric constant (1 Mc., 20°C)
	Strain point* (°C)	Annealing point† (°C)	Softening point§ (°C)					
Fused silica	1070	1140	1667	5.5	1.458	12.0	0.0002	3.78
96% silica glass	820	910	1500	8.0	1.458	9.7	0.0005	3.80
						11.2	0.00024	3.8
						11.7	0.00019	3.8
Soda-lime sheet glass	505	548	730	85	1.510	6.5	0.004	7.0
Soda-lime plate glass	510	553	735	87	1.510	6.7	0.011	7.4
Soda-lime container glass	505	548	730	85	1.520	7.0	0.011	7.6
Soda-lime bulb glass	470	510	696	92	1.512	6.4	0.009	7.2
Lead alkali silicate (elc.)	395	430	626	91	1.539	8.9	0.0016	6.6
		435	630	89	1.560	10.1	0.0016	6.66
High lead alkali silicate	390	430	580	91	1.639	11.8	0.0009	9.5
Alumino silicate	670	715	915	42	1.534	11.4	0.0037	6.3
	540	580	795	49	1.490	6.9	0.010	5.6
Low expansion borosilicate	520	565	820	32	1.474	8.1	0.0046	4.6
Low electrical loss Borosil	455	495	–	32	1.469	11.2	0.0006	4.0
Borosilicate for tungsten seal	485	525	755	36	1.487	8.8	0.0027	4.7
Borosilicate for Kovar seal	435	480	708	46	1.484	9.2	0.0026	5.1

* Strain point corresponds to a viscosity of $10^{14.5}$ poises and represents a temperature, usually 35.45°C below the annealing point.

† Annealing point corresponds to a viscosity of 10^{13} poises and represents a temperature at which internal strains are reduced to an acceptable limit in 15 minutes.

§ Softening point corresponds to a viscosity of $10^{7.5}$ to 10^8 poises.

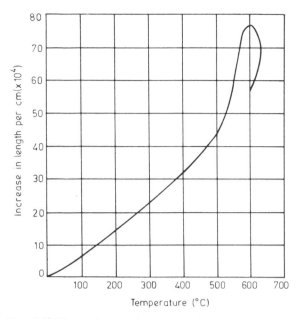

Fig. 4.18 Thermal expansion curve of a barium flint glass.

ition point) the rate of expansion increased rapidly. This rapid expansion
ued up to 580° or 590° C. With further heating the rate decreased, and an
ent contraction caused by settling or sagging began at 608° C. Similar curves
been obtained with other glasses, the actual temperatures being dependent
the composition of the glass. The influence of composition on the thermal
sion of some simple glasses is shown in Figure 4.19.

4.5   VISCOSITY OF GLASSES

viscosity of a glass-forming melt and its variation with temperature is of
mount importance in glass manufacture. When a shearing force is applied to
uid, a displacement results and, with continued application of the force, flow
s place. The ratio of force to displacement is a measure of the viscosity. For
viscosity the relation between flow and shearing force is a constant, and can
epresented by a straight line passing through the origin. If two parallel planes
rea A, and distance d apart, are subjected to a tangential force difference F, the
osity η is defined as

$$\eta = \frac{Fd}{Av}$$

ere v is the relative velocity of the two planes. The unit of dynamic viscosity in
modern SI system is the pascal second, a tenth of which has been commonly
own as a poise (P). But for common glasses the change of viscosity with

111

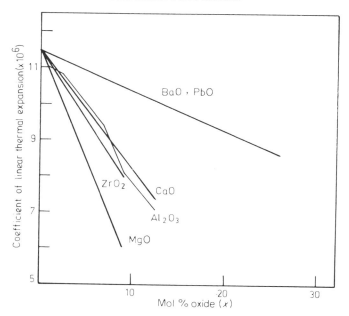

Fig. 4.19 Changes in thermal expansion of a sodium–silicate glass (75 SiO_2, $(25-X)Na_2O$, X MO, M_2O_3 or MO_2) when different oxides are substituted for Na_2O.

temperature is so large, that glass technologists commonly discuss and record log η. The fantastic changes of viscosity at various stages of a typical soda-lime–silicate glass manufacture are given in Table 4.7.

Table 4.7 shows that the viscosity at the feeder is 10 to 100 times that at the melting end, and that when the glass is removed from the mould a few seconds later, the viscosity is 100 000 times greater than at the melting end. Within the general limits set forth above, the ease with which glass can be worked into a variety of forms depends upon the length of the temperature interval between minimum and maximum viscosities at which it can be shaped. Usually, the longer the temperature range, the better the workmen likes it. He calls the glass 'Sweet' or 'Good natured'. Other glasses, 'Set up' more rapidly, in which case they are called 'Short'.

Table 4.7
Range of glass viscosities

Operation	Temperature (°C)	log η (poise)
Melting	1575	2.0
Automatic feeder	1300 – 1100	3.0 – 4.0
Gathering in mould	1000	4.5
Removal from mould	780	7.0
Annealing	580 – 555	13.0 – 13.5
Maximum service temperature	500 – 450	14.6 – 15.5

:o y also plays an important role in the melting conditions, refining
ic (removal of bubbles and seeds from the melt), upper temperature of
n devitrification rate of glasses. The viscosities of some commercially
ta silicate glasses are shown in Figure 4.20 as a function of temperature.

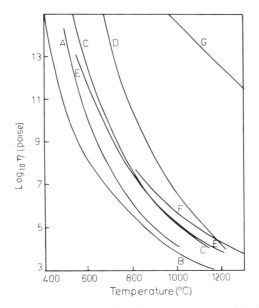

g .20 Viscosity–temperature relations for some commercial glasses.

(A) SiO_2 70.5, Al_2O_3 1.8, CaO 6.7, MgO 3.4, Na_2O 16.7 and K_2O 0.8 wt. % (lamp bulb).

(B) SiO_2 56.5, Al_2O_3 1.5, PbO 29.0, CaO 0.3, MgO 0.6, Na_2O 5.6 and K_2O 6.6 wt. % (vacuum seals).

(C) SiO_2 71.6, Al_2O_3 5.7 B_2O_3 11.0, CaO 3.6, MgO 0.6, Na_2O 3.5 and K_2O 3.9 wt. % (high voltage lamp bulb).

(D) SiO_2 54.5, Al_2O_3 21.1, B_2O_3 7.4, CaO 13.5 and BaO 3.5 wt. % (Hg-vapour lamp).

(E) SiO_2 71.0, Al_2O_3 7.4, B_2O_3 13.7, CaO 0.3, Na_2O 5.3 and K_2O 2.4 wt. % (high voltage lamp bulb).

(F) SiO_2 80.1, Al_2O_3 3.0, B_2O_3 12.0, CaO 0.2, Na_2O 3.9 and K_2O 0.3 wt. % (pyrex).

(G) SiO_2 100 wt. %. (fused silica).

'h :wo common methods frequently used to measure viscosity of glass and
is irming melts are the rotating cylinder method (at low viscosities, $\leqslant 10^8$
se and the fibre elongation method (at higher viscosities). In the former
th l the relative rate of rotation of two concentric cylinders is measured at a
is it torque, and the vicosity is inversely proportional to the rate of rotation.
is .n easily be shown as follows:

113

If the outer cylinder is rotated, some variable angular velocity ω is set up through the liquid. At any point in the liquid, say at radius r from the centre of the two cylinders, the linear velocity is ωr, and the linear tangential velocity gradient at this point is given by:

$$\frac{dv}{dr} = \frac{d}{dr}(\omega r) = \omega + r\frac{d\omega}{dr}$$

Only the second term of the above equation gives rise to viscous forces because the first term, the angular velocity ω, is to prevent any relative slipping in the uniformly rotating fluid.

Thus the velocity gradient due to viscosity is:

$$\frac{dv}{dr} = r\frac{d\omega}{dr}$$

The stress due to this velocity gradient, at radius r, is:

$$S = \eta 2\pi r l r\frac{d\omega}{dr}$$

where l is the effective length of the cylinder (a correction for end effects must normally be established experimentally).

If the torque, Γ on the inner cylinder is measured, this is given by:

$$\Gamma = r\eta\, 2\pi r l\, r\frac{d\omega}{dr} = 2\pi\eta l r^3\frac{d\omega}{dr}$$

The total torque is obtained by integration between radii R_1 and R_2 respectively.

$$\Gamma\int_{R_1}^{R_2}\frac{dr}{r^3} = 2\pi\eta l\int_{\Omega_1}^{\Omega_2} d\omega$$

or

$$\Gamma = 4\pi\eta l(\Omega_2 - \Omega_1)\Big/\left(\frac{1}{R^2_1} - \frac{1}{R^2_2}\right)$$

where Ω_1, Ω_2 are the angular velocities of the inner and outer cylinders respectively. Thus the viscosity in this rotating cylinder experiment may be derived from the simple definition:

$$S = \eta\frac{dv}{dx}$$

Usually this type of viscometer is calibrated with liquids of known viscosity.

Higher viscosities (10^9–10^{16} poise) are usually measured by the rate at which a glass fibre or rod elongates under a constant force. The viscosity is then given by:

$$\eta = \frac{Lmgf}{3\pi R^2 v} \tag{4.7}$$

114

is the length of the glass rod, R is its radius, m is the attached mass, g is
tional constant, v is an instrumental reading proportional to the rate of
, f is a calibration factor for the instrument.

ng viscosity of highly viscous materials like glass one has to be careful
problems:

ion of viscosity with time, and
ion of viscosity with the force applied.

riation of viscosity with time

s 1925, Stott, Irvine and Turner[20] reported that heat-treatment of
urs was required to reach a constant condition when the viscosity of
as high as $10^{16.5}$ poises. Lillie[21] made a systematic study of the
viscosity with time in the annealing range, and found that the time
to reach a constant viscosity characteristic of a given temperature
as the heat-treatment temperature was lowered. Some of his results are
Figure 4.21. The upper curve of Figure 4.21 is for a sample previously
77.8°C for 64 hours, a time sufficient for attainment of 'equilibrium'
and the lower curve is for a newly drawn fibre. The initial viscosity for
amples is different by a factor of more than ten, but they approach
final viscosity. It has been found that at higher temperatures less time is
and at lower temperatures longer times, until it becomes practically
e to reach the equilibrium viscosity.

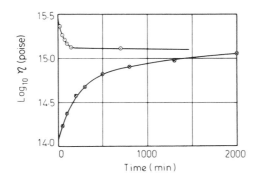

g. 21 Viscosity–time curves for two samples of a glass at 486.7°C.
The upper curve is for a sample previously heated at 477.8°C for 64 hours;
the lower curve is for a sample in the freshly drawn condition,
(after Lillie).

hange of viscosity with rate of shear

f rate of shear on viscosity of glass has been studied by many
tors[22]. The general finding is that the viscosity is not a function of
force or velocity of flow, implying that glass is a Newtonian liquid.

115

However, Bartenev [23] has reported the rate of flow of an alkali silicate glass at $655°C$ (in tension at low shear stress). Deviation from Newtonian behaviour was observed at shear stresses below $1 \, kgf \, cm^{-2}$. Thus it appears that there may be some elastic deformation in the viscous glass at low shear stresses before it flows.

4.5.3 Viscosity–temperature relationships below 10^9 poises

The variation of the viscosity with temperature of some simple liquids can be expressed by the relation:

$$\eta = A \exp (B/T) \tag{4.8}$$

where A and B are constants and T is the absolute temperature. This relation is very nearly adequate to express the variation of the viscosity of liquid GeO_2 and SiO_2 with temperature for about 10 decades of viscosity [24, 25], but it is inadequate for most other glass-forming inorganic liquids.

An empirical relation, Fulcher's equation:

$$\eta = A \exp\left(\frac{B}{T - T_0}\right) \tag{4.9}$$

(where T_0 is a constant) very closely expresses the experimental results, but Lillie [21] has demonstrated that it is not capable of fitting the best experimental data adequately. Over temperature intervals of a few hundred degrees equation (4.8) is sufficiently accurate for many measurements and the viscosity may be expressed approximately as the sum of two exponential terms:

$$\eta = A \exp\left(\frac{B}{RT}\right) + A_1 \exp\left(\frac{B^1}{RT}\right) \tag{4.10}$$

When plotting $\log \eta$ against $1/T$, the data can be fitted approximately by two straight lines; B and B^1 can thus be referred to as the 'high' and 'low' temperature activation energies. The use of such terms does not, however, imply that the behaviour can be so specifically interpreted.

The effect of small additions of Na_2O to GeO_2 is very marked and causes a considerable change in the variation of the viscosity with temperature. At the lowest concentration (0.045 mol %) shown in Figure 4.22 each sodium ion is separated by approximately ten germanium ions in all directions or, alternatively, only one Ge–O–Ge bond in a few hundred can possibly be affected by a sodium ion if the atoms remain in constant contact with their nearest neighbours. According to Kurkjian and Douglas [25], the large reduction in viscosity suggests perhaps that the sodium ions may travel around influencing many bonds. An estimate of the dwell time of the sodium ion, from diffusion data, suggests this may be about 10^{-12} s which is much shorter than the Maxwell relaxation time ($\sim 10^{-8}$ s).

The addition of small amounts of metal oxides to silica can reduce the viscosity markedly. Examples of this sharp variation are presented in Figure 4.23. At $1600°C$, for example, the addition of only 2.5 mol % of K_2O to SiO_2 causes the viscosity to drop from 2×10^7 to 2×10^3 poise. Effects of similar magnitude are shown by the other Groups I and II oxides. At higher concentrations of metal

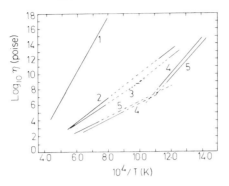

g. 4.22 Plots of log viscosity against $1/T$ for different glasses.
(1) SiO_2 (2) GeO_2 (3) 0.045 mol % Na_2O in GeO_2
(4) 20 mol % Na_2O in SiO_2 and (5) 24.5 mol % K_2O in SiO_2.

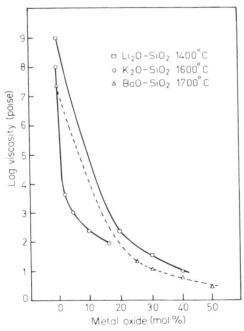

:. 4.23 Effects of metal oxides on the viscosity of binary liquid silicates[3].

the effects are less pronounced. For the alkaline-earth–silica systems
rgies of activation for flow change by less than 50 per cent over the
ition range 15–60 mol % metal oxide, whereas from 0–15 mol %, the
is over 100 per cent (Figure 4.24). In the Group I silicates the activation
with composition are illustrated in Figure 4.24. The large decrease of

117

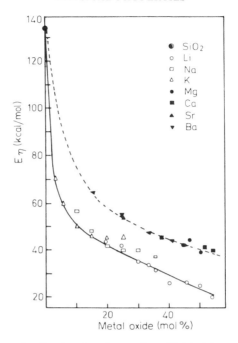

Fig. 4.24 Energy of activation for viscous flow (E_η) in binary liquid silicates [3].

activation energy from pure silica to about 10–20 mol % metal oxide, and the subsequent gradual decrease, indicate a rather abrupt and significant change in the structure of the melts as metal oxides are added to silica.

Although the presence of small amounts of alkali oxide is also effective in reducing the viscosity of B_2O_3, the magnitude is far less than that for the silicates. The corresponding ratios, $\eta_{B_2O_3}/\eta_{borate\ melt}$ containing 2.5 mol % metal oxide, vary from only 1.5 at 600°C to 6 at 1000°C. Considerable speculations have been made about the observed increase in the viscosity of liquid borates with increasing concentration of metal oxides; however, as pointed out by Mackenzie [26] no mention is made of the fact that measurements have sometimes been made at temperatures far below the liquidus. Thus, according to Mackenzie 'until the possibility of crystallization in the melt is unequivocally excluded by careful experiments, the validity of any model proposed in the past to explain variations of viscosity with composition at temperatures below the liquidus must be regarded with reservations'.

4.5.4 Viscoelastic behaviour or flow of glass in the transformation range

As discussed before when measuring viscosities above about 10^{10} poise, it is usual to handle the liquid in the form of a rod or bar. To introduce these measurements of very high viscosities, it is useful to look at viscosity from a rather different point of view – that of Maxwell. A substance is considered which shows both elastic and

haviour, a phenomenon typical of glass forming liquids when the
ty aches about 10^{10} poise. At this viscosity, it is quite possible to take a
o and break it in brittle fracture by applying a heavy load quickly.
ve if the same weight is applied slowly and carefully, the rod will extend
ll and be drawn out into a fibre. The material thus can be said to show a
is o stress typical both of a solid and a liquid. If some type of stress which
: drostatic, or a shear, or a uniaxial tension is applied to a mass of
a hen the relation between the stress, S and the strain, ε is given by the
o

$$S = M\varepsilon \qquad (4.11)$$

Λ s an elastic modulus. In a solid body this expression gives the complete
m etween stress and strain.
g e that a piece of glass is stressed so that it takes up a certain length. If it is
la ic and viscous a certain force is needed to stretch it to the desired length,
t t force is retained, the specimen will continue stretching because it is
is 'the force is reduced, the elastic part of the strain will be reduced and the
c be brought back to its original value. The rod will then flow viscously a
n e and so the force may be further reduced to restore the original strain
entually the desired length will be maintained but the force will
pe entirely.
is /hen the material is both elastic and viscous, equation (4.11) must be
er ated to allow for the variation of the stress with time, and an extra term
l allow for the disappearance of the stress:

$$\frac{dS}{dt} = M\frac{d\varepsilon}{dt} - \frac{S}{\tau}$$

e s the relaxation time.
vi ssume, as Maxwell did, that the force disappears at a rate proportional to
a constant strain, $d\varepsilon/dt = 0$, and the above equation becomes

$$\frac{dS}{dt} = -\frac{S}{\tau}$$

$$\frac{S}{S_0} = \exp\left(-\frac{t}{\tau}\right)$$

ii simple exponential decay of the stress.
t case of a constant rate of strain:

$$S = M\tau\frac{d\varepsilon}{dt} + C\exp\left(-\frac{t}{\tau}\right)$$

re is the stress required to keep the specimen stretching with a constant
ci and C is a constant. This expression means that this stress will start at a
ti value and then it will decay until the last term is zero; when the last term
d ppeared, the relation between stress and rate of strain will become:

$$S = M\tau\frac{d\varepsilon}{dt}$$

119

But this now corresponds to a steady state viscous extension and the viscosity coefficient is defined by:

$$S = \eta \frac{d\varepsilon}{dt}$$

Thus τ is equal to η/M.

When viscosity is measured by uniaxial stress, the coefficient of viscosity so determined is often called the coefficient of viscous traction, η_Y; it can be shown that to a reasonable approximation $\eta_Y = 3\eta_G$ where η_G is the coefficient of shear viscosity.

For a rod of length l under tension, F:

$$\eta = Fl/[\pi r^2 \, 3 \, (dl/dt)]$$

When a weight is hung on the glass rod there is an instantaneous elastic extension and then a creep, the rate of which gradually decreases to a constant value.

In the transformation range this changing rate of creep can be due to the slow change of configuration resulting in a changing viscosity or to delayed elastic behaviour. A volume viscosity should, however, also be recognized and, indeed, the slow isothermal changes of volume with time in the transformation range, which follow a sudden change of temperature can be expressed in terms of η_v, the volume viscosity. Summaries of the experimental evidence are available in Kurkjian and Douglas [27, 28].

4.6 SURFACE TENSION OF GLASS

A molecule in the interior of a liquid is completely surrounded by other molecules, and so, on average, it is attracted equally in all directions. A molecule in the surface, however, is subject to a net inwards attraction, because the number of molecules per unit volume is greater in the bulk of the liquid than in the vapour. The surface of a liquid will always, therefore, tend to contract to the smallest possible area. In order to extend the area of the surface it is obviously necessary to do work to bring the molecules from the bulk of the liquid into the surface against the inward attractive force. The work required to increase the area by $1 \, \text{cm}^2$ is called the *free surface energy*.

As a result of the tendency to contract, a surface behaves as if it were in a state of tension, and it is possible to ascribe a definite value to this surface tension, which is the same at every point and in all directions along the surface of the liquid. It is given the symbol γ, and may be defined as the force in dynes acting at right angles to any line of 1 cm length in the surface. The work done in extending the area of a surface by $1 \, \text{cm}^2$ is equal to the surface tension, which is the force per centimetre opposing the increase, multiplied by 1 cm, the distance through which the point of application of the force is moved.

When a liquid drop is placed on a smooth solid surface the shape it assumes depends upon the surface tension of the liquid, and under equilibrium conditions only two forces, namely surface tension and gravity, act on the drop. If the angle of contact between liquid and solid is θ, the equilibrium requires that:

$$\gamma_{sg} = \gamma_{ls} + \gamma_{lg} \cos \theta \tag{4.12}$$

where the subscripts s, l, and g represent solid, liquid and gas respectively.

tact angle thus depends on the three interfacial tensions, but whether it
 r less than 90° is governed by the relative magnitudes of γ_{sg} and γ_{sl}: if
 as tension (γ_{sg}) is greater than that for the solid–liquid interface (γ_{sl})
 must be positive, and θ less than 90°, but if the reverse is true then θ
 tween 90° and 180°. In the former instance, e.g. water on glass, the
 garded as wetting the solid, whereas in the latter, e.g. mercury on glass,
 is said not to wet the solid.

 liquid is brought in contact with a solid surface the degree of wetting
 pon the work of adhesion between liquid and solid (ω_{sl}) and can be
 d by the Dupre [29] relation:

$$\omega_{sl} = \gamma_{sg} + \gamma_{lg} - \gamma_{sl} \qquad (4.13)$$

 ; (4.12) and (4.13)

$$\omega_{sl} = \gamma_{lg}(1 + \cos\theta)$$

 wn as Young's equation.
 cept of a tension along a solid surface is however difficult to interpret
 nd it is better to deal with surface energy as suggested by Davies and
 ], so that the appropriate equation can be written as:

$$F_{sg} = F_{sl} + \gamma_{lg}\cos\theta$$

 ; the angle of contact, θ may be anomalously high due to adsorption of
 the solid–gas interface and then ω_{sl} will refer to the adhesion of the
 monolayer-covered solid surface. The presence of strongly adsorbed
 cally active gases may markedly affect θ, even when no visible chemical
 n takes place. For a $Li_2O.B_2O_3$ melt on platinum, the contact angle
 ogen is 85°. In less than one second after O_2 has been admitted, the
 ads and covers the metal surface. Conversely, the contact angle of
 licate on gold does not vary appreciably with the atmosphere[31].
 portant role played by surface tension in glass technology is often not
 d. To name a few, formation of cords and general homogeneity in
 fractory corrosion due to wetting, glass-to-metal seals, frothing of glass
 orption of gases, etc., may all be related to the surface tension.

etting of refractories and sand

 processes the phenomenon of formation of cords, as well as wetting
 rains by molten salts, is closely related to surface tension forces. In
 r case, those cords which have a relatively lower surface tension than the
 ing glass tend to spread out, become attenuated and disappear while the
 ing greater surface tension tend to shrink and sometimes take the form
 in molten glass. Silverman[32] showed that even a difference of a few
 can affect the persistence or disappearance of cords in glass. Similarly,
 f sand grains depends upon the surface tension of molten materials and,
 etting takes place in the earlier stages of melting, the batch-free time can
 lerably reduced. Further, a glass-forming melt may be an intimate
 of several different liquids. But if the surface tension of the embedded
 eater than its surrounding melt, the former will provide a minimum

possible surface for the process of homogenization due to diffusion. Jebsen-Marwedel [33] explained similar influences of surface tension in glass melting processes.

When molten glass comes into contact with a solid surface (refractory or metal) three independent surface tensions, corresponding to the three interfaces (liquid–gas, liquid–solid, solid–gas), determine whether or not the glass wets the solid and also its ability to penetrate into the pores of the refractory.

4.6.2 Refractory corrosion

The problem of refractory corrosion is of great importance in glass making and it seems that the life of a glass tank is the life of its flux line blocks. In other words a tank has to be put out when the glass has penetrated at the point of the worst attack which is of course the flux line.

The wetting process is important in flux line attack by the molten glass and further penetration of glass into the pores of the refractory is dependent upon the contact angle between the molten glass and surface of the refractory. However, the porosity of the refractory material is also important in the process of penetration because the size and shape of the pores also determine the manner in which penetration takes place. The penetration pressure of a liquid is given by the relation

$$P = \frac{2\gamma\cos\theta}{r}$$

where r is the radius of a uniform small pore and θ the contact angle.

The pores in refractories cannot be regarded as a number of fine capillaries as they have non-uniformity in diameter and shape. Therefore, in order to understand the mode of penetration it is necessary to consider the pores as having 'rough' surfaces and hence the apparent advancing contact angle [34] becomes important. Adams [35] indicated by geometric considerations how the penetration pressure is affected when a meniscus of liquid has to move past a constriction in a pore. If the advancing contact angle is 45° the penetrating pressure P will be $(2\gamma\cos 45°/r)$ before passing the constriction but if θ is greater than 45° the penetration pressure becomes negative just beyond the constriction of the pore and meniscus will not pass further.

4.6.3 Foaming

The possible effect of surface tension on the foaming behaviour of molten glasses may also be considered. Generally it can be said that a glass with a high surface tension would tend to destroy the foam by adopting the minimum possible surface area.

4.6.4 Improvement in surface quality

Surface tension plays an important role in the improvement of surface quality of different glass articles. For instance the rounding of cut or sharp ragged edges during operations such as fire-finishing and glazing can be readily observed. In the same way as surface tension helps remove sharp projections and edges, it also

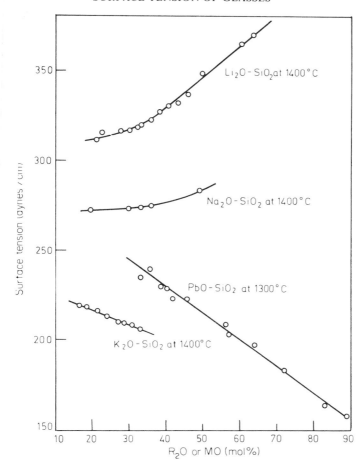

Fig. 4.25 Surface tension of some binary silicate melts.

ɛir formation; therefore it is difficult to produce sharp edges and designs
or pressed glass products.

trength of glasses

ιgth of a glass can be calculated by the Griffith's equation:

$$\text{Strength} = \sqrt{\left(\frac{2\varepsilon\gamma}{d}\right)}$$

is the modulus of elasticity, γ is the surface tension; and d is the crack
ι the surface. This suggests that suface tension values may have an
ιt effect on strength.

123

4.6.6   Methods of measurement of surface tension

Several methods are available for determining surface tension of liquids, but only a few of them are suitable for molten glasses. Even among the possible methods for the surface tension measurements of glasses few have been applied, chiefly because of the experimental difficulties caused by high temperatures and high viscosities.

The methods that have been successfully used for measuring surface tension of glasses are: (a) Drop weight method[36–38], (b) Fibre elongation method[39–44], (c) Pendent drop method[45, 46], (d) Sessile drop method[47–51], (e) Maximum bubble pressure method[41, 52–57] and (f) Pull-on-cylinder method[58]. The details of various methods can be found in the references cited.

Babcock[59] made surface tension measurements using a modified pull-on-cylinder method for a series of glasses. More extensive work on surface tension measurements was done by Shartsis and Smock[60] in the case of optical glasses and they found in general: glasses which are low in either silica or alkalis have positive temperature coefficients of surface tension; and the values observed were affected by the length of time the glasses were maintained at high temperature just before making measurements.

An extensive series of measurements of surface tension on binary alkali–silicate and alkali–borate melts has been made by Shartsis et al.[61]. Some of their results are shown in Figure 4.25. In binary silicates, surface tension increases with Li_2O and Na_2O, whereas increasing K_2O or PbO lowers it.

The surface tension of B_2O_3 is low compared to SiO_2 and increases with K_2O, Na_2O and Li_2O (Figure 4.26), the curves showing the change in surface tension with composition show a maximum with K_2O, Na_2O and to a lesser extent with Li_2O. The $PbO–B_2O_3$ melts show a flat portion over the region of liquid immiscibility.

4.7   ELECTRICAL PROPERTIES OF GLASS

The electrical properties of glass that have been extensively studied may be broadly divided into two groups: (a) D.C. conductivity, and (b) Dielectric relaxation. Electronic conduction in simple alkali–silicate, alkali–borate, and alkali–germanate glasses without dopant is extremely small. Electronic conduction in non-oxide chalcogenide glasses has been studied extensively by a number of workers and will not be discussed here. Interested readers are referred to volumes 2, 4, and 8–10 of *Journal of Non-Crystalline Solids*.

4.7.1   The D.C. conductivity in 'solid' glass

Although the phenomenon of ionic migration in solid glasses is usually considered to be similar to the corresponding phenomenon in ionic crystals, there are some important differences. The conductivity of glasses which are nominally alkali free, or which contain only a low percentage of alkali, are, in some cases, sensitive to trace constituents, notably where the latter provide the current carriers. But, as a general principle, it has been found that the mobility of any ion

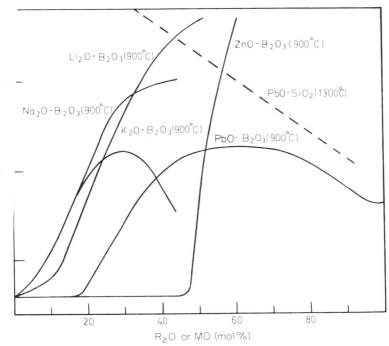

Fig. 4.26 Surface tension of some binary borate melts.

...s is not critically dependent on traces of other constituents.

...onductivity of a glass containing an alkali oxide as a major constituent ...at 10 mol %) is not critically affected by trace constituents. The situation is ...e in marked contrast to that in crystalline materials where minor ...ents have a profound influence on the conductivity.

...sic theory of D.C. conductivity in glass

...ventional model to describe D.C. conductivity of sodium silicate glasses ...in Figure 4.27(a)) consists of a sodium ion, say, in a potential well; the ion ...oing thermal vibration and having a finite and equal probability of ...inting the barrier and moving to any of the adjacent potential wells. Let us ...r for simplicity that the ions move in one dimension parallel to the x-axis, ...g from one potential well to the next over a potential barrier of height ...where N is Avogadro's number. The probability, assuming a Boltzmann ...ution of energy, that the ion will move either to the right or left is:

$$p = b \exp(-\Delta H/NkT)$$

b is the vibrational frequency of the ion in its well. On the application of a ...the system of energy barriers will be disturbed and may be represented by ...: 4.27(b), i.e. the effect of the field is to lower the potential barrier on one side ...well and raise it on the other by an equal amount thus favouring the motion

125

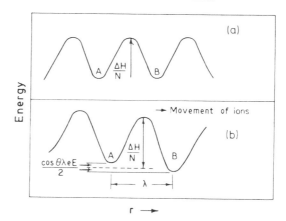

Fig. 4.27 (a) Schematic model of energy barriers in the absence of any applied electric field.
(b) Schematic diagram of energy barriers after applying an electric field.

of ions from left to right. If λ is the distance between the two neighbouring wells, the potential barrier separating each well to the right will be smaller by $1/2\,F\lambda$ (where $F = eE$, the force on the ion), while in the opposite direction it will be larger by the same amount. Thus, the probability of motion to the right is now

$$P_{+} = \tfrac{1}{2}b\exp - [(\Delta H/N) - \tfrac{1}{2}F\lambda]/kT$$

and in the opposite direction

$$P_{-} = \tfrac{1}{2}b\exp - [(\Delta H/N) + \tfrac{1}{2}F\lambda]/kT$$

The right-hand transitions of the ions will, therefore, be more frequent than the left-hand ones so that they will, on average, drift in the positive direction. The mean velocity of this drift motion is

$$\bar{v} = (P_{+} - P_{-})$$
$$= \tfrac{1}{2}p(\exp F\lambda/2kT - \exp - F\lambda/2kT)$$

or

$$\bar{v} = p\sinh\frac{F\lambda}{2kT}$$

where sin h is sin hyperbolic.

If the field strength is not too large so that $1/2\,F\lambda \ll kT$, the above equation may be approximated to

$$\bar{v} = \frac{\lambda^{2}p}{2kT}F \qquad (4.14)$$

126

ıer extreme of very large field strengths, the same equation may be

$$\bar{v} = C \cdot \exp F\lambda/2kT \qquad (4.15)$$

s a constant.

n temperature, $F\lambda$ is small compared with kT up to field strengths of V/cm. For weaker fields therefore where equation (4.14) holds, the ·ill be proportional to the field strength in agreement with Ohm's

·rent density i due to the drift of the ions (charge e) in the field is given by

$$i = ne\bar{v}$$

s the number of ions per cm^3. For moderate fields equation (4.14)

$$i = \frac{en\lambda^2 pF}{2kT} = \frac{e^2 n\lambda^2 pE}{2kT}$$

equation (4.15)

$$i = \frac{e^2 \lambda^2 nEb}{2kT} \exp -\Delta H/RT \, (Nk = R)$$

tivity ρ (ohms . cm) is given by E/i, therefore

$$\rho = \frac{2kT}{e^2 \lambda^2 nb} \exp \Delta H/RT$$

$$\log \rho = \log \frac{2kT}{e^2 \lambda^2 nb} + \frac{\Delta H}{RT} \qquad (4.16)$$

;ually called the activation energy for DC conductivity. It should be hat according to equation (4.16) the pre-exponential factor varies with ithm of the absolute temperature and therefore a plot of log ρ against .l absolute temperature should not give a straight line. The exponential ıce on $(1/T)$ predominates, however, and over a limited range of ure the deviations from linearity would be expected to be small and the n energy can be calculated from the slope of the line with reasonable

temperature on the D.C. conductivity of glass
:ndence of resistivity on temperature has been investigated by a number ːrs and their results generally fall into two groups:

some find that the resistivity is given by:

$$\log \rho = P + QT + RT^2 + \ldots$$

. Q, R etc. are constants. These results are quoted especially by those who de their measurements in or near the molten state of glass.

127

Group II: the results of the other group follow the relation:

$$\log \rho = A + \frac{B}{T} \tag{4.17}$$

where A and B are constants. This equation was first quoted by Rasch and Hindrichsen and is often referred to by their names. Comparing equation (4.16) with (4.17) it is clear that the constants A and B may be interpreted as

$$B = -\frac{\Delta H}{R}$$

$$A = \log \frac{2kT}{e^2 \lambda^2 nb}$$

In simple alkali–silicate or alkali–lime–silicate glasses the lithium glasses are found to be lower in resistivity than the sodium ones at equivalent concentration, and they are both lower in resistivity than the potassium glasses. Some typical results taken from the work of Taylor [62] are shown in Figure 4.28 for a variety of glasses. Even in this limited range of compositions there is a spread in resistivity, depending upon temperature, of some three to five decades. The activation energies vary from $16.4 \text{ kcal mol}^{-1}$ in D to $23.3 \text{ kcal mol}^{-1}$ in B.

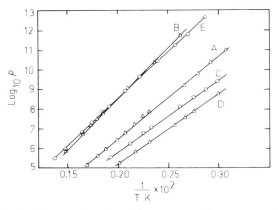

Fig. 4.28 Resistivity as a function of temperature for some typical glasses.
(A) 18 Na_2O, 10 CaO, 72 SiO_2
(B) 10 Na_2O, 20 CaO, 70 SiO_2
(C) 12 Na_2O, 88 SiO_2
(D) 24 Na_2O, 76 SiO_2
(E) Pyrex

The electrolytic nature of the conductivity of glass has been known since 1884 when Warburg passed a direct current through a test tube of glass which was filled with mercury or sodium amalgam and immersed in a bath containing a similar liquid to that inside the test tube; the bath temperature was generally about 500° C and the mercury or amalgam on the inside and outside of the tube served

des. With pure mercury electrodes the conductivity diminished to one-
th of its initial value within one hour and this was attributed to a
in the sodium concentration of the glass at the anode. When sodium
electrodes were used the decrease in conductivity was not observed and
erred that the glass was replenished with sodium from the amalgam. A
unt of literature has been accumulated since then mostly on the
aO–SiO$_2$ system. There is a good review up to 1963 in reference [63].
onductivity in glass is an intrinsic property in so far as it does not depend
constituents. However, it varies widely with changes in the major
nts even when the migrating ion remains unchanged. As a general rule
of composition affect primarily B, rather than A, in equation (4.17).
nerally observed in any glass system, containing only one alkali oxide,
ng the alkali content increases the conductivity at a given temperature and
e activation energy for conduction. This is true for single, binary and
nplex glasses provided the relative proportions of other constituents
constant and provided that there is at least 10 mol % alkali oxide present.
tions with less than 10 mol % alkali oxide may show a different effect.

inary alkali–silicate glasses

k of Seddon, Tippett and Turner [64] forms the basis for the great
of investigations that have been carried out on the Na$_2$O–SiO$_2$ system.
og resistivity against mol % Na$_2$O for a series of temperature using their
shown in Figure 4.29. Figure 4.30 shows the activation energy plotted

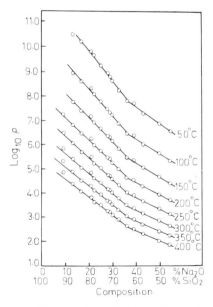

.29 Effect of composition on the resistivity of some binary sodium–silicate
glasses at different temperatures.

129

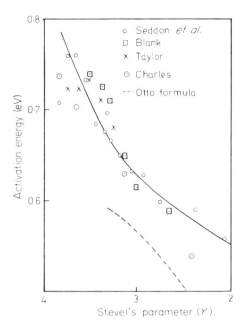

Fig. 4.30 Activation energy against Stevel's parameter (Y) for binary sodium–silicate glasses (after different workers).

against Stevel's parameter, Y which is the average number of 'bridging oxygens' linked to each silicon. Y is 4 for fused silica and 2 for the composition: $Na_2O.SiO_2$. For $Y < 2$, isolated silicate anions may form and crystallization takes place very rapidly.

Like most of the physical properties of glass the resistivity is rather sensitive to thermal history. Rebbeck, Mulligan and Ferguson [65] studied the resistivity (at $75°$ C) as a function of annealing time at $400°$ C of a typical soda–lime–silicate composition; initially there was a very rapid increase followed by a slow approach to a maximum resistivity, but even after 200 hours at $400°$ C appreciable changes still occurred. The resistivity of the well-annealed glass was found to be a factor of ten higher than that of the unannealed glass; and such results are quite typical.

For lithium–silicate glasses Charles[66] has shown that phase separation affects the D.C. conductivity very critically at low lithia content; the activation energy depends very much on the degree of heat treatment given and the values apply to the high silica matrix so they should not be plotted at the value calculated from the bulk composition. For high lithia content glasses, the heat treatment does not affect the D.C. conductivity and the activation energy changes very little with Li_2O content. However, the data of Kuznetsov [67] do not agree with these results, as they show a smooth decrease over a wide range of Li_2O concentrations; on the other hand the data of Blank [68], Otto [69] and Mazurin and Borisovskii [70] are similar to those of Charles.

130

)tassium, rubidium and caesium silicate glasses, the curves become
1 the order of increasing ionic radii. However, the data of Milberg
| for caesium silicate glasses show a discontinuous change of slope at
mol% Cs_2O and the activation energy was found to *increase* slightly
) content above this value. A similar curve has been given by Otto and
for thallous silicate glasses; furthermore discontinuous changes in the
asurements involving the Cs^+ and Tl^+ ions were reported at the same
ions as those where the activation energy changed slope.

literature there has been a considerable amount of discussion of 'kinks'
nductivity–composition plots of the binary silicates. These kinks are
hanges of slope at certain compositions and they are supposed to reflect
it changes in the 'structure'. Seddon, Tippett and Turner [64] represen-
conductivity data for each temperature by two straight lines intersecting
33 mol% Na_2O (Figure 4.29). However, smooth curves may also be
hrough the experimental points, although there is undoubtedly a
ced curvature between 30 and 40 mol% Na_2O. The composition
% Na_2O is particularly significant since there is on average one 'non-
oxygen' per silicon i.e. $Y = 3$. Below this some silicons have all four
bridging and the network would be expected to become extremely rigid.
verified the data of Seddon *et al.* and found that there is a change of slope
; ρ vs mol% Na_2O plot at about 33 mol% but not in the plot of log ρ vs n,
s the number of sodium ions per cm^3. Blank showed that the kink in the
mol% Na_2O graph arises from the values of ρ_0 are subject to
able scatter, as expected from the extrapolation of the log $\rho - 1/T$ plots
imental data, but show a linear increase with n.

from the data of Milberg *et al.* on caesium silicate glasses, and of Otto on
silicate glasses, which remain anomalous, it is considered that the
d kinks have arisen from an underestimate of experimental errors.
nore they have been reported at a wide range of compositions which does
it to an integral value of Y being involved or suggest that a single
al interpretation is not possible at present.

Mixed alkali effect

binary alkali–silicate or alkali–borate glass, one alkali oxide is pro-
ly substituted for another it is found that the resistivity does not vary
but goes through a very pronounced maximum when the two alkalis are
in roughly equimolecular amounts. A typical set of data is shown in
1.31 for a glass containing a total of 26 mol% sodium and lithium oxides.
quimolar composition the resistivity is greater than the resistivity of the
ıss by nearly ten thousand times. This is commonly known as the *mixed*
fect or poly-alkali effect by Russian workers. No satisfactory explanation
) be available for this behaviour although a number of hypotheses have
it forward by different workers.

general trend in resistivity of a simple sodium–silicate glass (18 wt%
on the substitution of silica by other commonly used oxides is shown in
4.32 after Fulda [72]. The most dramatic effect is produced by increasing
a concentration, or by substituting potash, which causes the resistivity to

131

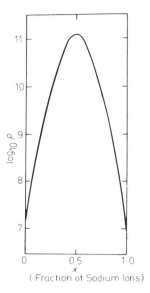

Fig. 4.31 Electrical resistivity of $(26-x)$ Li_2O, x Na_2O, 74 SiO_2 glasses.

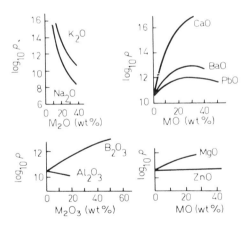

Fig. 4.32 Effect on resistance of replacing SiO_2 in the glass 18 Na_2O, 82 SiO_2 by various oxides (after Fulda [72]).

decrease very rapidly. All other oxides, except alumina, increase the resistivity at least initially, and this is especially marked in the case of calcium oxide. The comparison between the effects of boric oxide and alumina present an interesting exception to the generalization that the resistivity of glass is proportional to its viscosity. For, in the first place, the alumina glasses are more viscous yet less

:han the boric oxide glasses; and secondly, on increasing the B_2O_3
ition the viscosity decreases and resistivity increases whereas the
is true in the alumina series.

inary alkali–borate and alkali–germanate glasses

licate glasses in these systems the effect of alkali content can be studied
usly from 0 per cent upwards without any technical difficulty. In general,
ins can be distinguished; at low alkali contents, the effect of increasing
is small and may even lead to a *decrease* in conductivity and an *increase*
tion energy, while at higher alkali content, the conductivity increases
nd the activation energy decreases, with increasing alkali. The curves of
inst mol % R_2O either show a maximum at a low alkali content or else
sist of a nearly flat portion followed by a steep decrease. Figure 4.33
e data collected by Mazurin [70] for alkali–borate glasses and Figure 4.34
e data for sodium and potassium germanate glasses. A comparison of
4.33 and 4.34 with Figure 4.29 shows that the sodium borate and
te glasses are much more resistive than the silicate glasses containing the
il % Na_2O, while the resistivity of fused B_2O_3 is similar to that of fused
i that of fused GeO_2 is less.

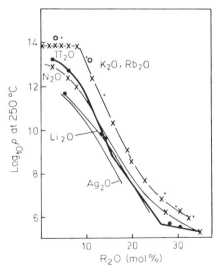

Fig. 4.33 Electrical resistivity of different binary borate glasses at 250°C.

D.C. conductivity in molten glass

iave been several investigations of the electrical conductivity of molten
silicates, but only a few workers have studied simple glasses of
atically varied compositions. Different workers disagree about which of

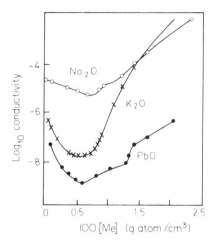

Fig. 4.34 Variation of the molar conductivity of germanate glasses with the volume concentrations of Na, K, and Pb ions in the system: Na_2O–GeO_2, K_2O–GeO_2, and PbO–GeO_2 at 300°C.

the three systems, Li_2O–SiO_2, Na_2O–SiO_2 or K_2O–SiO_2 is the most conducting. For a given mol % R_2O and fixed temperature, Endell and Helbrügge [73] found the order of conductivities: $K_2O > Li_2O > Na_2O$. Bockris et al. [74] reported: $Li_2O = Na_2O > K_2O$, whilst Urnes [75] found: $Li_2O > Na_2O > K_2O$. Tickle [76] has measured the electrical conductivity of R_2O–SiO_2 glasses (where R refers to Li, Na and K) over the temperature range 400–1400° C. His results are shown in Figures 4.35 and 4.36. The order of equivalent conductances is: $Li_2O > Na_2O > K_2O$ at high concentrations of R_2O, but below about 22 mol % R_2O the order is $Na_2O > Li_2O > K_2O$. For all melts, the plots of log (resistivity) against $1/T$ gave smooth curves.

4.7.6 The dielectric relaxation in glass

The static dielectric constant ε_s is usually defined as follows: A condenser of two parallel plates placed in vacuum at a small distance compared with its linear dimension and having $+\sigma$ and $-\sigma$ charges on the plates will produce an electric field given by:

$$D = 4\pi\sigma$$

If now the space between the plates be filled with a dielectric, the field will drop to a smaller value, say E

where

$$E = \frac{4\pi\sigma}{\varepsilon_s}$$

Thus

$$D/E = \varepsilon_s \quad \text{or} \quad D = E.\,\varepsilon_s \qquad (4.18)$$

134

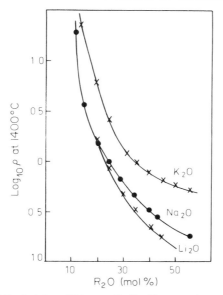

Fig. 4.35 Resistivity of binary alkali–silicate melts at 1400°C.

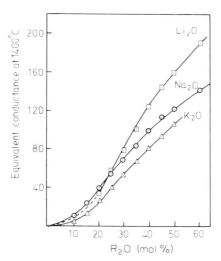

Fig. 4.36 Equivalent conductivity at 1400°C against mol% alkali oxide in binary alkali–silicate melts.

p in field strength may also have been achieved by reducing the surface of charge σ by the amount

$$P = \sigma - \frac{\sigma}{\varepsilon_s} = \left(\frac{\varepsilon_s - 1}{\varepsilon_s} \right) \sigma \tag{4.19}$$

135

where P is called the polarization of the dielectric. From equations (4.18) and (4.19)

$$D = E + 4\pi P$$

Let, in an alternating field, δ be the phase difference between D and E (and, therefore, between P and E). Using complex notation this can be expressed as:

$$E^* = E_0 \exp(i\omega t)$$

$$D^* = D_0 \exp i(\omega t - \delta)$$

where ω is the angular frequency of the applied field.

From equation (4.18) therefore

$$D^* = \varepsilon^* E^* \tag{4.20}$$

Hence

$$D^* = \frac{D_0}{E_0} \exp(-i\delta)$$

$$= \varepsilon_s \exp(-i\delta) = \varepsilon_s(\cos\delta - i\sin\delta) \tag{4.21}$$

The complex dielectric constant appearing in equation (4.20) is, in the complex form

$$\varepsilon^* = \varepsilon' - i\varepsilon'' \tag{4.22}$$

From equations (4.21) and (4.22)

$$\varepsilon' = \varepsilon_s \cos\delta \tag{4.23}$$

$$\varepsilon'' = \varepsilon_s \sin\delta \tag{4.24}$$

Combining equations (4.23) and (4.34)

$$\tan\delta = \frac{\varepsilon''}{\varepsilon'} \tag{4.25}$$

ε' is called the dielectric constant and ε'' is generally called the (dielectric) *loss factor*. The quantity $\tan\delta$ is some times called the *loss angle* or *dissipation factor* and $\sin\delta$ the *power factor*.

The D.C. conductivity of a glass depends upon the transport of charge over distances which are large on an atomic scale. In an alternating field, however, charged particles or dipoles which are restricted to much more limited movements will make a contribution to the dielectric properties. It can be shown that charged particles confined to oscillations between two adjacent potential minima give rise to a dielectric loss, ε'' with a maximum at a frequency $\omega_{max} = 1/\tau$, where the relaxation time, τ is related through a Boltzmann factor to the height of the potential barrier separating the two minima. This dielectric loss is equivalent to an 'A.C. conductivity' σ_{ac}, given by

$$\sigma_{ac} = \omega\varepsilon''$$

which is added to any 'D.C.' contribution. Similarly dipoles, which are limited to a process of alignment in the field, will make a contribution to the A.C. conductivity.

136

:tric properties of glass in an alternating field will depend therefore, not
e mobile ions which give rise to the D.C. conductivity but also on other
immobile ions or on dipoles etc., which may form part of the glassy
These various mechanisms may, according to temperature and other
s, overlap and add to each other, or they may occur in different parts of
:ncy spectrum. The general behaviour to be expected of a glass in an
g field has been represented by Stevels [77] in the form of a 'spectrum'
n δ) compounded from four mechanisms each having its characteristic
.

78] measured the dielectric constant and power factor of five com-
asses in the audio frequency range and from room temperature up to
e found the empirical relationship:

$$\tan \delta = K \exp (\alpha T)$$

he constants K and α depend upon the composition of the glass, and the
' of measurement, α decreasing with increasing frequency. Some typical
Moore and DeSilva [79] for $\tan \delta$ as a function of temperature at 1, 2
z, in a soda–lime–silica glass is shown in Figure 4.37. It will be noticed
at room temperature $\tan \delta$ is about 0.001 and this increase by a factor of
than 100° C. Such values are fairly representative of commercial glasses
g 10–20 per cent Na_2O. The variation of $\tan \delta$ with frequency at a
temperature, can likewise be represented by a simple empirical equation:

$$\tan \delta = B f^{-n}$$

and n are constants.

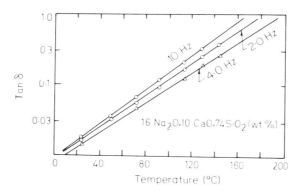

37 Typical behaviour of $\tan \delta$ (at audio-frequencies) with temperature for a
soda–lime–silica glass.

s [77] proposed a loss of spectrum to describe the general behaviour of
an electric field. This spectrum extending from 1 to 10^{14} Hz is shown in
.38. Stevels considers that this spectrum results from the sum of four
ting mechanisms, shown by the dashed lines in Figure 4.38, namely:

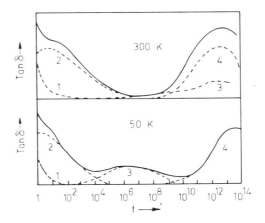

Fig. 4.38 Stevels' generalized loss spectrum for glass.

(a) conduction losses, (b) ionic relaxation losses, (c) vibration losses, and (d) deformation losses. Losses of types (a) and (b) are sometimes taken together and called migration losses.

(a) Conduction losses

Conduction losses are caused by those transitions which are represented by a continuous chain of ionic jumps leading to near transfer of charges in the direction of the field. These losses decrease very rapidly with increasing frequency of the applied field. As a general rule the conduction losses for frequencies higher than 50 Hz are negligible compared with other losses.

(b) Ionic relaxation losses

Due to the complete absence of long range order in glass, the modifying positive ions are situated in varied environments and thus confronted with varying heights of potential barriers illustrated in Figure 4.39. Limited movements such as b over the lower barriers give rise to dielectric polarization, whereas movements such as a over the higher barriers give continuous conduction.

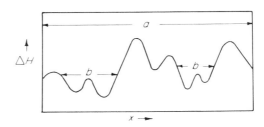

Fig. 4.39 Schematic diagram of randomly distributed energy barriers.

138

steady field is applied the equilibrium distribution of the ions amongst
s will be changed, giving rise to polarization of the medium. However,
zation will not be reached instantaneously. For the simplest model, as
Figure 4.39 where the heights of the b-type potential barriers are all
can be shown that the equilibrium polarization is approached
ally. Thus the frequency dependence of the complex dielectric constant
ternating field can be expressed as:

$$\varepsilon = \varepsilon_\infty + \frac{\varepsilon_s - \varepsilon_\infty}{1 - i\omega\tau}$$

s the static dielectric constant, ε_∞ is the dielectric constant at very high
', τ is the relaxation time. The above equation was first derived by Debye
ommonly known as the Debye equation. Writing $\varepsilon = \varepsilon' + i\varepsilon''$ and
g the real and the imaginary parts:

$$\varepsilon' = \varepsilon_\infty + \frac{\varepsilon_s - \varepsilon_\infty}{1 + \omega^2\tau^2}$$

$$\varepsilon'' = \frac{(\varepsilon_s - \varepsilon_\infty)\omega\tau}{1 + \omega^2\tau^2}$$

he power loss in the material. As discussed before this loss has a peak at
ency $\omega = 1/\tau$.
y be understood from Figure 4.39 and the discussion in Chapter 1 there
al relaxation times in glass, since the local structure differs considerably
e point to another in the volume of glass, with the result that the
n loss peaks are widely spread. In order to account for this fact one has to
e a distribution function $Y(\tau)d\tau$ which gives the contribution to the static
: constant from the ions which have relaxation times lying between τ and
The Debye equation then reduces to

$$\varepsilon = \varepsilon_\infty + \int\limits_0^\infty \frac{Y(\tau)d\tau}{1 - i\omega\tau}$$

h the above equation in its present form can describe the frequency
nce of the complex dielectric constant qualitatively, the knowledge of the
$Y(\tau)$ is essential for any detailed analysis. However, it is quite evident that
:s of this origin are small at $\omega \ll 1/\tau_m$ and at $\omega \gg 1/\tau_m$ have a maximum
$= 1/\tau_m$ where τ_m is the most probable value of the relaxation time.

· vibration losses
ions in the glass can vibrate around their positions of equilibrium at
:ies determined by their mass and by the restoring force in their potential
1enever the applied electric field alternates at a frequency near that of one
:onstituent ions they are excited to high resonant amplitudes. As the
; of the ions are damped due to energy exchange with the surroundings,
proportional to the square of the amplitude is absorbed from the applied
ing rise to high dielectric losses. This is the typical resonance absorption.

Due to the number of different kinds of ions in glass and due to the varied environments in the neighbourhood of the vibrating ions, the resonance absorption curve with frequency is not very sharp but is spread about the most probable frequency. For most of the glasses this most probable frequency usually lies in the infrared region, but due to wide spreading of the loss curve, substantial contributions from this type of loss may be expected even in the microwave region.

(d) The deformation losses

Stevels made some dielectric loss measurements at low temperatures. He observed some peaks in the loss temperature curves in the vicinity of 50 K in the frequency region of 1 MHz as shown in Figure 4.38. These peaks could not be accounted for by any of the loss mechanisms discussed before. He concluded that this might be caused by small displacements of the network-forming ions under a periodic electric field. These displacements, being of far more restricted nature, could be in the form of a small rotation or a small change in the bond angle or a similar kind of small deformation in the network (hence called deformation losses). This type of movement will be a relaxation phenomenon. As the activation energy involved in the process is very small, the time of relaxation will be small and the losses of this origin would show maxima in the frequency region of 10^{13} or 10^{14} Hz at room temperature. This type of loss will be generally masked by the vibration losses at room temperature and at high frequencies and cannot be distinguished experimentally. With a decrease of temperature, the relaxation time increases and hence deformation loss maximum shifts to medium frequency (1 MHz) at temperature near about 50 K.

The audio-frequency dielectric loss of glasses containing more than a few per cent of alkali oxides parallel very closely the behaviour of the D.C. conductivity. On increasing the proportion of the alkali oxide, $\tan \delta$ and the D.C. conductivity both increase rapidly. The addition of alkaline earth oxides CaO, SrO and BaO causes a dramatic decrease in the dielectric loss. According to Rinehart and Bonino [80] CdO and PbO behave very similarly to the alkaline earths and result in a steady decrease in $\tan \delta$. MgO and ZnO, however, produce a maximum in dielectric loss; this was suggested to be due to the intermediate nature of these oxides i.e. Mg^{2+} and Zn^{2+} can enter the glass network either as a network-former or as a network-modifier. The same authors also found that B_2O_3 and Al_2O_3 produce opposite effects on the loss factor; B_2O_3 produces a general decrease in $\tan \delta$ whereas Al_2O_3, after a small initial decrease, rapidly increases $\tan \delta$. The addition of SnO_2 and ZrO_2 had rather similar effect to Al_2O_3. The poly-alkali effect in silicate and borate glasses is markedly prominent in $\tan \delta$–frequency plots.

4.8   STRENGTH OF GLASSES

A. K. Varshneya
Alfred University, USA

Strength of a material is the magnitude of the applied stress which causes spontaneous fracture. Ordinarily glass articles break easily. The strength of such articles varies from as low as ~ 2000 psi (15 MPa) to as much as $\sim 20,000$ psi

.). Glass fibres normally break around 50 000 psi although in a freshly
te it is not uncommon to observe strengths exceeding 500 000 psi. The
ıl strength, σ_{th}, of the glass based upon the arguments of the energy
o break the bonds and create two new surfaces can be shown [82] to be
$E \gamma/a)^{1/2}$ where $\gamma =$ surface energy, and $a =$ interatomic spacing. Using
ılues for the parameters,

$$\sigma_{th} = \sim 2 \times 10^6 \text{ psi } (1.4 \times 10^4 \text{ MPa}).$$

red experimentally observed strengths, their rather large variation with
men size and its pristineness were explained by Griffith [83] by
g that glasses were riddled with *surface flaws*. The stress at the tip of
ws was considerably higher than the applied stress. By finding the
energy maximum in forming a narrow ellipsoidal crack of length L
or axis) when plotted against L, Griffith showed that the breaking
σ_f of the glass was given by

$$\sigma_f = (2E\gamma/\pi L)^{1/2} \qquad (4.26)$$

little as a micron sized flaw can reduce the observed strength of the glass
or of 100.
workers noted [84, 85] the fact that glass strengths decreased greatly in
of various atmospheres compared to those in vacuum. These were
to be consistent with the notion that γ varied in various atmospheres, the
)eing in vacuum. However, glass strength has been shown to decrease
reased temperatures, increased time of loading and cycling.
)w generally accepted that, as long as the applied stress in glass is greater
imiting amount called the *fatigue limit*, the flaws are always growing.
eous fracture occurs when the velocity of crack growth reaches a critical
)out half that of longitudinal acoustic waves in the glass). These concepts
ıre mechanics were advanced greatly by Weiderhorn and co-workers
to explain the strength of glass.
ıderstand fracture mechanism, it is useful to start from Irwin's definition
he *stress intensity factor K*. For a crack of length L (semimajor axis) in an
y wide plate, the stress intensity factor at the tip of the crack is defined as
$L)^{1/2}$ where σ is the applied stress. Clearly, if one starts with a defined
nder a fixed load in a specimen, the value of K increases as the crack
:s (i.e., L increases). Wiederhorn [86] plotted the observed crack velocity
K_1 (said as 'K one', and implying mode I application where the crack
ıtes in a plane normal to the direction of the applied stress) in soda lime
ınder various conditions of humidity. This is shown in Figure 4.40. Three
t regions of behaviour are observed: region I where the crack velocity
; upon the relative humidity (RH) as well as K_1, region II where the crack
is independent of K_1 but depends upon RH, and region III where the
elocity is independent of RH but increases rapidly with K_1. When the
as advanced to region III, the velocities are so high that a fracture is
nt: the value of K_1 at the instant of fracture is termed the critical stress
y factor K_{1C} or simply, the *fracture toughness* and is thought to be a
ıl constant. (Although K_{1C} does not depend upon RH, it may depend upon
ıvironments, and the speed at which a test is carried out). The behaviour.

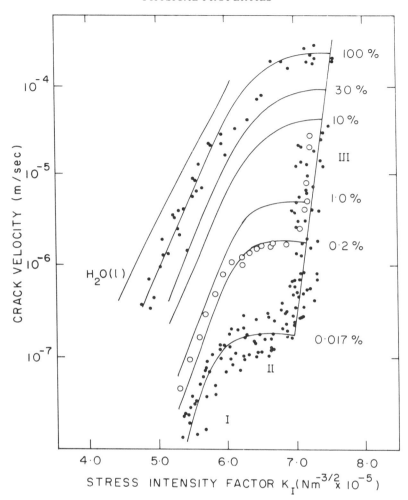

Fig. 4.40 Crack-velocity K_I curves for soda-lime glass tested in N_2 gas of varying relative humidity (after [86]).

in region I (region of subcritical crack growth) for some glass compositions of commercial interest is shown in Figure 4.41. The soda lime and the borosilicate data appear to become vertical at low values of K_I, apparently indicating a minimum K_I below which measurable crack propagation does not occur. This minimum K_I is called the *fatigue limit*. In Figure 4.42, the effect of temperature on the subcritical crack velocities for soda lime glass is shown. The velocity increases with temperature reflecting the common observation of 'easier failure' at higher temperatures.

The apparent exponential relationship between the crack velocity and K_I in the subcritical crackgrowth region is understood in terms of the *stress corrosion*

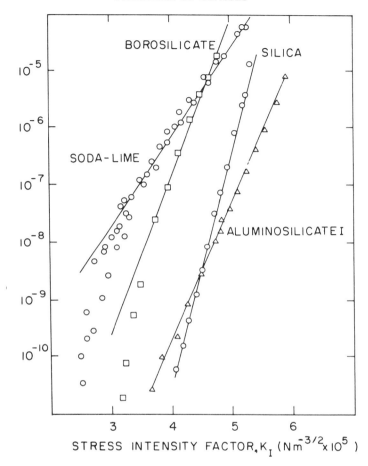

1.41 Effect of glass composition on crack velocity-K_I behaviour of various glasses tested in water at room temperature (after [87]).

ʒ mechanism advanced by Hillig and Charles [91]. Hillig and Charles :d that the rate v of corrosion of a flaw is a thermally activated process ʰe activation energy Q, in general, decreases with increasing stress, i.e.,

$$v = A \exp(-Q/RT) \qquad (4.27)$$

he total activation energy Q is thought to consist of Q_a, the energy to : a crack and γ, the surface energy necessary to create newly fractured i.e. $Q = Q_a + \gamma$. The stress dependence of Q_a is written as

$$Q_a = Q_0 + \sigma (dQ/d\sigma)_{\sigma=0}$$

$)_0$ is a constant.

143

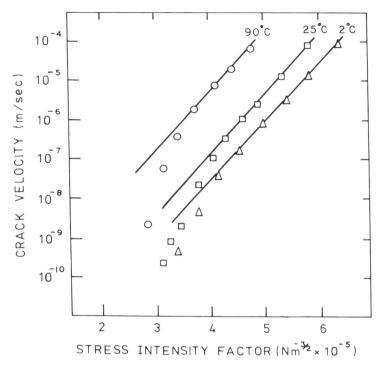

Fig. 4.42 Effect of temperature on crack velocity behaviour of soda-lime in water. (The fit was made to data above the point at which velocity decrease occurred). After [87].

Thus

$$v = A \exp-[Q_0 + \gamma + \sigma(dQ/d\sigma)_{\sigma=0}]/RT$$

or
$$v = v_0 \exp-[\gamma + \sigma(dQ/d\sigma)_{\sigma=0}]/RT \qquad (4.28)$$

Note that $dQ/d\sigma$ is negative, hence Q_a decreases with increasing σ, and v increases with σ. The surface energy γ itself is a function of $1/r$ where r is the local radius of curvature. Equation 4.28 predicts a different velocity of corrosion along a flaw. The schematics of the corrosion of a flaw at various levels of σ are shown in Figure 4.43. In the absence of stresses, the rate of corrosion is equal along the sides as well as at the tip of the flaw, with the result that blunting of the crack tip occurs ($(L/\rho)^{1/2}$ decreases, see Figure 4.43 (a)). In the presence of stresses, σ is the highest at the tip (by stress multiplication), hence the corrosion velocity is also the highest at the tip. Along the side of the flaw, σ is low and hence a low corrosion rate is predicted. This implies that when the stress is high (Figure 4.43(b)) the flaw tip will get corroded faster leading to a sharper curvature ρ at the tip. In turn, this will cause a higher σ at the tip (a larger value for $(L/\rho)^{1/2}$) and consequently a higher V. At lower values of σ (Figure 4.43 (c)), the value of γ competes with $\sigma \, (dQ/d\sigma)_{\sigma=0}$ in Equation 4.28. Hence, the corrosion rate along

144

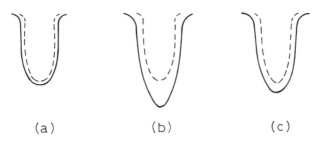

(a)   (b)   (c)

43 Effect of stress corrosion on the geometry of the crack tip. (a) With no
applied stress the tip blunts; (b) with larger applied stress the crack tip
sharpens; (c) with a stress equal to the static fatigue limit the stress
concentration factor $(L/\rho)^{1/2}$ remains constant.

is comparable to that at the tip such that the crack does not get much
$(L/\rho)^{1/2}$ hardly increases). One can readily imagine a fatigue limit where
enough that $(L/\rho)^{1/2}$ remains essentially constant, and no measurable
owth occurs during a prolonged experiment. Equation 4.28 also explains
nential relationship, and the temperature dependence between the crack
(corrosion rate) and the stress (and hence the stress intensity factor)
d by Wiederhorn.
ternate equation suggested for the $V - K_1$ dependence is

$$V = A\, K_1^n \qquad (4.29)$$

of n measured for several glasses vary between ~ 10 and ~ 40.
t from the crack velocity measurement experiments, researchers have
ted two other classes of strength tests which appear to yield fracture
ics parameters essentially similar to those obtained from the $V - K_1$
These are (i) *static fatigue*, and (ii) *dynamic fatigue* experiments. Static
experiments are carried out by placing the sample under constant load
asuring the time to failure t_f. Pioneering experiments of this type were
out by Mould and Southwick. [92] Figure 4.44(a) shows their raw data (a
applied stress σ vs. log t_f) for a variety of abrasion conditions. When the
are normalized against the inert strength σ_N (strength in liquid N_2) and
horizontally by log $t_{0.5}$ (which is the time to failure for achieving $0.5\sigma_N$),
different curves merge to form a 'universal fatigue curve' shown in Figure
. The postulate of a universal fatigue curve can be predicted from Hillig
narles model.
ne dynamic fatigue experiment, one increases the load on the specimen at a
nt rate and measures the stress to failure σ_f as a function of the loading rate
ilts of a typical experiment are shown in Figure 4.45. It may be shown [93]
racture mechanics that the slope of the log σ_f vs. log $\dot{\sigma}$ line should
$n + 1$).
time prediction under fixed loading conditions is, of course, straight-
rd using a static fatigue curve (Figure 4.44(b)). The generation of a static
e curve, however, can be quite tedious. Often it suffices to consider $t_f \propto \sigma^{-n}$
allows a conservative estimate of t_f at some service stress after determining

145

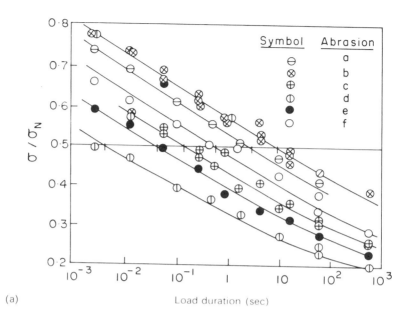

(a)

Load duration (sec)

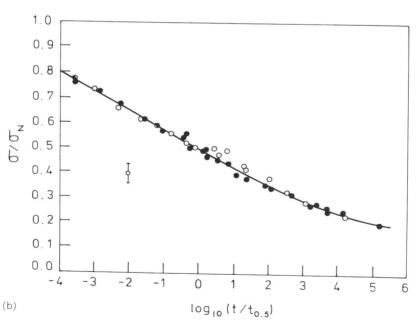

(b)

$\log_{10}(t/t_{0.5})$

Fig. 4.44 (a) Fracture time as a function of reduced strength (σ/σ_N) for different surface treatments of soda-lime glass; (b) 'Universal fatigue' curve. After [92].

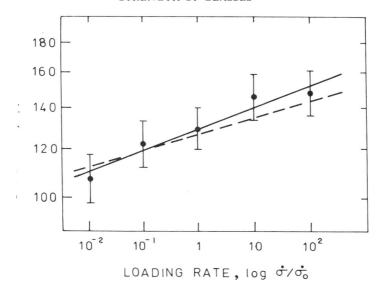

45 Equivalence of strength behaviour of a SiO_2–TiO_2 glass as determined by stressing rate experiments (solid line) or crack velocity measurements (dashed line). After [89].

num time to failure at two values of proof stresses (generally higher than
wn maximum service stress). The use of $V - K_I$ curves for lifetime
in is as follows. A numerical integration scheme is set up whereby one
s K_I for a given σ (and making some assumption about flaw geometries),
; corresponding value of V which is written as $\Delta L/\Delta t$; the ΔL is used to
te K_I and hence a new value of V is obtained; all the small increments of
immed. Time to failure, of course, is the integrated time when $K_I \rightarrow K_{IC}$.
ss products, a random distribution of the flaw sizes, mostly on the
can be assumed. Under given conditions of the loading, the flaw which
e largest $\sigma\sqrt{L}$ will cause the failure. The strength is then that of the
link in a chain: the strength distribution can be approximated by
statistics. The cumulative probability of failure F in a large enough
ion upon the application of stress σ is given in its simplest form by

$$F = 1 - \exp - [\sigma/\sigma_0]^m.$$

$_0$ is a normalizing stress and the exponent m is called the *Weibull modulus*
generally about 5 to 15 for glasses. Plots of the density function
$= (dF/d\sigma)$, and F vs. σ are shown in Figure 4.46. These figures can be
to estimate the percentage of cracks in a given glass processing operation
:ressing conditions change. For instance, if m is assumed $= 10$ as an
, then a glass plant operating at 2% cracks-in-process level (i.e., $F = 0.02$)
xperience a 60% jump in the crack level (to $F \simeq 0.0324$) if the stress level
:d only by 5%, say by a 5% increase in the thermal expansion coefficient
lass.

147

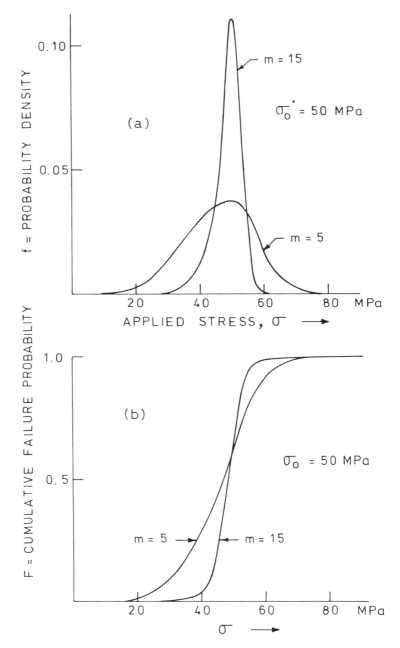

Fig. 4.46 Weibull distribution.

eresting to note that the key to strength of glass lies in the chemical
of glass. Glass is weak in those media which corrode a given glass
asic solutions, and even water, are good examples of such media for
ies. The kinetics of ion exchange, e.g., of Na^+ and H^+, and of bridging
ond breakup play a pivotal role in determining the strength of glass.

astic properties of glass

mmercial glasses act as mechanically rigid solids at ordinary tempera-
is implies that upon the application of a stress, glass undergoes
eous deformation to the fullest extent and these deformations are fully
d at the instant the applied stresses are removed. Observations of creep
ed recovery in glass are exceptions to this generality, and are discussed

astic properties of rigid solids are expressed in terms of four moduli of
which relate the applied stresses to observed strains, and are constants
deformations (Hooke's law for linear elasticity):

g's modulus $E =$ (uniaxial stress/axial strain).

lk modulus $K = \left(\dfrac{\text{hydrostatic stress}}{\text{fractional volume change}} \right)$.

ar modulus $G = \left(\dfrac{\text{shear stress}}{\text{shear strain}} \right)$.

Poisson ratio $v = \left(\dfrac{-\text{transverse strain}}{\text{axial strain}} \right)$.

iprocal of the bulk modulus is called compressibility, and is generally
1 as β. Appropriate units for the moduli of elasticity are psi in English
iPa in SI units, dynes/cm^2 and Kbars in c.g.s. units. (For convenience, one
memorize $1\,GPa = 10^{10}$ dynes/$cm^2 = 10$ Kbars $\simeq 1.45 \times 10^5$ psi). The
n ratio is unitless. For homogeneous (properties do not vary from point to
and isotropic (properties at any point are identical in all directions)
als, such as glasses, the four elastic constants are interrelated to yield only
dependent moduli. These relations are given by

$$K = E/3\,(1 - 2v)$$
$$\text{and } G = E/2\,(1 + v)$$

Young's modulus of glass can be readily measured by stretching
ately gripped specimens or by beam bending using universal tensile testing
nes. By far the most favoured techniques, however, employ acoustic
ment. The reader is referred to the book by Schreiber, Anderson and Soga
which is one of many on this topic.
cedures have been developed [94–95] to calculate the elastic moduli of
as a function of its chemical composition. Simple additivity factors
shed by Winklemann and Schott [94] and by Clarke and Turner [96] are
n in Table 4.8. To estimate E (Kbars), one simply multiplies the weight

149

Table 4.8
Factors for calculating Young's modulus from chemical composition.

	Winkelmann and Schott (94)			Clarke and Turner (96)
	A	B	C	
SiO_2	6.9	6.9	6.9	3.9
B_2O_3		5.9	2.5	
Na_2O	6.0	9.8	6.9	10.8
K_2O	3.9	6.9	2.9	
CaO	6.9	6.9		23.5
BaO		6.9	2.9	
ZnO	5.1	9.8		
PbO	4.5		5.4	
MgO		3.9	2.9	29.4
Al_2O_3	17.6	14.7	12.7	11.8
Fe_2O_3				11.8
P_2O_5			6.9	
AS_2O_5	3.9	3.9	3.9	

Note: Factors under A are for glasses free from B_2O_3, P_2O_5, BaO, MgO; under B, for glasses free from PbO, P_2O_5; under C, for borosilicates, lead borosilicates, and phosphates.

Reproduced from reference [1].

Table 4.9
Values of elastic moduli for some glasses.

Glass type	Young's modulus (GPa)	Shear modulus (GPa)	Poisson ratio	Knoop hardness (Kg/mm²)
Fused silica	70	30	0.16	550
Soda-lime silica (GE type 008)	69	29	0.22	460
Corning type 7740 borosilicate	63	26	0.20	480
Corning type 7052 borosilicate	57	23.4	0.22	375
Corning type 1720 aluminosilicate	85	34	0.24	600
GE type 001 potash soda lead silicate	62	25.6	0.21	320

of the component oxides with the corresponding additivity factor and various contributions. Direct calculations of E utilizing the dissociation the component oxides and a 'packing density' have been made by ia and Mackenzie [97]. Measured values of the elastic moduli for iss compositions are shown in Table 4.9. The effect of replacing Na_2O ixides in a $18\,Na_2O$, $82\,SiO_2$ (wt%) glass is shown in Figure 4.47 (after and Thomas [98]). Addition of most alkaline earths, particularly of eases the Young's modulus. Additions of B_2O_3 appear to show an s behaviour, commonly referred to as 'the B_2O_3 anomaly'.

ients to measure the compressibility have clearly shown [99–101] that e application of a critical pressure, all glasses undergo an irreversible

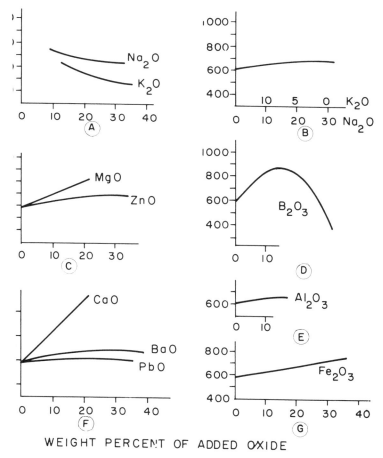

WEIGHT PERCENT OF ADDED OXIDE

The effect of Young's Modulus, in a glass containing 18% Na_2O and 32% SiO_2 by the indicated percentages by weight of other oxides. After [97].

compaction (departure from Hooke's law). This densification is explained in terms of the attainment of new equilibrium values of the bond angles such that the interstitial free volume is reduced. A similar argument based upon bond angle changes in the glass network is put forward [102] to explain the observed increase of E with extensional strain in some glasses, notably in fused silica. (In 'normal' materials one expects E to vary as $1/a^4$ where $a =$ interatomic spacing).

Hardness or microhardness generally refers to a material's resistance to plastic deformation. The tests can be carried out using several available indentation tests, such as the square-pyramid diamond (Vicker's indentation) or the elongated-pyramid diamond (Knoop indentation). Microhardness tests in glasses are simple; but they must be carried out with care. As little as 25 g load on the indentor can sometimes give rise to radial cracking from the corners of the indentation in some glasses. Because cracking represents absorption of energy (to produce new surfaces), it is obvious that the apparent values of the microhardness can vary greatly with loading conditions. Available experimental data clearly suggest [103] that there is some irreversible densification of glass under the indentation. There is evidence that plastic flow also occurs [104] in some glasses where the material has in fact been pushed aside or folded. A result of such plastic flow is the appearance of residual stresses [105] in the glass. Typical values of microhardness for some compositions are also shown in Table 4.9.

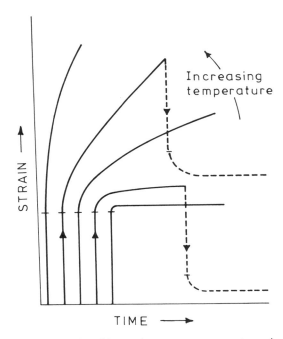

Fig. 4.48 Strain–time relationships under constant stress at varying temperatures (illustrated on shifted scales).

ependence of the mechanical behaviour is often termed delayed
or elastic after-effect). In glasses, the delayed response arises in part due
is flow (because the structure of glasses is thought to be liquid-like) and
ə to a rearrangement of network modifying cations under the influence
The viscous flow in glass increases with temperature, and hence the
havior can be sketched as shown in Figure 4.48. Upon the application
i, there is an instantaneous elastic strain, followed by a curved portion
strain) and then a seemingly linear region (Newtonian) of creep. Upon
val of the stress, the elastic strain is instantaneously recovered, and a
of the delayed strain is recovered with time. The balance, which
s of the creep portion plus a portion of the delayed strain, is due to

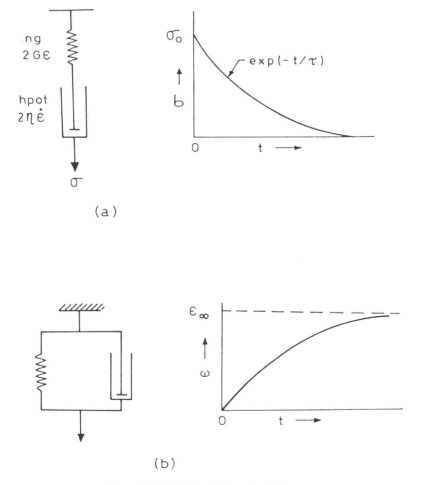

(a)

(b)

Fig. 4.49 Models of viscoelasticity.

153

viscous flow and is never recovered. The recoverable portion of the delayed strain (due to elasticity) and the irrecoverable portion of the delayed strain (due to viscosity) together are called viscoelasticity, and can be modelled using springs (purely elastic) and dashpots (purely viscous). The two simplest of these are shown in Figure 4.49. For a spring and a dashpot in series, called the Maxwell element, it is easy to show that

$$\sigma_t = \sigma_0 \exp(-t/\tau)$$

where

$$\sigma_t = \text{stress at time } t, \text{ and } \tau = \text{relaxation time} = \frac{\text{shear modulus G}}{\text{viscosity } \eta}.$$

This equation is useful to model stress relaxation under constant strain in glass. For the Voigt element, consisting of the spring and the dashpot in parallels, the release of strain ε under constant load with time is given by $\varepsilon_t = \varepsilon_\infty (1 - e^{-t/\tau})$. Here, τ is called the retardation time. The observations of delayed elasticity in the vicinity of the glass transition region are quite understandable, hence. However, according to Murgatroyd and Sykes [105], delayed strain was roughly $\sim 1.2\%$ of the total strain in a sheet glass loaded at ambient temperatures where the relaxation time was expected to be too long for any observable viscous flow to occur. Perhaps, such observations can only be explained in terms of the motion of network modifiers under the influence of applied stresses. A safer statement, however, is to say that relaxation mechanisms in glasses at temperatures far below the glass transition are not well understood.

REFERENCES

[1] Morey, G. W. (1954), *Properties of Glass*, Reinhold, New York.
[2] Napolitano, A. (1965), *J. Amer. Ceram. Soc.*, **48**, 613.
[3] References to different trivial models of B_2O_3 See: Mackenzie, J. D. (1960), *Modern Aspects of Vitreous State* Vol. 1, Butterworths, London.
[4] Lewis, G. N. and Randall, M. (1961), *Thermodynamics* (Revised by K. S. Pitzer and L. Brewer), McGraw-Hill Book Co. Inc., New York, pp. 207–208.
[5] Callow, R. J. (1952), *Trans. Soc. Glass Technol.*, **36**, 137.
[6] Morey, G. W. and Merwin, H. E. (1932), *J. Amer. Opt. Soc.*, **22**, 632.
[7] Bockris, J. O'M., Tomlinson, J. W. and White, J. L. (1956), *Trans. Farad. Soc.*, **52**, 299.
[8] Riebling, E. F. (1967), *J. Amer. Ceram. Soc.*, **50**, 46.
[9] Bray, P. J. and O'Keefe, J. G. (1963), *Phys. Chem. Glasses*, **4**, 37.
[10] Shartsis, L., Capps, W. and Spinner, S. (1953), *J. Amer. Ceram. Soc.*, **36**, 35.
[11] Riebling, E. F. (1963), *J. Chem. Phys.*, **39**, 3022.
[12] Ivanov, A. O. and Yevstropov, K. S. (1962), *Dokl. Akad. Nauk. USSR*, **145**, 797.
[13] Murthy, M. K. and Ip, J. (1964), *Nature*, No. 4916, p. 285.
[14] Riebling, E. F. (1966), *J. Chem. Phys.*, **44**, 2857.
[15] Batsanov, S. S. (1966), *Refractometry and Chemical Structure*, Van Nostrand Rheinhold, New York.
[16] Megaw, H. D. (1939), *Z. Krist.*, **100**, 58.
[17] Nenglein, F. A. (1925), *Z. Physik. Chem.*, **115**, 91, **117**, 285.
[18] Austin, J. B. and Pierce, R. H. H. (1933), *J. Amer. Chem. Soc.*, **55**, 661.
[19] Peters, C. G. and Cragoe, C. H. (1920), *Bur. Standards Scientific Papers*, **393**.

V. H., Irvine, E. and Turner, D. (1952), *Proc. Roy. Soc. (London)*, **A, 108**, 154.

H. R. (1933), *J. Amer. Ceram. Soc.*, **16**, 619.

ferences see Morey, G. W. (1954), *Properties of Glass*, Reinhold, New York, p.

nev, G. B. (1965), in *Physics of Non Crystalline Solids* (ed. J. A. Prins), North nd Pub. Co., Amsterdam, p. 461.

ı, J. F., Hasapis, A. A. and Wholley, J. W., Jr. (1960), *Phys. Chem. Glasses*, 1,

jian, C. R. and Douglas, R. W. (1960), *Phys. Chem. Glasses*, **1**, 19.

enzie, J. D. (1960), *Modern Aspects of Vitreous State*, Vol. 1, Butterworths, on, pp. 188–218.

las, R. W. IV Int. Congress on Rheology, pp. 1–27.

jian, C. R. and Douglas, R. W. (1958), Compt. Rend. 31st Congr. Intern. ıie. Ind., Leige, pp. 1–58.

e, A. (1869), *Théorie Méchanique de la Chaleur*, p. 69.

ew and Rideal (1961), *Interfacial Phenomena*, Academic Press, New York, p.

son and Taylor (1938), *J. Amer. Ceram. Soc.*, **21**, 193, 205.

rman, W. B. (1942), *J. Amer. Ceram. Soc.*, **25**, 168.

n-Marwedel, H. (1937), *J. Soc. Glass Technol.*, **21**, 436.

ie, A. D. (1948), *Contact Angles*, Faraday Soc., **3**, 11.

ns, N. K. (1948), *Dis. Farad. Soc.*, **3**, 5.

ıcke, G. (1968), *Pogg. Ann.*, **135**, 621.

tson, E. W. (1911), *Ind. Eng. Chem.*, **3**, 631.

enier, A. and Gilard, P. (1924; 1925), *Bull. Soc. Chem. Belg.*, **33**, 119; **34**, 27.

mann, G. and Rabe, H. (1927), *Z. Anorg. u. Allgem. Chem.*, **162**, 17.

kel, G. (1934), *Glastech. Ber.* **12**, 413.

peler, G. (1937), *J. Soc. Glass Technol.*, **21**, 53.

ey, J. (1938), *J. Soc. Glass Technol.*, **22**, 38.

ıotin, M. V. and Melikova, I. G. (1952), *Steklo Keram.*, **9**, 4.

e, H. R. (1952), *J. Amer. Ceram. Soc.* **35**, 150.

ies, J. K. and Bartell, F. E. (1948), *Anal. Chem.*, **20**, 1182.

gery, W. D. (1959). *J. Amer. Ceram. Soc.*, **42**, 6.

fson, B. S. and Taylor, N. W. (1938), *J. Amer. Ceram. Soc*, **21**, 193.

hforth, F. and Adams, J. C. (1883). *An attempt to correlate the theories of ıllary action*. Cambridge University Press.

rthingham, A. M. (1885), *Phil. Mag.*, **20**, 51.

guson, A. (1914), *Phil., Mag.*, **28**, 403.

rsey, N. E. (1928), *J. Washington Acad. Sci.*, **18**, 505.

ger, F. M. (1917), *Z. Anorg. Allgem. Chem.*, **101**, 210.

malee, C. W. and Lyon, K. C. (1937), *J. Soc. Glass Technol.*, **21**, 44.

rödinger, E. (1915), *Ann. Physik.*, **46**, 413.

rsey, N. E. (1926), *Bur. Stand. Sc. Paper*, No. 540, **21**, 33.

malee, C. W. and Harman, C. G. (1937), *J. Amer. Ceram. Soc.*, **20**, 224.

rett, L. R. and Thomas, A. G. (1959), *J. Soc. Glass Technol.*, **43**, 179.

rrison, W. and Moore, G. (1938), *J. Res. Nat. Bur. Stand.*, **21**, 337.

bcock, C. L. (1940), *J. Amer. Ceram. Soc.*, **23**, 12.

ırtsis, L. and Smock, A. W. (1947; 1951), *J. Res. Nat. Bur. Stand.*

ırtsis, L. and Spinner, S. (1948), *J. Res. Nat. Bur. Stand.*, **40**, 61.

ylor, H. E. (1956), *Trans. Farad. Soc.*, **52**, 873.

ven, A. E. (1963), *Progress in Ceramic Science*, Vol. 3, Pergamon Press, pp. –196.

ıddon, E., Tippet, E. and Turner, W. E. S. (1932), *J. Soc. Glass Technol.*, **16**, 450.

155

[65] Rebbeck, J. W., Mulligan, M. J. and Ferguson, J. B. (1925), *J. Amer. Ceram. Soc.*, **8**, 329.
[66] Charles, R. J. (1963; 1966), *J. Amer. Ceram. Soc.*, **46**, 236; **49**, 55.
[67] Kuznetsov, A. Ya. (1959), *Zuhr Fiz. Khim.*, **33**, 1492. (*J. Phys. Chem. Moscow.*, **33**, 20).
[68] Blank, K. (1966), *Glastech. Ber.*, **39**, 489.
[69] Otto, K. (1966), *Phys. Chem. Glasses*, **7**, 29.
[70] Mazurin, O. V. and Borisovskii, E. S. (1957), *Zh. Tech. Fiz.*, **27**, 275. *Soviet Physics–Technical Physics* (English Translation), **2**, 243.
[71] Milberg, M. E., Otto, K. and Kushida, T. (1966), *Phys. Chem. Glasses*, **7**, 14. Otto, K. and Milberg, M. E. (1967), *J. Amer. Ceram. Soc.*, **50**, 513.
[72] Fulda, M. (1927), *Sprechsaal*, **60**, 769, 789, 810, 831, 853.
[73] Endell, K. and Helbrugge, H. (1942), *Glastech. Ber.*, 277.
[74] Bockris, J. O'M., Kitchener, J. A., Ignatowicz, S. and Tomlinson, J. W. (1952), *Trans. Faraday Soc.*, **48**, 75.
[75] Urnes S. (1959), *Glass Ind.*, 237.
[76] Tickle, R. E. (1967), *Phys. Chem. Glasses*, **8**, 101 and 113.
[77] Stevels, J. M. (1957), *Handbuck der Physik* XX, Springer-Verlag, p. 372.
[78] Strutt, M. J. (1931), *Arch. Electrotech.*, **25**, 715.
[79] Moore, H. and DeSilva, R. C. (1952), *J. Soc. Glass Technol.*, **36**, 5.
[80] Rinehart, D. W. and Bonino, J. (1959), *J. Amer. Ceram. Soc.*, **42**, 107.
[81] Malkin *et al.* (1967), *Inorg. Materials*, **3**, 123, (translated from Russian).
[82] Orowan, E. (1945–6), *Inst. of Engineers and Shipbuilders in Scotland*, **89**, 165.
[83] Griffith, A. A. (1920), *Phil. Trans. Roy. Soc.*, Series A, **221**, 163–198.
[84] Milligan, L. H. (1929), *J. Soc. Glass Technol.* **13**, 351.
[85] Baker, T. C. and Preston, F. W. (1946), *J. Appl. Phys.* **17**, 170–8.
[86] Wiederhorn, S. M. (1967). *J. Am. Ceram. Soc.* **50**, 407–14.
[87] Wiederhorn, S. M. and Bolz, C. H. (1970), *J. Am. Ceram. Soc.* **53**, 543–8.
[88] Wiederhorn, S. M. and Johnson, H. (1973). *J. Am. Ceram. Soc.* **56**, 192–7.
[89] Weiderhorn, S. M., Evans, A. G., Fuller, E. R., and Johnson, H. (1974), *J. Am. Ceram. Soc.*, **57**, 319–23.
[90] Irwin, G. R. (1958), *Encyclopedia of Physics.* Vol. VI pp. 551–90, Springer-Verlag, Berlin.
[91] Hillig, W. B. and Charles, R. J. (1965), in *High Strength Materials* (V. F. Zackay, ed.) pp. 682–705. John Wiley, New York.
[92] Mould, R. E. and Southwick, R. D. (1959), *J. Am. Ceram. Soc.* **42**, 542–7. ibid **42**, 582–92.
[93] Freiman, S. W. (1980), in *Glass Science and Technology* (D. R. Uhlmann and N. J. Kreidl, eds) p. 57. Academic Press, New York.
[93a] Schreiber, E., Anderson, O. L., and Soga, N. (1973), in *Elastic Constants and Their Measurement*, McGraw-Hill Book Co., New York.
[94] Winklemann, A. and Schott, O. (1894), *Ann. Physik Chem.* **51**, 697; ibid (1897) **61**, 105.
[95] Clarke, J. R. and Turner, W. E. S. (1919), *J. Soc. Glass Technol.* **3**, 260–66.
[96] Makishima, A. and Mackenzie, J. D. (1973), *J. Non-Cryst. Sol.* **12** (1), 35–45.
[97] Gehlhoff, G. and Thomas, M. (1926), *Z. Tech. Physik.*, **7**, 105.
[98] Bridgman, P. W. and Simon, I. (1953), *J. Appl. Phys.* **24**, 405–13.
[99] Cohen, H. M. and Roy, R. (1965), *Phys. Chem. Glasses* **6**, 149–61.
[100] Mizouchi, N. and Cooper, A. R. (1971), *Mater. Sci. Res.* **5**, 461–76.
[101] Mallinder, F. P. and Proctor, B. A. (1964), *Phys. Chem. Glasses* **5**, 91–103.
[102] Ernsberger, F. M. (1968), *J. Amer. Ceram. Soc.* **51**, (10) 545–7.
[103] Peter, K. W. (1970), *J. Non-Cryst. Solids*, **5**(2), 103–15.
[104] Lawn, B. R. and Swain, M. V. (1975), *J. Mater. Sci.* **10**, 113–222.
[105] Murgatroyd, J. B. and Sykes, R. F. R. (1947), *J. Soc. Glass Technol.*, **31**, 47.

CHAPTER 5

Batch Melting Reactions

P. Hrma
Case Western Reserve University, USA

5.1 INTRODUCTION

~ry of glass batch melting reactions has not yet been satisfactorily
:d. Available phase equilibria of ceramic systems [1–5] cover only a
1umber of simple batches, and do not include even as basic a system as
iO$_2$–CO$_2$. Unlike chemical engineering, which connects laboratory
:ments with industrial reactors, glass processing theory has not yet
ed kinetic models of batch melting reactions that would allow us to
'aboratory crucible melts with the processes that occur in glass melting
s. Reaction mechanisms are even less understood. There are no sys-
studies concerned with spatial distribution of reactants and products in a
g batch, which is a granular material, and how the species are transported
come the distances that separate them from each other. Nevertheless, a
nount of literature on the subject exists. Unlike reactions in a beaker,
nelting chemistry is affected by the shape of the whole batch body, and the
to glass conversion process spans a wide temperature range.
version of batch to glass can be described as a process of three main stages:
st stage is characterized by the absence of any melt. In this stage, all free
iost of the bonded water is removed. If water or steam are present, batch
onents may undergo hydrothermal reactions. Most of the crystalline
ions occur during this stage. Also, organic materials burn, react with
nts, or decompose. Solid state reactions result in development of new
alline phases. Evolution of gases, such as water, carbon dioxide, nitrogen, or
∙n is typical.
 the second stage, the melting reactions approach equilibrium in the
∙nce of melt. More crystalline compounds precipitate and are eventually
∙lved in the melt. Inorganic salts, if they were present in the batch, melt and
∙mpose or partially dissolve in the glass melt. All gases except refining gases
liberated. At the end of this stage, the mixture consists of a melt with
∙ended refractory particles and gas bubbles.
n the third stage, the remains of refractory particles are dissolved and bubbles
removed. Dissolution of particles and bubble removal, or fining, are often
It with as two separate processes. However, since bubbles are nucleated on
d surfaces, it is convenient to think about grain dissolving and fining as one
cess.

157

The next section summarizes experimental methods. Then the three melting processes (solid state reactions, reactions in the presence of melt, and refining) are discussed, the first two in detail. As regards the third stage, only the aspect of dissolution of remaining solid particles is dealt with in detail because refining reactions fall into a different category not covered by this chapter.

5.2 EXPERIMENTAL TECHNIQUES

From the point of view of the heat treatment schedule, the methods for studying batch melting reactions can be divided into two groups, isothermal and non-isothermal. Non-isothermal techniques cannot provide accurate kinetic data because the processes are too complex and numerous; their mechanisms change with temperature and several reactions overlap. Data obtained from isothermal techniques are easy to handle and apply to a wide range of temperature histories. However, it is practically impossible to perform isothermal measurements at high temperatures because the initial heating rate of even small batches to the set temperature is slower than the conversion rate, and so the beginning of the process is not under a proper control.

Differential thermal analysis [6] and differential scanning calorimetry, both of them measuring reaction heat, thermogravimetry [7], measuring mass loss due to gas release, X-ray diffraction and Raman spectroscopy [8, 9], identifying crystalline and glassy phases, classical chemical analysis, optical and electron microscopy, hot stage microscopy [10], and evolved gas analysis [11] (gas chromatography and mass spectrometry) have been employed in the studies of batch melting reactions. Other special techniques, such as measurement of electrical conductivity to trace the development of melt [12], dripping of melt drained from batch resting on a porous support to analyse the composition of primary melts, dissolution of large bodies of silica in the melt prepared from the rest of the batch components [13, 14], or direct observation of foam evolution [15], were occasionally used. Visual investigation of incompletely molten batches, or sections through them [16], also provide valuable and unique information about melting in crucibles or industrial furnaces.

5.3 FIRST STAGE OF MELTING

This stage includes all processes a batch undergoes before the first melt appears. In batches with silica as a major component, the most important processes belonging to this stage are hydrothermal [17–19] and solid state reactions that convert solid silica into silicates, which are much easier to melt. Other processes, not so directly connected with melting, are water release [11], crystalline inversions [6, 7, 20, 21], and oxidation of carbon and organics [22, 23].

5.3.1 Solid state reactions

As the term suggests, solid state reactions between batch constituents occur at temperatures lower than that at which the first melt appears in the batch and involve only solid and gaseous phases. Almost all materials originally present in batches, such as silica and carbonates, can be converted into new crystalline

s by solid state reactions. For example, sodium carbonate and silica
ɔdium metasilicate,

$$Na_2CO_3 + SiO_2 = Na_2O . SiO_2 + CO_2$$

further react with silica to form sodium disilicate,

$$Na_2O . SiO_2 + SiO_2 = Na_2O . 2SiO_2$$

ɔdium carbonate to form sodium orthosilicate:

$$Na_2O . SiO_2 + Na_2CO_3 = 2Na_2O . SiO_2 + CO_2$$

te reactions may be beneficial for glass making, because the new
ɪds easily melt without producing gases. However, the melt viscosity is
tely high because all refractory constituents that usually increase
are dissolved, and so homogenization, removal of small trapped
and dissolution of some residual refractory particles that have not been
d, become more difficult. Süsser [24] argues that presintering also
water that would otherwise remain dissolved in glass and lower its
/ during refining.
state reactions can be controlled by heat transfer, volume or surface
n, or other processes. In small samples, such as those used for thermal
ɜ, heat transfer is usually rapid and does not affect the degree of
ion. Efficient heat transfer to large batches [25] can be accomplished by
ing sand before mixing, by percolation of hot gases, or by exothermic
ns, such as between sodium hydroxide and sand [26, 27] or nitrates and
ɜs.
ɪng density and degree of homogeneity determine the contact area
n reactants and diffusion distances, and thus influence both the reaction
nd the final degree of conversion. Batch treatments affecting packing
y and preventing demixing, such as batch pelletizing or briquetting, are
ɔre effective in the early stage of melting, before the first melt appears
ɔ]. The degree of homogeneity [31] depends on particle size of batch
ials, their agglomeration (particles of the same kind tend to form clusters),
ɜgregation (particles of different kind tend to separate to different ends of
ɪtch body).
ɜ degree to which solid state reactions convert batch materials into solid
ɪcts depends on the rate of heating. Although there are numerous solid state
ions between batch components which can proceed with a large variety of
ɪanisms according to their geometrical configuration, compactness, pres-
or absence of water, heat treatment, and so forth, these reactions are
ɪvely unimportant if the batch is heated rapidly because the time is not
ɜient for them to progress to a significant extent before the first melt occurs.
ɪever, if the batch particles are fine and well mixed and if there is sufficient
for the processes to progress, solid state reactions may consume large
ions of silica and decompose most of the carbonates. Batch preheating,
eacting, or presintering provide favourable conditions for both solid state
ɜtions and reactions in which low viscosity molten batch salts are present,
ɪe high viscosity silicate melts do not form or form in a limited amount. These
ɪtments can be beneficial for some batches that are otherwise difficult to melt.

5.3.2 Reaction kinetics

The kinetics of solid state reactions are usually fitted by simple models. The powder reaction models assume that spherical grains of the reactant A are uniformly distributed in space and equal in size, and are embedded in a continuum of the reactant B. The degree of conversion is defined as $\xi = (c - c_e)/(c_i - c_e)$, where c is the concentration of the reactant A and the subscripts i and e denote the initial and equilibrium values. The product grows from the grain surface. The reaction rate can be controlled by nucleation of the product, volume diffusion through the product layer, surface diffusion along the grain surfaces, the rate of removal of the liberated gas, or other mechanisms. This creates a great diversity in reaction kinetics even at temperatures at which batch is free of melt.

If there is no net volume change due to reaction and nucleation is fast, the initial stage of conversion can be described by the Jander equation [32]

$$[1 - (1 - \xi)^{1/3}]^2 = k_1 \Delta C D R_0^{-2} t, \tag{5.1}$$

where k_1 is a constant, ΔC the concentration difference of the diffusing species over the product layer, D the diffusion coefficient, R_0 the original effective grain radius, and t the time. Ginstling and Brounstein [33] derived an equation valid in a wider range of ξ:

$$1 - 2/3\xi - (1 - \xi)^{2/3} = k_1 \Delta C D R_0^{-2} t. \tag{5.2}$$

For small values of ξ, the expressions on the left hand side of equations (1) and (2) become identical and equal to $1/9\,\xi^2$; consequently, k_1 is the same constant in both equations.

If the particles are polydisperse and spatially randomly distributed, the problem can still be analytically approached in many special cases, most of which can be expressed by a general equation

$$-\ln(1 - \xi) = k_m t^m, \tag{5.3}$$

where m depends on the reaction mechanism, rate of nucleation and geometry of nuclei.

Equations such as (5.1) to (5.3) usually do not fit experimental data at short reaction times and at the final stages. Initially, other processes, such as nucleation of the product, are rate-controlling. Since the solid state reactions are slow, the initial stage may last from a few minutes to several hours. At the end of the reaction, the reaction products from neighbouring particles touch and overlap each other, and so the assumptions regarding particle geometry are no longer valid. Also, imperfect mixing results in slower reaction at the end of the process. Another factor that can affect the rate of glass batch reactions is continuous shrinkage of batch due to sintering of sodium carbonate that begins below 500°C and amounts 5 to 12 vol.% [10].

In experimental studies, the time change of ξ at $T = $ const. is usually fitted by an appropriate kinetic equation. As a final result, the value of the activation energy is obtained by plotting $\ln(k_1 \Delta C D)$ vs. T^{-1}, where T is the absolute temperature. If the plot is linear, then $k_1 \Delta C D = k_2 \exp(-B/T)$, where B is the characteristic temperature associated with the activation energy. If ΔC can be determined from a phase diagram as a function of temperature, the activation energy for D can be obtained provided that k_1 is temperature independent.

160

ssible to predict non-isothermal behaviour from isothermal data by equation (5.2) in a more general form

$$1 - 2/3\xi - (1 - \xi)^{2/3} = R_0^{-2}k_2 \int_0^t \exp(-B/T(t))dt. \qquad (5.4)$$

perature history, $T = T(t)$, is known and the reaction mechanism does ge over the temperature range in question, equation (5.4) can be d to obtain the desired $\xi(t)$ function and thus isothermal and non-al data can be transformed to each other. Probably no such attempt has made with glass batches.

inetic studies

ntly studied glass batch is a mixture of silica sand and sodium carbonate. est eutectic, that between silica and sodium disilicate at 790°C, sets the mperature limit for the first stage reactions in this system. This lower ture limit is given by the time at which a measurable amount of product generated.

of the early studies of the reaction between soda ash and silica was carried Pole and Taylor [34], who mixed two moles of quartz powder with one sodium carbonate, both with particle size 75–100 μm, and heat treated ture isothermally in the temperature range 700 to 760°C (below 700°C the conversion was too slow to be measured). They assumed that the reaction t was sodium metasilicate and obtained the degree of conversion, sed as the fraction of total silica reacted, from the amount of carbon e evolved. Since only $\xi \leq 0.1$ was reached at the end of the experiment, on (5.1) was applicable. The resulting reaction rate was expressed as $R_0^{-2} = k_3 \exp(-B/T)$, where $k_3 = 1.8 \times 10^{15} \, \text{s}^{-1}$ and $B = 5.2 \times 10^4$ K; he average silica grain radius was $R_0 \approx 44 \, \mu$m, $k_2 = k_3 R_0^2 = 3.5 \times 10^6 \, \text{m}^2/\text{s}$. rington et al. [35] heat-treated a mixture with 20 mol% of sodium nate at temperatures from 680 to 780°C. They used a coarser grained re with 100 μm sodium carbonate particles and 150–180 μm silica particles finer grained mixture with the corresponding particle sizes 60 μm and 5 μm (only one measurement was taken with quartz 37–44 μm particle size). eaction product was determined by X-ray analysis as sodium metasilicate; e, only up to one fourth of the total silica could react, that is, $\xi = 0.25$ was the mum degree of conversion. Apart from the initial period, they found tion (5.2) fitting their data up to 90% of sodium carbonate conversion. The um metasilicate crystals were oriented, but when vitreous silica was used ad of quartz, the reaction was faster and the metasilicate crystals showed no erred orientation.

he investigators [35] obtained different activation energies for the coarser and r grained mixtures ($B = 4.0 \times 10^4$ K and 2.8×10^4 K, respectively), both lower that obtained by Pole and Taylor [34]. However, calculating $k_1 \Delta CD$ from r original data, and plotting $\ln(k_1 \Delta CD)$ vs. $1/T$ (Figure 5.1), one can see that lata agree with each other very well. From the combined plot one can obtain $CD = k_2 \exp(-B/T)$ with $k_2 = 0.50 \, \text{m}^2/\text{s}$ and $B = 3.6 \times 10^4$ K (notice the large

161

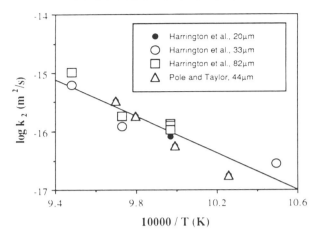

Fig. 5.1 Temperature dependence of the reaction rate constant for the Na_2CO_3–SiO_2 mixture.

difference in the value of k_2 between this value and that of Pole and Taylor). This example shows that a narrow temperature span combined with a relatively small number of measurements may result in a low accuracy of k_2 and B. Extrapolation from such reaction rate data outside the temperature range of measurement would lead to large errors. Therefore, extreme caution is needed.

Thermodynamic calculations indicate that under 1 atm. (0.1 MP) of CO_2, sodium metasilicate can form from 270°C upwards. Wilburn and Thomasson [7], who used non-isothermal techniques, indeed observed commencement of this reaction at 300°C if fine material ($< 40\ \mu$m) were used (the mixture contained 15 mol% sodium carbonate, the heating rate was 10°C/min). The mass loss rate reached maximum at 600°C, then decreased to a minimum at about 720°C and increased to a second sharp maximum at 790°C (Figure 5.2). It has been suggested that the minimum at 720°C is due to formation of disilicate from metasilicate and silica. This disilicate layer separates silica from sodium carbonate, whereas metasilicate did not show such an effect.

According to the combined data by Pole and Taylor [34] and Harrington *et al.* [35], shown in Figure 5.1, the time for 1% conversion of quartz with grain radius 1 μm is 5 min. at 550°C and 89 min. at 500°C, too slow to obtain any measurable mass loss at temperatures below 500°C, even with very fine grains. The fact that Wilburn and Thomasson [7] were able to measure mass loss at 300°C indicates that finely grained quartz is more reactive than coarse quartz. This indicates that extrapolation of reaction rate data obtained with medium grain sized materials to small grain sizes is not justified.

In batches with high silica content, there is not enough sodium carbonate to cover all the available surface of silica. However, the models based on the assumption that silica grains are embedded in sodium carbonate continuum may still be applicable if sodium oxide diffuses rapidly enough along silica grain surfaces. Vidal [10] observed under hot stage microscope formation of a viscous film coating on silica grains at temperatures above 600°C, and attributed this to

162

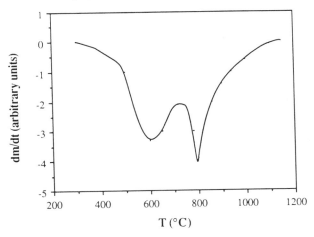

.2 Rate of mass change against temperature for quartz–sodium carbonate mixtures (15% Na_2CO_3, both grained below 40 μm) heated at 10 K/min [7].

diffusion. Ott *et al.* [36] also claim that this mechanism operates in carbonate–silica batches above 700°C, but is absent in potassium ite–silica batches.

n-isothermally treated batches, the solid state reactions may not occur to ectable extent if the temperature increase rate is rapid and the specific area of the reactants is small (the particles are large). For example, when ;h and quartz of the grain size 150–420 μm were used, an appreciable n started at a relatively high temperature of 700°C and the peak rate was l only after the first melt appeared [7]. Similar observation is valid for um silicate batches. Using Raman spectroscopy, Verweij *et al.* [8] found otassium disilicate was formed by solid state reaction after 1 hour ent at constant temperature. No such reaction was observed by Lindig *et*] who heated the glass at 10°C/min.

nerous solid state reactions can occur in other batches. If limestone and n carbonate are present together, they form double carbonate, $\iota(CO_3)_2$, 400–500°C [37]. Wollastonite forms in the mixture of silica and nly above 1400°C [21], but it can form at 850–950°C if Na_2SO_4 is present melting temperature of Na_2SO_4 is 880°C). SiO_2–K_2CO_3–PbO and K_2CO_3–$CaMg(CO_3)_2$ batches were studied by Lindig *et al.* [37] ehrmann and Frischat [17], Na_2CO_3–$BaCO_3$–SiO_2 batches by Krol and n [39], and batches containing zinc oxide by Russell and Ott [40].

5.4   SECOND STAGE OF MELTING

Formation of liquid phase

second stage of melting is characterized by the presence of unreacted ial batch constituents, intermediate crystalline reaction products, and melt.

There are two kinds of melt that can be produced in the batch: molten inorganic salts and glass forming melts. Hence, three basic situations can occur: (i) presence of inorganic molten salts without glass forming melt, (ii) presence of glass forming melt without inorganic molten salt, (iii) presence of both melts. Since these two types of melts are partly mutually soluble, we can also treat these as the low viscosity and high viscosity liquid phase. The viscosity of inorganic salts is of the order 0.1 poise (10^{-2} Pas), whereas viscosity of molten silicates and borates is about six orders of magnitude higher. Table 5.1 lists several liquid phases that can occur in soda-lime-silica glass batches during heating.

Formation of melt is usually a heat transfer controlled process that proceeds in batches more easily than processes involving diffusion. Liquid assists reactions of solids by dissolving them and transferring ions more rapidly. Therefore, when the first melt occurs, the reaction rate increases dramatically [41, 42].

When liquid is present, reaction kinetics become more complex than those in solid state mixtures because a larger number of factors affects the transfer mechanism [43, 44]. Some primary melts readily wet solid phases, until their surfaces are completely coated. Precipitating solid products reduce the amount of melt and a large number of tiny precipitated crystals hold the melt and obstruct its mobility. Reduced fraction of melt means thinner liquid films separating solid grains of different composition, and so reaction rates are enhanced due to shorter diffusion distances. If gaseous products evolve, liquid is stirred, which enhances diffusion, but gas may also separate liquid from solid, thus inhibiting reaction. Finally, a combination of steep concentration gradients within the liquid phase and a large gas–liquid interface area introduce surface tension gradients that drive additional convection.

As the amount of melt increases, batch volume initially decreases because of sintering and pulling solid particles together by melt bridges. Heat conductivity of the batch is increased due to melt bridges between solid particles, but is reduced again when foam develops.

Unless the batch size is very small, or the batch is percolated by other gases, or melting proceeds in a fluidized state, the reaction atmosphere is that of the

Table 5.1
Primary melts in soda-lime-silica batches

320°C	$NaNO_3$ (m)
624°C	Na_2SO_4–$NaCl$ (e)
630°C	Na_2CO_3–$NaCl$ (e)
725°C	$Na_2O \cdot 3CaO \cdot 6SiO_2$–$SiO_2$–$Na_2O \cdot 2SiO_2$ (e)
725°C	$Na_2CO_3 + CaCO_3$ solid solutions–$Na_2Ca(CO_3)_2$ (e)
790°C	SiO_2–$Na_2O \cdot 2SiO_2$ (e)
800°C	$NaCl$ (m)
821°C	$Na_2O \cdot 2SiO_2$–$Na_2O \cdot SiO_2$–$2Na_2O \cdot CaO \cdot 3SiO_2$ (e)
830°C	Na_2CO_3–Na_2SO_4 (e)
840°C	$Na_2O \cdot 2SiO_2$–$Na_2O \cdot SiO_2$ (e)
850°C	Na_2CO_3 (m)
880°C	Na_2SO_4 (m)

(m) melting, (e) eutectic

gases. These gases can initially freely escape through open pores. When
lation limit is exceeded with a further increase of the amount of melt,
rosity becomes closed. If gases still evolve and melt viscosity is high, the
acquires a foamy structure with a volume that can many times exceed its
value [45]. Eventually, the foam collapses, all solid particles except some
y grains are dissolved, and the process enters its final stage.

ffect of minor additions

est temperature at which melt occurs can be reduced by minor additions
Ginstling and Fradkina [46] describe a scenario at which such a minor
of melting agent operates. Consider the binary phase diagram of sodium
te and an inorganic salt (the melting agent) that does not react with silica
5.3). At a constant temperature between the eutectic temperature and the
elting temperature of either salt, the equilibrium mixture will consist of
lium carbonate and a melt containing all the inert salt (present only in a
antity), saturated with sodium carbonate (point A in Figure 5.3). If silica

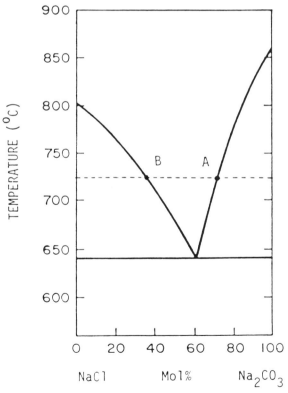

Fig. 5.3 NaCl–Na$_2$CO$_3$ phase diagram.

particles are present, both solid and dissolved sodium carbonate react with silica producing crystalline sodium silicates. At the same time, solid carbonate dissolves in the melt to compensate for that which was lost from the melt due to reaction with silica. The overall reaction rate depends on the fraction of the wetted silica surface. The amount and composition of melt remain constant at this stage.

After the solid carbonate disappears, the melt becomes impoverished in carbonate until the point of saturation by the inert salt (point B in Figure 5.3) is reached. Then the melt composition remains constant due to crystallization of the inert salt. When all carbonate is finally consumed, the mixture consists of unreacted silica, crystals of the inert salt, and crystalline silicates.

The rate-controlling process is most likely diffusion of sodium carbonate through the melt. Very little is known about the spatial configuration of melt, silica grains and different crystalline phases during the process. It is certainly a dynamic situation, because evolving carbon dioxide continuously disturbs the mixture.

Kröger and Marwan [47] found that additions of 4 wt.% of K_2CO_3 or Li_2CO_3 increased the CO_2 evolution rate from soda-lime-silica batches, Li_2CO_3 having the strongest effect. Boron oxide had a similar, although somewhat milder, effect than alkali carbonates. Addition of water did not show any effect by itself, but had an indirect effect by dissolving sodium carbonate and wetting sand grains with the saturated solution. The increase of reaction rate was unaffected by drying prior to melting.

Thomasson and Wilburn [20] showed by thermoanalysis (heating rate 10°C/min) that if the sodium carbonate–silica batch (with 150–200 μm grains of both components) contained 2 mol% NaCl, the temperature of the maximum mass loss rate decreased by 55°C (from 845°C without additions, see Figure 5.4);

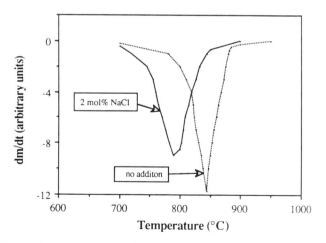

Fig. 5.4 Rate of mass change against temperature for quartz–sodium carbonate mixtures (15% Na_2CO_3, both materials of grain size 150–200 μm; heating rate 10 K/min) with and without addition of NaCl [20].

as used, the temperature drop was 85°C. In a batch with less than 50 μm
ie drop (from 790°C without additives) was even larger: 135°C in the
IaCl and 90°C in the case of NaF. The reversal of order of the effects of
.d NaF on batches with fine and coarse silica cannot be taken too
the reproducibility of TGA results was not reported by the authors.
[48] studied alkali aluminosilicate glass batches and observed similar
.used by CaF_2 and mixtures of $NaCl + Na_2SO_4$ (eutectic melt at 624°C)
Cl + Na_2SO_4 + CaF_2; Na_2SO_4 alone had no effect. Krämer [11], who
.lved gas analysis, found that if 2% of Na_2O was introduced to window
tch as nitrate, the temperature of the maximum CO_2 evolution rate (at
ing rate 5°C/min) decreased by 150°C.

eaction of silica with solid soda ash in the presence of melt

odium carbonate–silica sand mixture without additions, the first melt
at 790°C (the eutectic between silica and sodium disilicate). This melt
iolve a small amount of carbonate [49] (approx. 0.4%, but much less
eratures higher than 790°C).
inciple, any of the three possible crystalline compounds, that is, sodium
e, sodium metasilicate, and sodium orthosilicate, can be formed, but
dynamically disilicate is the most and orthosilicate is the least stable.
)dium carbonate (solid or liquid) and silica dissolve in the silicate melt,
he crystalline silicates precipitate or dissolve and carbon dioxide is
l.
e equilibrium product is a pure melt, the most probable isothermal
n path goes through the stage when only melt and remains of silica grains
sent. However, it may not be so if very fine silica is used. According to
ature and soda–silica ratio, the final equilibrium may contain a small
1 of melt or no melt at all. Kröger [41] followed formation and
earance of melt in sodium carbonate–silica batches at a constant tempera-
' measuring the electrical conductivity, and found that in the metasilicate
sition (with both components of 150–250 μm grain sizes), the maximum
:tivity (roughly corresponding to the maximum fraction of melt) decreased
'0 min at 850°C to 12 min at 920°C.
riya and Sakaino [50] studied carbon dioxide evolution and silica
ition in mixtures of sodium carbonate and silica with 20, 33 and 50 mol %
O_3 under isothermal conditions at the temperature range 762 to 833°C,
is below the melting temperature of sodium carbonate. They ground
n carbonate to obtain as fine powder as possible while the radius of silica
varied between 6 and 850 μm. The final equilibria of mixtures correspon-
, a mixture of sodium disilicate and silica, pure sodium disilicate, and pure
n metasilicate. Although these investigators generated a wealth of useful
they did not interpret them with regard to the possible reaction path.
:ver, a number of useful inferences can be made by examining their results
y.
measuring the degree of conversion of each batch constituent indepen-
y, Moriya and Sakaino [50] showed that in the metasilicate composition
onstituents were initially consumed in equimolar proportions. At the

167

advanced stage, when sodium carbonate approached its full decomposition, some silica remained unreacted. However, at the lowest temperature and low conversion degrees there was a slight opposite tendency, that is, a larger amount of SiO_2 was consumed than that corresponding to metasilicate. These results indicate that the main reaction product was sodium metasilicate and only small fractions of sodium orthosilicate, sodium disilicate and eutectic melt were formed.

For batches with SiO_2 in excess to metasilicate, silica conversion data are not given. However, it seems logical to assume that the beginning of the reaction path was the same for all mixtures and that the rate of reaction was dependent on the silica surface area per unit mass of sodium carbonate, $s = 3m_{SiO_2}/(R_0\rho_{SiO_2}m_{NaCO_3})$, where m_{SiO_2} and m_{NaCO_3} are the mass of silica and sodium carbonate in the batch, and ρ_{SiO_2} is silica density. This is confirmed by Figure 5.5 in which data for two silica grain sizes ($R_0 = 66$ and $134\,\mu m$) and three compositions are used.

As shown in Figure 5.6, the conversion rate strongly depended on the total mass of the batch. This could be due to the effect of carbon dioxide partial pressure: small batches, which were exposed to air with low p_{CO_2}, reacted rapidly, whereas large batches, which were under higher p_{CO_2}, reacted slowly. This effect of p_{CO_2} is probably responsible also for the unexpected result that $t \sim R_0^{0.45}$ (Figure 5.7), where t is the reaction time to attain a constant degree of conversion; according to equation (5.2), which is based on the assumption that $p_{CO_2} = const.$, $t \sim R_0^2$. The difficulties of degasing fine batches would cause a higher p_{CO_2}, and thus the reaction time would be prolonged. It would be interesting to analyze Moriya and Sakaino's [50] data using equation (5.3) to establish the temperature dependence of the reaction rate constant and compare the results with those by Pole and Taylor [34] and Harrington et al. [35].

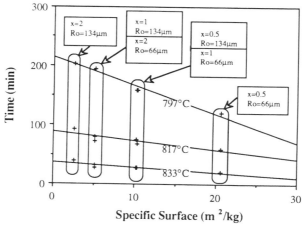

Fig. 5.5 The time to reach the sodium carbonate conversion degree $\xi = 0.915$ as a function of the initial quartz surface area per unit mass of Na_2CO_3; x is the Na_2O/SiO_2 molar ratio and R_0 the initial average quartz grain radius (based on data from [50]).

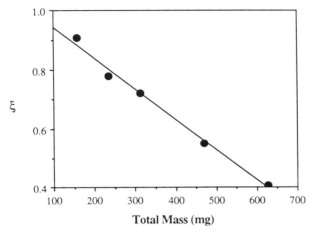

.6 The sodium carbonate conversion degree at $t = 24$ min ($T = 833°C$) as a function of the total batch mass; the Na_2O/SiO_2 molar ratio was 2, and the initial average quartz grain radius was 134 μm (based on data from [50]).

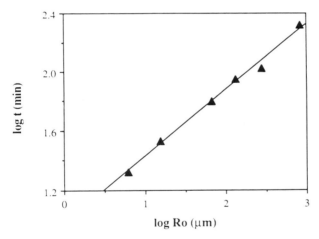

. 5.7 The time to reach the sodium carbonate conversion degree $\xi = 0.8$ ($T = 833°C$) as a function of the initial quartz grain radius; the Na_2O/SiO_2 molar ratio was 2 and the batch mass 313 μg (based on data from [50]).

cording to Kröger [41, 42], the reaction rate of a sodium carbonate–silica ❚ suddenly increases in the temperature region from 835 to 845°C, well ❚ the melting point of sodium carbonate. Kröger and Ziegler [51] studied ❚eaction between sodium carbonate and sodium disilicate and found a ❚en increase in the reaction rate at the temperature interval 830 to 840°C. As

Table 5.1 indicates, the increase is caused by generation of glass forming melt, which may dissolve sodium carbonate; this would decrease its viscosity. Thus, although the silica content of glass forming melt is high, its presence significantly accelerates the reactions.

In summary, silica surface per unit mass of the mixture, partial pressure of the gas produced, parameters affecting this partial pressure, such as batch size and compactness, and the melt fraction in the mixture, determine reaction kinetics. The melt fraction can monotonously increase or decrease after reaching a maximum if the conditions are isothermal. Multiple crystalline products can form.

5.4.4 Reaction of silica with molten soda ash

Since glass forming melt appears in Na_2CO_3–SiO_2 mixture without minor additions at temperatures lower than 850°C (the melting temperature of sodium carbonate), when sodium carbonate melts, a second liquid phase is produced. Approximately at the same temperature, the eutectic melt between sodium metasilicate and sodium disilicate can appear (see the Na_2O–SiO_2 phase diagram in Figure 5.8). Above 874°C, crystalline sodium disilicate can no longer exist and a relatively wide compositional range of thermodynamically stable silicate melt is possible (above 1020°C, the eutectic between sodium orthosilicate and sodium metasilicate, another silicate may occur until the orthosilicate melts at 1090°). Vidal [10] observed under hot stage microscopy that the silicate melts readily dissolve in the carbonate melt.

The reactions proceed in two main stages [43]. In the first stage, liquid sodium carbonate wets silica grains and the mixture reacts vigorously evolving carbon dioxide. In the second stage, no free sodium carbonate is present and the mixture of silicate melt, silica and crystalline sodium silicates react with each other approaching equilibrium that depends on temperature and soda–silica ratio of the mixture. Since the silicate melt wets and separates individual solid components, the reactions are controlled by diffusion in the melt. The key process is dissolution of silica grains because silica is usually the solid component to disappear last, if the final product at a given temperature is a melt. Reaction kinetics of sodium carbonate–silica mixture at a constant temperature above the melting point of sodium carbonate were discussed by Hrma [43] and experimentally investigated by Bobalbhai and Hrma [44].

There is a large variety of possibilities for the rate controlling step in the first stage of melting in laboratory crucibles, such as flow of carbonate melt towards the grains, from which it tends to be separated by the evolving gas, or heat transfer from the crucible walls to the rapidly reacting mixture. As with the solid sodium carbonate, the reaction rate increases as the silica surface per unit mass of sodium carbonate increases. Hence, at a fixed content of silica, the batch with finer silica grains reacts more rapidly and at a fixed silica grain size the batch with higher content of silica reacts more rapidly. The experimental results show that at the end of the first stage one mole of silica has been consumed for two moles of sodium oxide and thus the amount of solid silica at the beginning of the second stage is determined [43]. This, however, is not true for more complex batches, for

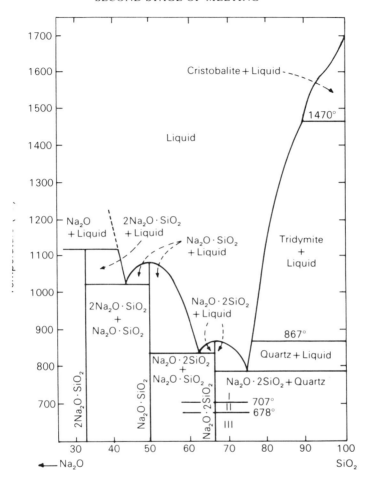

Fig. 5.8 Na_2O–SiO_2 phase diagram.

ie fraction of silica dissolved during decarbonation depends on silica
e and melting temperature [52].

be assumed that during the second stage the crystalline products are
d from silica grains by a layer of the melt of thickness δ, over which the
oxide concentration drops from that which is at equilibrium with the
ne product to that at equilibrium with silica. Naturally, sodium oxide
ransported by diffusion towards silica grains and these grains will shrink
s according to the equation

$$dr/dt = \beta D \Delta f/\delta, \qquad (5.5)$$

is the fraction of silica surface wetted by the melt, D is the diffusion
nt, and Δf is the difference between the mass fraction of sodium oxide at

171

the interface of melt with sodium metasilicate and that at the interface melt with silica. As the amount of silicate melt increases with time, β increases from 0 to 1. At the same time, δ may also increase. It certainly increases at the end of the process when the crystalline reaction products tend to dissolve completely. The amount of melt depends on how much of the crystalline products have precipitated.

The situation is too complex to predict the effect of temperature, total silica content, and initial silica grain size on dissolution rate and dissolution time. Experiments show that temperature has a smaller effect than one would expect because the melt fraction increases as temperature increases and thus δ increases, which, by (5.5), compensates for the increase of D. Of course, the first stage of dissolution affects the total time to dissolve because it determines the size of the residual grains with which the second stage begins.

If temperature is continuously increasing (at the rate about $10°C/min$), most of the carbon dioxide evolves when the primary melt is present, usually above $700°C$ (or $500°C$ if wet batch is used) with a maximum at $850°C$, the melting temperature of sodium carbonate [7].

Reactions of silica with some other sodium containing materials, such as NaCl, NaF, $NaNO_3$, Na_2SO_4, and NaOH, were reported by Thomasson and Wilburn [20], Abe et al. [53], Pugh [26], Cable and Siddiqui [27], Conroy et al. [54], and Höland and Heide [55]. Reaction of silica with potash was studied by Verweij et al. [8, 56, 57], Lindig et al. [37], Ott et al. [36], and Abou-el-Azm and Moore [58].

There are several differences between sodium carbonate and potassium carbonate behaviour in reaction with silica [36]. Sodium ions readily spread over silica grains, whereas potassium carbonate tends to drill into silica grains. Solubility of potassium carbonate in potassium silicate melt seems to be higher than solubility of sodium carbonate in sodium silicate melt. Potassium carbonate reacts more rapidly with silica than sodium carbonate and a mixture of both reacts more rapidly than either of them separately.

5.4.5 Reactions in complex batches

Soda-lime-silica and soda-lime-dolomite-silica batches [21] melt in a similar way to soda-silica batches. The difference is that sodium carbonate reacts with both limestone and dolomite to form double carbonate, $Na_2Ca(CO_3)_2$, at about $500°C$ (if the mixture is heated at the rate of $10°C/min.$; double carbonate is thermodynamically stable above $335°C$, Figure 5.9). Since commercial glasses have higher molar content of soda than lime, this reaction consumes all lime from limestone and dolomite, from which magnesium carbonate is set free to decompose at $630°C$. At $725°C$ (the Na_2CO_3–$CaCO_3$ eutectic temperature [4]), a melt is formed containing $62 \text{ mol}\%$ Na_2CO_3; at $813°C$, double carbonate melts incongruently. Unreacted sodium carbonate, double carbonate, and eutectic melt react with silica, producing CO_2. If a small amount of batch is heated at the rate $5°C/min$ and CO_2 is immediately removed, CO_2 is evolved in the temperature range from 600 to $1100°C$ with maximum at $785°C$ [11].

According to Wilburn et al. [21], the intermediate reaction products formed during $10°C/minute$ heating are sodium metasilicate and sodium disilicate.

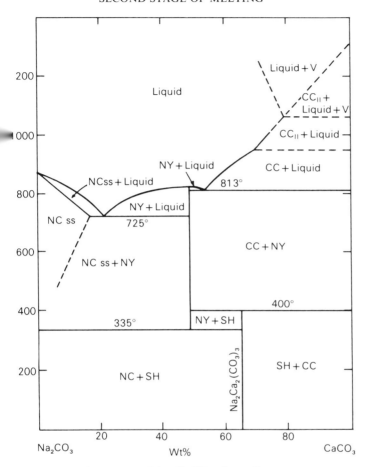

Fig. 5.9 Na_2CO_3–$CaCO_3$ phase diagram.

ger [42], who studied isothermal melts, arrived at the same conclusion; he reasoned that ternary compounds $2Na_2O.CaO.3SiO_2$ and $Na_2O.$ $O.3SiO_2$ form at later stages. Mukerji et al. [59] claimed to find that the product from isothermal melts held at 727–754°C was $Na_2O.CaO.SiO_2$; ever, according to the Shadid and Glasser's [4] phase diagram, this tance is not included among the four ternary compounds of the O–CaO–SiO_2 systems, namely $Na_2O.CaO.5SiO_2$, $Na_2O.3CaO.6SiO_2$ itrite), $2Na_2O.CaO.3SiO_2$, and $Na_2O.2CaO.3SiO_2$.

ilburn et al. [21] paid attention to the development of silicate melt that ns glass during cooling. They detected the first glass forming liquid phase in ples cooled from temperatures 900°C or higher. The estimated fraction of olved silica (of the total silica in the batch) was 0.4 at 900°C, 0.8 at 1000°C, 0.85 at 1100°C. Since the rate of temperature increase was constant, these

173

numbers reflect the rate of dissolution of silica grains; the decrease in dissolution rate at temperatures above 1000°C is remarkable, but not contrary to expectation.

Bader [60] undertook a systematic study of the effect of grain size (in the range 10 to 500 μm) of individual components of soda lime batches on the melting reactions using thermoanalytical techniques; the heating rate was 5 and 10°C/min. Grain size of soda ash had no effect. The extent to which sodium carbonate reacted with silica to form sodium metasilicate decreased with increasing silica grain size until it dropped to zero at 80 μm. The same was true about the reaction of sodium carbonate with sodium metasilicate and sodium disilicate to form a sodium-rich melt. Extent to which sodium carbonate was consumed to produce sodium disilicate and silica-rich melt showed a flat maximum at 30–80 μm. The extent to which double carbonate reacted with silica grew with growing silica grain size; as the silica grain size approached 500 μm double carbonate became the almost exclusive compound reacting with silica. The extent of formation of double carbonate decreased with growing lime and dolomite grain size, dropping to zero at 100 μm of lime and 400 μm of dolomite. Large limestone grains (> 100 μm) decomposed to CaO and CO_2; then soda ash reacted with silica to form melt rather than with crystalline sodium silicates.

Höland and Heide [55] used very fine silica powder obtained from $SiCl_4$ with the result that most of CO_2 was released by 600°C in soda-silica batches. Sodium carbonate grains were <0.3 mm. In soda-lime-silica batches with silica grains <0.1 mm, the CO_2 release temperature increased with increased grain size of either of the carbonates. The quality of the resulting glass, based on the number of undissolved sand grains in the bulk and surface of the melt and on the number of seeds, increased as the temperature of the end of CO_2 release decreased. This temperature was affected by the grain size of raw materials (maximum silica grain size was 0.2–0.3 mm).

An interesting observation about the effect of heating rate on batch melting reactions is due to Riedel [61], who placed a soda ash sphere between a lime cube and a quartz cube (the bodies were about 5 mm in size) and observed their interaction using a hot stage microscope. If the heating rate was 10°C/min, sodium carbonate reacted with silica and shrank away from lime; at a high heating rate, achieved by placing the materials into the furnace preheated at 1100°C, sodium carbonate reacted preferably with lime.

Both finer grains and slower heating rates allow the batch reactions to come closer to equilibrium. Therefore, grain size decrease and slower heating rate have a similar effect on batch reactions. For example, sodium metasilicate forms when the heating rate is very slow or silica is very fine, whereas sodium disilicate and double carbonate precipitate when heating is fast or silica particles are large. In an extreme case when all carbon dioxide is removed before liquid phase can trap it in the form of bubbles, better quality glass results even without refining agents [55]. To achieve similar results with coarse silica and fast heating, fining agents must be added; they remove trapped bubbles and accelerate dissolution of residual silica grains.

Decarbonation of batches is affected by addition of fine-grained refractory raw materials other than silica. Bader [48] showed that introduction of Al_2O_3 into an alkali aluminosilicate glass batch as aluminum hydroxide instead of by

hifted the CO_2 evolution temperature range down by about 30°C.
id Cable [52] observed that 1% Al_2O_3 increased dissolution rate of a
in a carbonate melt at 1000°C almost ten times.
um carbonate-dolomite-quartz batches were studied by Lindig et al.
ches containing lead oxide by Rosenkrands and Simmingskold [63],
[36], and Lindig et al. [37], a mixture of $15Na_2CO_3$, $10BaCO_3$, and
oy Krol and Janssen [9, 39, 64], Abou-el-Azm and Moore [58, 65]
i reaction extent in SiO_2–B_2O_3 mixtures as a function of time and
ure, Stoch and Pater [66] used thermal analysis and X-ray diffraction
borosilicate batches during temperature increase, and Bader [60]
the evolved gas analysis to B_2O_3–SiO_2 batches containing sodium
te and sodium nitrate.
i is a common constituent of batches. It can be viewed as a high viscosity
led to the batch. According to Bader [67], cullet brings about a decrease
artial pressure of the evolved gas caused by its easier escape through the
t also dilutes the granular part of the batch, thus reducing the specific
surface area between reactants.

5.5 THIRD STAGE OF MELTING

ird stage of melting is strongly affected by the processes the batch has
gone during the first two stages. When the stage of vigorous melting
ins is over, the mixture becomes a fluid with suspended refractory grains
ibbles. In this stage several processes occur simultaneously: dissolution of
al refractory grains, evolution of oxygen from redox reactions, and
ion of fining gases; they result in removal of solid and gaseous inclusions,
gation, and foaming. These processes are described only briefly below.
vival of a refractory constituent in the glass melt depends on the solubility
diffusion coefficient, hydrodynamic conditions, and melting history. Hy-
namic conditions and concentration distribution of the dissolved re-
ory component in the melt affect the concentration layer thickness around
grains (see equation 5.5) and thus control the rate at which the particles
lve. The concentration distribution of melt components depends on the
ing history [51]. If the history is such that the melt is homogeneous,
entration layers will be extremely thin and dissolution will proceed rapidly.
a the other hand, a grain is surrounded by an almost saturated melt and the
centration layer is spread over a long distance, dissolution will be extremely
/. Either of these situations can occur in glass melting.
i most commercial batches, silica grains are large and dissolve slowly;
efore, they are the last solid particles to be dissolved. Their presence at high
iperatures is beneficial because solid grains provide sites for heterogeneous
leation of fining bubbles [68, 69].
The situation when a silica grain is surrounded by an almost saturated melt is
ially a result of grain accumulation by segregation. Vierneusel et al. [70]
inted out that the melt rich in alkali and alkaline earth oxides has higher
nsity than quartz and cristobalite and so buoyancy may be responsible for
gregation. Segregation can be reduced if particles are small initially, or dissolve
rly to a substantially smaller size, or dissolve rapidly enough so that their

lifetime is short and thus there is not enough time for them to segregate. However, if initially small particles agglomerate [71], the agglomerates behave like large particles. Also, any convection that brings down the rising particles, will reduce segregation.

Minor ingredients producing bubbles (refining agents) at high temperatures, deliberately introduced into batches, oversaturate melt with gases, such as O_2 or SO_2, that diffuse into small carbon dioxide bubbles or nitrogen bubbles from air trapped within the melt. These small bubbles grow and quickly leave the melt [69]. Refining agents themselves produce bubbles (nucleated on solid surfaces) that stir and effectively homogenize the melt and provide surfaces into which other oversaturated gases can sink. The remaining small bubbles from minor additions easily dissolve on cooling and so the resulting glass is not seedy.

A classical refining agent is arsenic combined with sodium nitrate. According to Krämer [11], sodium carbonate is involved in their initial interaction, which is quite complex and can be described by the overall scheme

$$As_2O_3 + 24\,NaNO_3 + 10\,SiO_2 + Na_2CO_3 \rightarrow$$
$$2\,Na_3AsO_4 + 10\,Na_2SiO_3 + 24\,NO_2 + 5\,O_2 + CO_2.$$

Oxygen is evolved at temperatures above $1200°C$ (with a maximum at $1450°C$) by the reaction:

$$2\,Na_3AsO_4 + 3\,SiO_2 \rightarrow 3\,Na_2SiO_3 + As_2O_3 + O_2.$$

Any multivalent oxide, if introduced at higher oxidation state into batch, is reduced as temperature increases and produces oxygen [72]. If the evolution of oxygen or another gas is too vigorous and melt viscosity relatively high, foam occurs. In commercial glass batches containing carbonates, the carbonate foam is produced at $1000-1500°C$. If sulfate is added to the batch, the sulfate foam occurs at $1400-1500°C$. In nuclear waste glasses, in which batch composition control is limited, it is difficult to avoid gas evolution in the viscosity region 10^3 to 10^5 poise (10^2 to 10^4 Pas) in which foam has maximum persistence. In fact, several redox reactions of nuclear waste constituents evolve oxygen in this viscosity region and cause severe foaming [15, 22, 45].

REFERENCES

[1] Levin, E. M., Robbins, C. R. and McMurdie, H. F. (1964), in *Phase Diagrams for Ceramists*, The American Ceramic Society.

[2] Levin, E. M., Robbins, C. R. and McMurdie, H. F. (1969), in *Phase Diagrams for Ceramists*, The American Ceramic Society.

[3] Levin, E. M. and McMurdie, H. F. (1975), in *Phase Diagrams for Ceramists*, The American Ceramic Society.

[4] Roth, R. S., Negas, T. and Cook, L. P. (1981), in *Phase Diagrams for Ceramists*, Vol. 4, The American Ceramic Society.

[5] Roth, R. S., Negas, T. and Cook, L. P. (1983), in *Phase Diagrams for Ceramists*, Vol. 5, The American Ceramic Society.

[6] Wilburn, F. W., Thomasson, C. V. and Cole, H. (1959), in *Symposium on Glass Melting*, Union Scientifique Continental du Verre, Charleroi Belgium.

[7] Wilburn, F. W. and Thomasson, C. V. (1958), *J. Soc. Glass. Tech.*, **42**, 158.

[8] Verweij, H., van den Boom, H. and Breemer, R. E. (1978), *J. Amer. Ceram. Soc.* **61**, 118.

REFERENCES

D. M. (1982), *Rivista della Staz. Sper. Vetro*, **5**, 194.
, A. (1963), *Glastech. Ber.*, **36**, 305.
ner, F. (1980), *Glastech. Ber.*, **53**, 177.
ger, C. (1952), *Glastech. Ber.* **25**, 307.
e, M. and Martlew, D. (1984), *Glass Technol.*, **25**, 24.
e, M. and Martlew, D. (1984), *Glass Technol.*, **25**, 139.
tong, C. and Hrma, P. (1988), *J. Amer. Ceram. Soc.*, **71**, 323.
ting, J. A. and Bieler, B. H. (1969), *Amer. Ceram. Soc. Bull.*, **48**, 781.
etzki, K. H. (1982), *Silikattechnik.*, **33**, 149.
rmann, E. and Frischat, G. (1986), *J. Amer. Ceram. Soc.*, **69**, C84.
iams, J. A. (1978), in *39th Ann. Conf. on Glass Problems*, Columbus, Ohio.
masson, C. V. and Wilburn, F. W. (1960), *Phys. Chem. Glasses*, **1**, 52.
burn, F. W., Metcalfe, S. A. and Warburon, R. S. (1965), *Glass Technol.*, **6**, 107.
dinec, M. J. (1986), *J. Non-Cryst. Solids*, **84**, 206.
kford, D. F. and Diemer, R. B. (1986), *J. Non-Cryst. Solids*, **84**, 276.
ser, V. (1963), *Veda vyzk.*, **8**, 33.
oper, A. R. (1979), *Riv. Staz. Sper. Vetro.*, **9**, 219.
gh, A. C. P. (1968), *Glastek. Tidskr.*, **23**, 95.
ble, M. and Siddiqui, M. Q. (1981), *Glass Technol.*, **21**, 193.
sta, P. (1977), *Glastech. Ber.*, **50**, 10.
sta, P. (1977), *Glastech. Ber.*, **50**, 301.
mamoto, J. and Komatsu, E. (1968), *Glass Ind.*, **49**, 491.
ooper, A. (1973), *Glasteck. Tidskr.*, **28**, 27.
nder, W. (1927), *Z. anorg. allgem. Chem.*, **163**, 1.
instling, A. M. and Brounstein, B. I. (1950), *J. Appl. Chem. USSR*, **23**, 1327.
ole, G. R. and Taylor, N. W. (1935), *J. Amer. Ceram. Soc.*, **18**, 325.
arrington, R. V., Hutchins, III, J. R. and Sherman, J. D. (1962), in *Advances in Glass 'echnology*, Plenum Press, New York, p. 75.
Ott, W. R., McLaren, M. G. and Harsell, W. B. (1972), *Glass Technol.*, **13**, 154.
indig, M., Gehrmann, E. and Frischat, G. (1985), *Glastech. Ber.*, **58**, 127.
aylor, T. D. and Rowan, K. C. (1983), *Commn. Amer. Ceram. Soc.*, **66**, C-227.
Krol, D. M. and Janssen, R. K. (1982), *J. de Physique*, **43**, C9.
Russell III, H. H. and Ott, W. R. (1980), *Glass Technol.*, **21**, 237.
Kröger, C. (1952), *Glastech. Ber.*, **25**, 307.
Kröger, C. (1957), *Glastech. Ber.*, **30**, 42.
Hrma, P. (1985), *J. Amer. Ceram. Soc.*, **68**, 337.
Bodalbhai, L. and Hrma, P. (1986), *Glass Technol.*, **17**, 72.
Bickford, D. F., Hrma, P., Bowan, S. W. and Smith, P. K. (in press) *Amer. Ceram. Soc. Bull.*
Gistling, A. M. and Fradkina, T. P. (1952), *J. Appl. Chem. USSR*, **25**, 1325.
Kröger, C. and Marwan, F. (1956), *Glastech. Ber.*, **29**, 275.
Bader, E. (1978), *Silikattechnik*, **29**, 84.
Pearce, M. L. (1964), *J. Amer. Ceram. Soc.*, **47**, 342.
Moriya, T. and Sakaino, T. (1955), *Bull. Tokyo Inst. Tech.*, **B(2)**, 13.
Kröger, C. and Ziegler, G. (1953), *Glastech. Ber.*, **26**, 346.
Hrma, P., Barton, J. and Tolt, T. L. (1986), *J. Non-Cryst. Solids*, **84**, 370.
Abe, O., Utsunomiya, T. and Hoshino, Y. (1983), *Chem. Soc. Japan*, **56**, 428.
Conroy, A. R., Manring, W. H. and Bauer, W. C. (1966), *Glass Ind.*, **46**, 84.
Höland, W. and Heide, K. (1981), *Silikattechnik*, **32**, 344.
Verweij, H., van den Boom, H. and Breemer, R. E. (1977), *J. Amer. Ceram. Soc.*, **60**, 529.
Verweij, H. (1981), *J. Amer. Ceram. Soc.*, **64**, 493.
Abou-El-Azm, A. M. and Moore, H. (1953), *J. Soc. Glass Technol.*, **37**, 129.

[59] Nandi, A. K. and Mukerji, J. (1977), in *Proc. XIth Int. Cong. Glass*, Prague, Vol. 4, p. 177.

[59a] Mukerji, J., Nandi, A. K. and Sharma, K. D. (1980), *Amer. Ceram. Soc. Bull.*, **59**, 790.

[60] Bader, E. (1983), *Silikattechnik*, **34**, 7.

[61] Riedel, L. (1962), *Glastech. Ber.*, **35**, 53.

[62] Cable, M. (1974), *Glastek. Tidskr.*, **29**, 11.

[63] Rosenkrands, B. and Simmingskold, B. (1962), *Glass Technol.*, **3**, 46.

[64] Krol, D. M. and Janssen, R. K. (1983), *Glastech. Ber.*, **56K**, 1.

[65] Abou–El–Azm, A. M. and Moore, H. (1953), *J. Soc. Glass Technol.*, **37**, 155.

[66] Stoch, Z. and Pater, W. (1980), *Silikattechnik*, **31**, 364.

[67] Bader, E. (1979), *Silikattechnik*, **30**, 112.

[68] Němec, L. and Žlutický, J. (1979), *Sklar Keram.*, **29**, 353.

[69] Němec, L. (1974), *Glass Technol.*, **15**, 153.

[70] Vierneusel, U., Goerk, H. and Schüller, K. H. (1981), *Glastech. Ber.*, **54**, 332.

[71] Jebsen-Marwedel, H. (1957), *Glastech Ber.*, **30**, 122.

[72] Schreiber, H. D., Balaz, G. B., Carpenter, B. E., Kirkley, J. E., Minnix, L. M. and Jamison, P. L. (1984), *J. Amer. Ceram. Soc.*, **67**, C106.

CHAPTER 6

Chemical Durability of Glass

lern container is kept full of water for a year, it will lose 30 mg of material to
er, but this will fall to 3 mg if the bottle had first been treated with sulphur
, and to only 1 mg if the bottle were to be made from chemically resistant
icate glass.

Cookson Monthly Bull. (1988) No. 566.

ı 'chemical durability' has been used conventionally to express the
: offered by a glass towards attack by aqueous solutions and atmos-
ents. There is no absolute or explicit measure of chemical durability and
e usually graded relative to one another after subjecting them to similar
ntal conditions; the nature of the experiment usually determines the
ırder. Interferometry and weight loss measurements on the attacked
well as alkalimetric titration, pH and electrical conductivity measure-
ı the extracts have been in common use in the past. In recent years a
analysis of the leached solutions for all the glass components and a
analysis of the leached glass surface have provided useful information
? various factors involved in the decomposition of glass. For example,
in the near corroded surface of glass $(1-20 \text{ Å})$ can be monitored with
lectron spectroscopy (AES)[1]. Electron spectroscopy for chemical
[ESCA)[2], ion-scattering spectroscopy (ISS)[3] or secondary-ion mass
:opy (SIMS)[4]. Coupling these methods with ion-beam milling yields
letailed compositional profiles of the intermediate glass surface
0 Å)[5]. Measurement of the average composition of the deeper surface
$-10\,000 \text{ Å})$ can easily be made with electron-microprobe analysis (EMP),
lispersive X-ray analysis (EDX) in the scanning electron microscope
ır infrared reflection spectroscopy (IRRS)[6].
us durability tests have been devised to compare the rates of attack on
glass compositions under standard conditions. These tests are usually
ı carefully[7] prepared graded glass grains or on a manufactured glass
such as a bottle. The choice would normally depend upon whether
tion is required on the intrinsic durability of a certain composition, or on
ınished product will stand up to attack.
hemical durability of a formed glass article can be improved by lowering
li content of the surface of the article before use. Such a reduction can be
l by heating the articles in an atmosphere of SO_2 [8] or by adding various
ls to the glass surface which react with alkali diffusing to the surface at
nperature ($\sim 650°C$), forming an alkali compound such as Na_2SO_4 [9].
phate can then be rinsed off thus reducing the concentration of alkali at
s surface.

6.1 MECHANISM OF REACTIONS OF GLASSES WITH AQUEOUS SOLUTION

When a piece of ordinary glass is brought into contact with an aqueous solution, alkali ions are extracted into the solution in preference to silica and an alkali-deficient leached layer is formed on the surface of the virgin glass. The formation of this layer usually reduces the rate of alkali extraction by forming a barrier through which further alkali ions must diffuse before they can be brought into solution. Recent X-ray diffraction analysis using pair function and disorder distribution function analysis has shown that the silica-rich films formed during the corrosion process are more closely equivalent to vitreous SiO_2 produced from the molten state than a hydrated silica structure [10]. The thickness of the silica-rich films and probably their compactness also vary with the composition of the glass and, for the same glass, depends on the test conditions, namely, time, temperature and pH of the solution; under identical conditions of corrosion, a poorly durable glass usually produces a thicker film than a durable glass.

According to Charles[11] the corrosion of alkali–silicate and alkali–lime–silicate glasses by aqueous solutions can be described in terms of three chemical reactions:

(a) The penetration of a 'proton' from water into the glassy network, replacing an alkali ion into solution:

$$\equiv Si - OR + H_2O \rightleftharpoons \; \equiv Si - OH + R^+ + OH^- \qquad (6.1)$$

(b) The hydroxyl ion in solution disrupts siloxane bonds in glass:

$$\equiv Si - O - Si \equiv \; + OH^- \rightleftharpoons \; \equiv Si - OH + \; \equiv Si - O^- \qquad (6.2)$$

(c) The non-bridging oxygen formed in reaction (6.2) interacts with a further molecule of water producing a hydroxyl ion, which is free to repeat reaction (6.2) over again:

$$\equiv Si - O^- + H_2O \rightleftharpoons \; \equiv Si - OH + OH^- \qquad (6.3)$$

Penetration of a bare proton, as suggested in reaction (6.1), is energetically improbable for the hydration energy of H^+ to H_3O^+ is very large and negative (~ -367 kcal/mol). From a recent infrared study on corroded thin films of glass, Scholze [12] has reported the existence of free water molecules inside the leached layer: the ratio of entrant protons to H_2O molecules inside the leached layer has been found to change with the temperature of leaching and the nature of the replaceable alkali ion, but it is apparently independent of the alkali content of the glass. However it is not clear from Scholze's study that the H_2O molecules he estimated with infrared spectroscopy in the leached layer have really diffused from the solution phase as H_2O or H_3O^+ species or formed *in situ* inside the leached layer due to an auto-condensation reaction of the type:

$$\equiv Si - OH + HO - Si \equiv \; \rightleftharpoons \; \equiv Si - O - Si \equiv \; + H_2O \qquad (6.4)$$

It should be pointed out that this type of condensation reaction is well known on a hydrated silica surface [13] and indeed it is the reaction which converts soluble silicic acid to insoluble silica in the gravimetric estimation of silica in silicate materials.

and Isard [14] studied the extraction of alkali from a com-
~ilicate glass (SiO$_2$–69.9 %, Al$_2$O$_3$–2.6 %, CaO–5.4 %, MgO–3.6 %,
~ %) and found that the amount of sodium removed from the glass
~the action of distilled water varied as the square root of time. From this
~n they concluded that the rate-controlling process involved is one of
~on and the rate of extraction of the sodium should be related to the
~onductivity of the glass. They assumed that, below the softening
~re of the glass, the atoms of the silica network are in fixed positions
~e Na$^+$ ions can move from one site to a neighbouring site if they
~definite energy E, and that the passage of electric current through the
~s place exclusively by migration of Na$^+$ ions. During the diffusion of
to the glass surface, the electrical neutrality of the glass must be
~d by the contra-diffusion of other ions, otherwise an electric double
ld be set up at the glass surface which would prevent further removal of
~ns. The amount of sodium that could be removed before the double
~ped the process was estimated and found to be of the order of 100 times
the amount of sodium actually extracted. On the assumption of simple
~, the following relation between the quantity of Na$^+$ ions removed Q, the
coefficient D, and the electrical conductivity, σ of the glass can be

$$Q = \frac{2N_0}{\sqrt{\pi}} \cdot \sqrt{(Dt)} \qquad (6.5)$$

$$\frac{Q}{\sqrt{t}} = \frac{2N_0}{\sqrt{\pi}} \cdot \frac{\sigma KT}{N_0 \varepsilon^2} \qquad (6.6)$$

$\checkmark_0$ is the initial concentration of Na$^+$ in the glass, t is the time, K is
~ann's constant, T is the absolute temperature, and ε is the electronic charge.

Table 6.1 Diffusion coefficients of alkali ions in glass

~ass composition (mol %)	Temperature (°C)	D_R (cm^2/s)	D_C (cm^2/s)	D_R/D_C
), 85 SiO$_2$	68.4	2.959×10^{-12}	5.146×10^{-14}	57.5
), 5 CaO. 80 SiO$_2$	70.5	2.954×10^{-14}	1.115×10^{-15}	26.5
~, 85 SiO$_2$	49.7	7.31×10^{-12}	1.433×10^{-16}	64300
~, 5 CaO, 80 SiO$_2$	70.9	1.80×10^{-12}	6.087×10^{-17}	29.571

~iglas and Isard observed that when the experimental values are inserted in
~on (6.6), the equation balanced except for a factor of about 3. However,
~r work [15] has shown a marked difference between the diffusion
~ient D_R calculated from the leaching experiments and the diffusion
~cient D_C estimated from the electrical conductivity measurements. Some
~l results are given in Table 6.1 which shows that D_R/D_C varies considerably
the glass composition. This discrepancy may be due to any one or a
~ination of the following factors:

(a) The diffusion medium in the leaching experiments is different from that of the conductivity experiments;
(b) No account was taken in the above treatment of the amount of silica removed from the glass surface during leaching;
(c) It has been assumed that the diffusion coefficient is not concentration dependent and this is not necessarily true;
(d) In this deduction it has been assumed that all the Na^+ ions in the glass take part in the leaching and in the conductivity and this also may not be true.

Numerous other workers have also studied the extraction of alkali from different types of glasses as a function of time, and nearly all the available data can be summarized into two main forms of rate equations:

6.1.1 Rate of alkali extraction

The rate of alkali extraction varies linearly with the square root of time at short times and at low temperatures and then linearly with time at sufficiently long times and/or at high temperatures. As an example the results of water-leaching of $15Na_2O$, $85SiO_2$ glass at various temperatures are shown in Figures 6.1. Figure 6.1(a) shows a plot of the weight in milligrams of Na_2O extracted from one gram of glass against the square root of time for the early stages of the extraction while Figure 6.1(b) represents a plot of the weight of Na_2O extracted against the time for longer periods of extraction. This sort of extraction behaviour can be represented by an empirical relationship of the following form

$$Q = a\sqrt{t} + bt \tag{6.7}$$

where t is time, and a and b are empirical constants.

$$\frac{d \log Q}{d \log t} = \left(\frac{a}{2\sqrt{t}} + b\right)\left(\frac{1}{(a/\sqrt{t}) + b}\right) \tag{6.8}$$

The gradient of (6.8) has the limiting values of $1/2$ as $t \to 0$ and 1 as $t \to \infty$ and increases slowly with time at intermediate values. Over limited times approximately linear plots of $\log Q$ vs. $\log t$ would be expected, and the slopes of these plots would be expected to vary between $1/2$ and 1 as time and temperature increase.

In the case of leaching of silicate glasses by aqueous solutions, sodium and silicon (in the form of soluble silicate groups) are extracted simultaneously. Hlavac et al. [16] have developed a mathematical model with which it is possible to explain the whole course of the decomposition of glass in aqueous solutions including the period of constant rate. They assumed that during corrosion of glass in aqueous solution, the following two processes occur simultaneously:

Process I: Exchange of glass cations for H_3O^+ cations from the solution controlled by the diffusion of the ions involved through the thus formed leached layer.

Process II: Dissolution of the leached layer controlled by a surface reaction.

Process II continuously diminishes the thickness of the leached layer, thus increasing the concentration gradient and affecting also Process I.

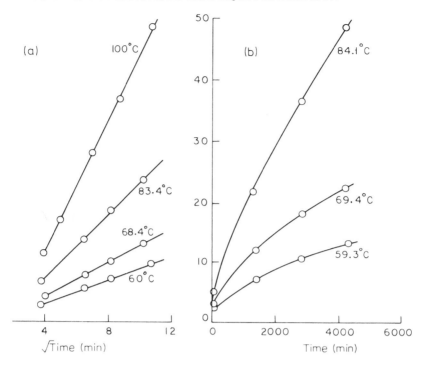

Fig. 6.1 (a) Short-time water-leaching of 15 Na$_2$O, 85 SiO$_2$ glass.
(b) Long-time water-leaching of 15 Na$_2$O, 85 SiO$_2$ glass [20].

: a mathematical formalism to such a complex process, they made the simplifying assumptions:

ar concentration gradient in the leached layer whose thickness is thus letermined. It should be stressed here that this assumption is not correct ise the concentration gradient in the leached layer has been found to be inear [17].

ate of process II is constant.

sion products are being removed from the solution and therefore do ffect the further course of corrosion.

osion takes place on a planar interface or on particles whose radii are iently large in relation to the rate of the process.

oresence of only one kind of alkali ion is considered. With the above lifying assumptions, the derivation is based on the relation

$$\frac{dQ}{dt} = D \frac{C_A}{x(t)} \cdot F \qquad (6.9)$$

is the apparent diffusion coefficient, C_A the concentration of alkali ion in nal glass, x the thickness of the leached layer, and F the reaction area.

183

The total amount of alkali transferred into solution in time t by diffusion according to (6.9) consists of two parts: the amount that corresponds to a fully destroyed portion of glass, Q_2, and the amount that corresponds to the thickness of the remaining layers, Q_{AV}:

$$Q_{AV} = \frac{F C_A x(t)}{2} \qquad (6.10)$$

and

$$Q_2 = B F C_A t \qquad (6.11)$$

where B is a constant giving the rate of heterogeneous reactions as the rate of progress to the depth. The total amount, Q is then given by:

$$Q = Q_2 + Q_{AV} = F C_A \left[\frac{x(t)}{2} + Bt \right] \qquad (6.12)$$

By differentiating (6.12) and substituting (6.9), the velocity equation for x is obtained:

$$\frac{dQ}{dt} = F C_A \left[\frac{1}{2} \frac{dx}{dt} + B \right] \qquad (6.13)$$

and

$$\frac{dx}{dt} = 2 \left[\frac{D}{x} - B \right] = \frac{A}{x} - 2B \qquad (6.14)$$

where $A = 2D$

With the initial condition $t = 0$, $x = 0$, the integration of (6.14) leads to the relation

$$t = \frac{1}{(2B)^2} \left[A \ln \frac{1}{1 - (2Bx/A)} - 2Bx \right] = f(x) \qquad (6.15)$$

For $t \rightarrow \infty$ the thickness of the layer tends to the value of $A/2B$.

Hlavac et al. [16] have shown that with two adjustible parameters A and B, (6.15) describes the leaching behaviour of alkali oxides from binary and ternary silicate glasses satisfactorily.

Lyle [18] has reported an extensive amount of leaching data in which plots of $\log Q$ against $\log t$ can be represented by straight lines over two decades of $\log t$; the slope of the lines varied between 1.107 to 0.500.

6.2 FACTORS AFFECTING CHEMICAL DURABILITY MEASUREMENTS OF A GLASS

Factors affecting chemical durability measurements of glass are:

(a) Weight of glass grains used and the surface area exposed,
(b) Ratio of the weight of the glass to the volume of the leaching solution,
(c) Nature of the leaching solution and the frequency of replenishing it,
(d) Temperature of leaching.

·face area

ea is an important factor and the amounts of various constituents
ɣ a glass under certain conditions are proportional to the surface area
n most of the investigation it is assumed that carefully prepared grains
te size range have a surface area which is proportional to their weight.
:y of this assumption can be checked from the results of an experiment
ie weight of glass grains is varied while other factors are kept constant.
s strongly support the assumption adopted.

tio of the surface area of the glass to volume of the leaching solution

and Turner[19] found that the percentage weight loss of an
₂O–SiO$_2$ glass increases about twice with a fourfold increase in the
f grains tested in a given volume of solution. This suggests that the
ıf material extracted from a silicate glass varies with the ratio of surface
ɪss to the volume of the leaching solution. An investigation was carried
ımy [20] to determine the effect of variation of this ratio on the amount
ividual constituents extracted from binary alkali oxide silicate glasses.
:hing experiments were carried out on a 25 K$_2$O, 75 SiO$_2$ glass. In these
ıts the volume of the leaching solution was kept at 100 ml while the
the glass grains was increased systematically. The quantity of alkali and
acted after various time intervals was determined in every case. The
these experiments are given in Table 6.2.

Table 6.2
Results of leaching a 25 K$_2$O, 75 SiO$_2$ glass at 40° C

Volume of the leaching solution	Weight of grains	Approximate ratio of surface area to volume of solution ($cm^2\ ml^{-1}$)	mg oxide extracted per gram glass		Final pH	Mean final pH
			K_2O	SiO_2		
			13.33	3.11	11.8	
			30.00	7.07	11.9	
100 ml	3.00 g	1.5	53.33	18.75	11.9	11.72
			87.66	52.62	11.6	
			126.00	95.72	11.4	
			15.5	3.21	11.6	
			32.0	6.04	11.7	
100 ml	2.00 g	1.0	54.5	12.36	11.7	11.62
			84.5	29.39	11.6	
			121.5	63.46	11.5	
			15.0	2.63	11.4	
			31.75	5.30	11.5	
100 ml	1.00 g	0.5	54.75	9.75	11.5	11.56
			83.25	20.45	11.5	
			115.25	40.60	11.4	

The quantity of silica extracted per gram of glass after a given time increases as the ratio of the surface area of the glass to the volume of the leaching solution increases. The ratio of alkali to silica in the extracts is higher than in the glass. Final pH of the leaching solution increases slightly when the ratio of the surface area of the glass to the volume of the solution is increased. As will be shown later the rate of silica extraction from a glass increases with the pH of the solution above a pH of about 9. The rise in the amount of silica extracted from this glass, when the ratio of the surface area to the volume of the solution is increased, can be attributed to the accompanying increase in the pH of the solution.

The quantity of alkali removed in a given time did not vary appreciably with the ratio of the surface area of the glass to volume of the solution. The results show that the pH of the solution increases when the ratio of the surface area of the glass to the volume of the solution is increased. The increase in the pH would be expected to suppress the exchange of alkali ions of the glass with protons from the solution. However, increasing the pH of the solution also favours the dissolution of silica and this has the opposite effect on the removal of alkali from the glass because it causes alkali to pass into solution through breakdown of the silica network and reduces the thickness of leached layer. It is, therefore, probable that the apparent independence of alkali extraction on the ratio of the surface area of the glass to volume of the solution is due to these two factors counter-balancing one another.

Limitations on the ratio of the surface area of glass to the volume of the leaching solution to be used in any experiment are usually imposed by the sensitivity of the analytical methods as well as the size of the sample to be used. Yet the choice of this ratio remains arbitrary over a relatively wide range depending on the details of the experiment.

6.2.3 Nature of the leaching solution and the frequency of replenishing it.

An experiment was carried out by Shamy [20] to study the effect of frequency of replenishing the leaching solution on the amount of alkali and silica extracted from a glass under certain conditions. 2.00 grams of grains of $25\,K_2O$, $75\,SiO_2$ glass were leached in water having an initial pH of about 6.0 at $40°C$. The experiment continued for 45 minutes during which the leaching solution was renewed 18 times. The experiment was repeated under nearly identical conditions while the number of replenishing times was decreased to 8, 4 and 2. The results show a marked increase in the amount of silica extracted as the number of replenishing times is decreased. This can be attributed to the accompanying rise in the pH of the attacking solution.

Silica is severely attacked by fluorine, hydrofluoric acid and alkaline solutions. The resistance of silica towards attack appears to depend on the form in which silica exists. Quartz was found to be less susceptible to attack by alkaline solutions than fused silica. The depolymerization reactions of silica can be represented by:

$$\equiv Si\text{-}O\text{-}Si \equiv\ +\ Na^+OH^- \rightarrow\ \equiv Si\text{-}OH +\ \equiv Si\text{-}O^-Na^+ \qquad (6.16)$$

$$\equiv Si\text{-}O\text{-}Si \equiv\ +\ H^+F^- \rightarrow\ \equiv SiOH +\ \equiv SiF \qquad (6.17)$$

$$\equiv Si\text{-}O\text{-}Si \equiv\ +\ Na^+F^- \rightarrow\ \equiv Si\text{-}O^-Na^+ +\ \equiv SiF \qquad (6.18)$$

.ions imply that the essential step in the depolymerization process is
ʒ of a siloxane bond Si–O–Si. The siloxane bond, though strong, is
y be represented as $(Si^{\delta^+}\text{--------}O^{\delta^-})$. The incremental positive charge
n atom makes it susceptible to attack by nucleophilic reagents such a
'⁻ ions. The high rate of depolymerization appears to be specific to
)ntaining these two particular ions. Budd [21] has suggested that
16) proceeds through the nucleophilic attack on the silicon atom
ɔ the equation:

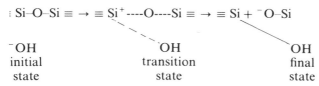

≡ Si–O–Si ≡ → ≡ Si⁺----O----Si ≡ → ≡ Si + ⁻O–Si

⁻OH   OH   OH
initial   transition   final
state   state   state

   reaction (6.17), it was suggested that it proceeds through the
ıs nucleophilic and electrophilic attack on the network silicon and
ms respectively according to:

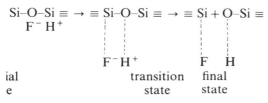

Si–O–Si ≡ → ≡ Si–O–Si ≡ → ≡ Si + O–Si ≡
F⁻ H⁺

F⁻H⁺   F   H
ial   transition   final
e   state   state

ted out that 'the bridging oxygen atoms' of the network, although
:lectrophilic attack, are rarely affected because hydrogen ions are not
 powerful to cause disruption of the O–Si bond. However, if the
c attack on the silicon atom is proceeding simultaneously, then it
)ossible for electrophilic attack on the oxygen atom to occur and this
ce the nucleophilic attack on silicon. This type of reaction would occur
reagent comprises both a strong nucleophilic agent and a strong
ic agent. Normally this is a stringent requirement, for instance H⁺ and
ot exist together in any substantial quantity (for K_w at 25°C must be
ese conditions, however, are met in the case of hydrofluoric acid. The
n is a moderately strong nucleophilic reagent. In the presence of H⁺
aneous nucleophilic and electrophilic attack on the network can occur
ıner represented by equation (6.19).
h acids like hydrochloric and hydroiodic satisfy the requirements for a
)us nucleophilic–electrophilic attack, as put forward by Budd, better
)fluoric acid, yet their degrading effect on silica is almost negligible. It
t the problem of the depolymerization of silica requires a more
:ed atomistic interpretation than the simple one presented by Budd.

mperature of leaching

ity of alkali extracted from a glass in a given period of time increases
ʌsing temperature. For most silicate glasses the quantity leached in a

187

given time is nearly double for every 8° to 15° C rise in temperature depending on the composition of the glass and the type of alkali ion. Some workers have attempted to express the temperature-dependence of alkali extraction in terms of the Arrhenius equation:

$$A = Be^{-E/RT} \qquad (6.20)$$

Where A is the specific reaction rate changing with temperature, B is a constant, R the gas constant, T the absolute temperature, and E the activation energy, defined as the minimum energy which a system must acquire for a reaction to occur. Apparent activation energies were obtained for the process of alkali removal from the glass and these were different from those obtained from electrical conductivity measurements. In cases where equation (6.20) applies it is difficult to express the results in terms of the Arrhenius equation because the two reaction constants a and K in the relation $Q = Kt^a$ both vary with the temperature. In glass, alkali extraction is always associated with pH changes and these depend not only on the quantity of alkali released but also on the quantity of silica so that it is not possible to eliminate the effect of temperature on either alkali or silica extraction. Great care should, therefore, be taken before any physical meaning is ascribed to such apparent activation energies.

6.3   EFFECT OF GLASS COMPOSITION

The rate of alkali extraction from glass by aqueous solutions is largely determined by the composition of the glass. Generally, the rate decreases with decreasing alkali content of the glass, with decreasing ionic radius of the alkali ion or when part of the silica is replaced by almost any other divalent oxide.

Dubrovo and Shmidt [22] made a systematic study of the reaction of vitreous sodium silicates with water and with hydrochloric acid solutions. Nine glasses having Na_2O/SiO_2 ratios equal to 1:1, 1:1.4, 1:1.7, 1:2, 1:2.3, 1:2.6, 1:3, 1:4 and 1:6, were studied. Glasses in the form of discs having a known surface area were subjected to the action of water and hydrochloric acid at two or three different temperatures. Samples of the leaching solution were removed after certain periods of time; these were analysed for soda and silica. The amounts of alkali and silicic acid found in the solution were calculated as mols of the corresponding oxides (Na_2O and SiO_2) going into solution from one centimetre square of surface and were denoted by n_{Na_2O} and n_{SiO_2}. In order to characterize the process of interaction the authors also used some other quantities. If the composition of the glass is expressed in the form $Na_2O.mSiO_2$ then the amount of SiO_2 in the layer of the glass that has undergone reaction corresponding to n_{Na_2O} would be $m.n_{Na_2O}$ and the fraction of this, a, which has passed from this layer to the solution will be $(n_{SiO_2})/(m.n_{Na_2O})$. Thus a is the fraction of silicic acid which has passed into solution from the layer of the glass that has undergone reaction. Its value gives an idea of the nature of the process that is taking place. When $a = 1$ the components are passing into solution in the same proportion as those in which they occur in the glass i.e. dissolution is occurring. On the other hand when $a = 0$ leaching is occurring with the result that a layer of silicic acid remains on the surface of the glass.

clusions of Dubrovo and Shmidt can be summarized as follows:
process of interaction of sodium silicates with water can be divided
tages: exchange of sodium ions of the glass for hydrogen ions of the
esulting in the formation of a residual layer of silicic acid, which
ith the silica of the original glass comprises a protective film on the
d reaction of the protective layer with the alkaline solution that has
ed, resulting in removal of silicic acid from the surface by the solution.
he case of sodium silicates having a low silica content, a kinetic
m is established between the primary and secondary reactions, so that
n of the glass appears to take place. In the case of sodium silicates of
content, the main process is the leaching of Na_2O from the glass.
en sodium silicates are treated with hydrochloric acid solutions,
the solution of SiO_2 lags behind that of Na_2O to a greater extent than
se of water treatments. In the case of glasses having a silica content
an that of the disilicate, no SiO_2 could be detected in the solution.
relation between the logarithms of the amounts of components
nto solution and the molecular percentage of silica in the glass is
ed over the range of compositions examined by smooth curves or by
ines.
ith rise in temperature the amounts of the components passing into
increase considerably, and at the same time the boundary of the region of
ilicates moves in the direction of glasses of higher original silica content.
kali removed by hydrochloric acid treatment is greater than that
d by water and does not vary with the concentration of the acid in the
om 1.0 N to 0.01 N HCl i.e. in the pH range from about 0 to 2.

Effect of lime in silicate glasses

n of CaO to a binary alkali silicate glass is known to increase its durability.
pical results of a series of $(25 - x)Na_2O$, $x CaO$, $75 SiO_2$ (where $x = 0$ to
% at 2.5 mol% intervals) are shown in Figures 6.2 and 6.3. It can be seen
igure 6.2 that extraction of Na_2O decreases sharply as it is partially
d by equivalent amount of CaO in the glass. Increasing the CaO content
to 10 mol% (Figure 6.2(a) and (b)) caused a rapid decrease in soda
ion. However, with an addition of more than 10 per cent CaO, the Na_2O
ion is virtually the same and the rate of change of soda extraction with lime
t is negligible.

tion of CaO:

e lime extraction from these glasses in the buffer solution of pH = 1 is less
an 0.1 p.p.m. i.e. below the usual detection limit.
O passes into the solution at pH = 10, but the amount of extraction is very
all. Figure 6.3 shows that the extraction of lime increases with increasing
aO content of the glass.

ction of SiO_2:

lica extraction at pH = 1 is extremely small, and below the limit of
easurement ($\sim$ 0.1 p.p.m.).

189

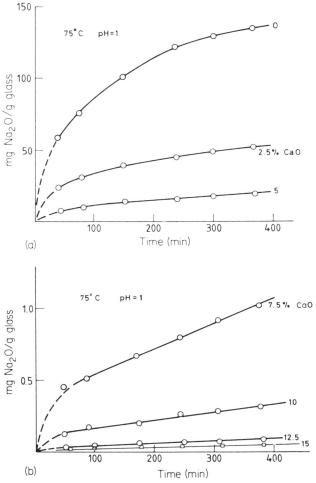

Fig. 6.2 Effect of CaO on soda extraction from Na_2O–CaO–SiO_2 glasses at 75°C. (pH of the leach solution = 1)

(ii) At pH = 10, the silica extraction is very high for low lime glasses (1.57 mg from the glass containing 2.5 mol % CaO).

Under identical conditions, the silica extraction goes through a minimum around 10 mol % CaO, and increases again with further increase of CaO in the glass. Similar behaviour has also been reported by Budd [23] in a different series of Na_2O–CaO–SiO_2 glasses. This enhanced silica extraction with high CaO glasses may be either due to microphase separation and/or the activity of CaO in these glasses, containing more than 10 mol % CaO, may be disproportionately higher than in other glasses of this series.

Gastev [24] has calculated the depth to which various oxides are leached from a series of glasses of the composition 18 % Na_2O, (82 − x) % SiO_2, x MO where M

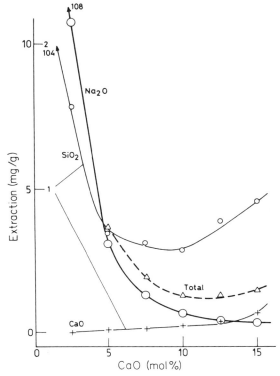

5.3 Effect of CaO in glass on the extraction of soda, lime and silica from a
(25-x) Na$_2$O, x CaO, 75 SiO$_2$ glass at 75°C after 5 hours of leaching. (pH of
the leach solution = 10)

nts an alkaline-earth metal. The depth to which CaO (and oxides of other
e-earth metals) was leached was found to increase gradually as its content
sed in the glass, but remained less than the depth of leaching of soda and
The results obtained by Das [25] also suggest that in 15 Na$_2$O, 5 MO,
$_2$ glasses the amount of MO extracted by water from a glass is less than
ed if it had been removed as a result of network breakdown of the silica
ure. Possible explanations for this phenomenon are that calcium released
: glass through network breakdown is partly retained by absorption on the
e, or that silica is preferentially extracted from the bulk of the leached layer
d on the surface of the glass leaving it with an alkaline-earth oxide/silica
higher than that of the original glass.

5.4 EFFECT OF pH OF THE SOLUTION ON CHEMICAL
DURABILITY OF GLASS

chemical durability of silicate glasses is critically dependent on the pH and
ature of the attacking solution. When an alkali–silicate glass is placed in pure

191

water, the water instantaneously becomes a solution of alkali oxide and silica. The pH of the solution of this type depends on the concentration of alkali (activity, strictly) as well as on the relative ratio of alkali oxide to silica. Both the concentration and the ratio do change with time and the pH of the attacking solution would also be expected to change (for example see Table 6.2).

The effect of the different pH values of the solutions on the decomposition of simple glasses, and the rate at which the constituents of the glass go into solution, has been studied by a number of investigators [26] and their results showed that all silicate glasses become particularly susceptible to decomposition above pH $\sim$ 9–10. This may be easily seen in Table 6.3 where the thermodynamic equilibrium constants for the formation of different silicate and silico–fluoride ions are given.

Table 6.3

Standard free energy and formation constants of different silicate anions in aqueous solution at 25 °C

Reaction	ΔG° $cals\,mol^{-1}$	$log\,K$
SiO_2 (quartz) $+ H_2O$ (liq) $\rightleftharpoons H_2SiO_3$	$+\ 7\,090$	$-\ 5.198 =$ $log\,a_{[H_2SiO_3]}$
$H_2SiO_3 \rightleftharpoons HSiO_3^- + H^+$	$+\ 13\,640$	$-\ 10.000 =$ $log\,a_{[HSiO_3^-]} - pH$
$H_2SiO_3 \rightleftharpoons SiO_3^{2-} + 2H^+$	$+\ 30\,000$	$-\ 21.994 =$ $log\,a_{[SiO_3^{2-}]} - 2pH$
SiO_2 (glass) $+ 6HF$ (liq) $\rightleftharpoons SiF_6^{2-} + 2H_2O + 2H^+$	$-\ 11\,020$	$+\ 8.079 =$ $log\,a_{[SiF_6^{2-}]} - 2pH$

The formation of silicate and silico-fluoride ions are pH-dependent; activity of $HSiO_3^-$ becomes important around pH $\sim$ 10 and thus even vitreous silica will be attacked at pH 10. This thermodynamic prediction has been verified by Shamy, Lewins and Douglas [27] and some of their results are shown in Figure 6.4.

In the presence of HF or fluoride ions, silico–fluoride anions will be formed. Although this reaction is also pH-dependent, the equilibrium constant is so large and positive that almost quantitative reaction will take place in all practically attainable acid and alkaline media. Indeed the very corrosive action of fluorides on silicate glasses is well known.

Before going any further in discussing the thermodynamic stability of different glasses in aqueous solutions, let us first define the term 'stability'. A system is stable if it does not change with time. There are two types of stability in vogue: thermodynamic stability and kinetic stability. In common usage, these are sometimes intermixed in a confusing way.

In thermodynamic stability the system is in equilibrium corresponding to minimum possible free energy, the system is stable in the strict sense, that is none

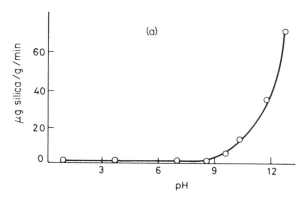

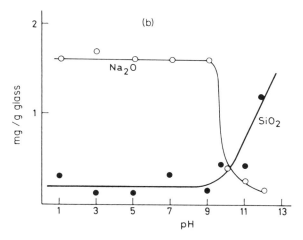

. 6.4(a) Effect of pH on the rate of extraction of silica from fused silica powder at 80°C.

(b) Effect of pH on the extraction of soda and silica from a Na_2O, $3SiO_2$ glass at 35°C.

e conceivable changes in the system can occur spontaneously. On the other
1, in the case of apparent or kinetic stability, the system is not in a state of
ibrium, some changes can occur spontaneously but at an immeasurably slow
   The best possible example of kinetic stability is the existence of glass itself.
here may be some argument as to whether the chemical durability of glass is a
ter of thermodynamic equilibrium (true stability) or apparent stability
olving a high activation barrier for the various diffusion processes involved
ing decomposition of glass in aqueous solutions.
he work of Morey and co-workers [28] showed that vitreous silica or quartz
ins continuously rotated in water at a speed of 0.5 rev min^{-1} produces less

than 1 p.p.m. soluble silica even after one year. At the same time the use of glass electrodes for measuring H^+ ion activities and other ionic species in solution is a fairly common practice. As will be shown later a major part of the observed potential with glass electrodes originates from the differences of chemical potential of the species in the solution and those exchanged on the glass surface. A reproducible steady potential with glass electrodes is possible with the establishment of ion exchange equilibria, at least at the surface of the glass. Since with conventional glass electrodes steady potentials are obtained within a short time, a kinetic barrier does not seem to play any important part, at least on the surface of the glass.

In reality the durability of glass may be expressed as a function of both thermodynamic and kinetic stability of its component oxides:

Durability $= f$ (kinetic stability) $\times f$ (thermodynamic stability).

The relative influence of either of these two factors on durability will depend on the nature of test. If the test is carried out at low temperature, the system will have small thermal energy to overcome the activation barrier, and the kinetic part will be predominant. On the other hand, if the surface area of the glass sample exposed to the corroding medium is large, and/or the experiment is carried out at a relatively high temperature, the thermodynamic part, will be more important. As, for example, Figure 6.5 shows, the rate of Na_2O extracted (mg Na_2O extracted

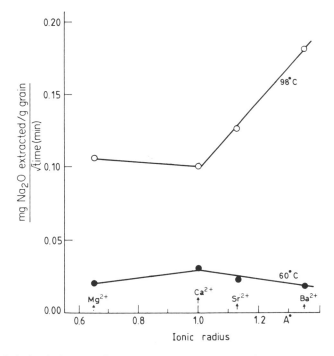

Fig. 6.5 Relative influence of different alkaline-earth metal oxides on the rate of extraction of soda from $15\,Na_2O$, $5\,MN$, $80\,SiO_2$ glasses.

^{2} of glass surface/(time in minutes)$^{1/2}$) by distilled water from a series
), 5 MO, 80 SiO$_2$ glasses (where M refers to Mg, Ca, Sr and Ba) at two
mperatures. At the higher temperature 98° C, the glass containing BaO
able, and the durability increases with decreasing ionic size of M^{2+}. As
en from Table 6.4 this behaviour is in accordance with the thermodyn-
lity of different binary alkaline earth silicates in water. However, the
ehaviour of the same glasses at 60° C indicates the apparent stability of
containing glass compared with even the MgO-containing glass. This is
due to the fact that at lower temperatures the diffusion of the large Ba^{2+}
mes energetically restricted while the smaller Mg^{2+} ion can move
he leached layer relatively easily and thus can be extracted, opening up a
he leaching of Na$^+$ from the glass.

Table 6.4
nge equilibrium constant of different crystalline binary silicates in water at 25° C

Reaction	$\Delta G°$ (Cals)	log K (at 25° C)
(cryst) + 2H$^+$ (aq) $\rightleftharpoons$ H$_2$SiO$_3$ + 2Li$^+$ (aq)	− 22 740	16.67
$_3$(cryst) + 2H$^+$ (aq) $\rightleftharpoons$ H$_2$SiO$_3$ + 2Na$^+$ (aq)	− 28 815	21.125
$_3$(cryst) + 2H$^+$ (aq) $\rightleftharpoons$ H$_2$SiO$_3$ + 2K$^+$ (aq)	− 41 735	30.598
$_3$(cryst) + 2H$^+$ (aq) $\rightleftharpoons$ H$_2$SiO$_3$ + 2Rb$^+$ (aq)	− 46 500	34.091
$_3$(cryst) + 2H$^+$ (aq) $\rightleftharpoons$ H$_2$SiO$_3$ + 2CS$^+$ (aq)	− 46 820	34.325
$_3$(cryst) + 2H$^+$ (aq) $\rightleftharpoons$ H$_2$SiO$_3$ + Mg^{2+} (aq)	− 13 888	10.182
$_3$(cryst) + 2H$^+$ (aq) $\rightleftharpoons$ H$_2$SiO$_3$ + Ca^{2+} (aq)	− 16 116	11.815
$_3$(cryst) + 2H$^+$ (aq) $\rightleftharpoons$ H$_2$SiO$_3$ + Sr^{2+} (aq)	− 24 400	17.889
$_3$(cryst) + 2H$^+$ (aq) $\rightleftharpoons$ H$_2$SiO$_3$ + Ba^{2+} (aq)	− 30 570	22.412

om the above discussion it is clear that a large increase of either
nodynamic or kinetic stability will make the glass more durable. In the
ng case, it can be argued that in the case of thermodynamic stability or
ibrium, the chemical potentials of the species on the glass surface and those
lution are equal and thus no net mass transfer will take place and the glass
be durable in the strict sense.
onsider a chemical reaction:

$$aA + bB \rightleftharpoons cC + dD$$

re the small letters denote the number of mols of that component undergoing
nge.

$$K = \frac{a_{C}{}^{c} \cdot a_{D}{}^{d}}{a_{A}{}^{a} \cdot a_{B}{}^{b}}$$

$$= \frac{(C^c)(D^d)}{(A^a)(B^b)} \cdot \frac{\gamma_C^c \cdot \gamma_D^d}{\gamma_A^a \cdot \gamma_B^b} \tag{6.21}$$

195

a denotes activity, ( ) denotes concentration (usually mols per litre), and γ denotes activity coefficient, γ is unity at the standard state (pure solid, pure liquid, or solution at extreme dilution).

The relationship between the equilibrium constant K and the free energy change of the reaction is

$$\Delta G = \Delta G^\circ + RT \ln K \qquad (6.22)$$

Where the reactants and the products are at their standard states, $\Delta G = 0$ and therefore

$$\Delta G^\circ = - RT \ln K \qquad (6.23)$$

Where ΔG° is the standard free energy change of a reaction, and is the sum of the free energies of formation of the products in their standard states, minus the free energies of formation of the reactants in their standard states.

i.e.
$$\Delta G^\circ = \sum G^\circ_{\text{products}} - \sum G^\circ_{\text{reactants}}$$

In (6.23) R is the gas constant, T the absolute temperature; at 298.15 K therefore

$$\Delta G^\circ = - 1.987 \text{ cal deg}^{-1} \times 298.15 \times 2.303 \log K$$
$$= 1364 \log K$$

or

$$\log K = - \frac{\Delta G^\circ}{1364} \qquad (6.24)$$

6.5 SOLUBILITY OF SILICA IN AQUEOUS SOLUTIONS

The relative insolubility of silica in water is one of the main factors in the corrosion of glass. When silica (quartz) is brought into contact with water at ordinary temperatures the value of the equilibrium solubility is very low (~ 6 p.p.m. for quartz), but it is the extremely slow rate of hydration that is responsible for this glass having a high resistance to the attacking water.

In principle the thermodynamic stability of a glass may be considered to be the stability of its component oxides which in turn is a function of activity (chemical potential) (for $\mu_i = \mu_{0_i} + RT \ln a_i$) of that particular oxide in glass and the equilibrium constants of hydration, ionization and complexation. With the available thermodynamic data it is possible to calculate the various energy changes being associated with these processes, and therefrom the stability of the glass under various conditions can be judged. Standard chemical potentials of some selected species relevant to common glass compositions are given in Table 6.5. Since this information is commonly available only for 25° C, and one atmosphere pressure, the following discussion will refer to that temperature and pressure only.

Table 6.5.
d free energies of formation, the standard heats of formation, and the standard entropies of some selected species at 25 °C

	Description	State	$\Delta G°$ (k cal)	$\Delta H°$ (k cal)	$S°$ (cal deg^{-1})
		gas	0	0	31.211
		aq.	0	0	0
		gas	− 54.636	− 57.798	45.106
		liq.	− 56.690	− 68.317	16.716
		aq.	− 37.595	− 54.957	− 2.519
	metal	cryst.	0	0	6.70
		aq.	− 70.22	− 66.554	3.4
		cryst.	− 133.9	− 142.4	9.06
		cryst.	− 106.1	− 116.45	12.0
		aq.	− 107.82	− 121.51	0.9
		aq.	− 101.57	− 106.577	16.6
		aq.	− 317.78	− 350.01	10.0
		aq.	− 96.63	− 115.926	38.4
₃		aq.	− 266.66	− 294.74	− 5.9
₂SiO₂		cryst.		− 586.8	28.5
SiO₂		cryst.		− 381.2	19.2
SiO₂		cryst.		− 543.2	29.0
Al₂O₃, 2SiO₂	B-eucryptite	cryst.		− 984.49	49.6
Al₂O₃, 4SiO₂	spodumene	cryst.		− 1638.36	61.9
Al₂O₃, 4SiO₂	spondumene	cryst.		− 1624.96	73.8
n					
	metal	cryst.	0	0	12.2
		aq.	− 62.589	− 57.279	14.4
)		cryst.	− 90.0	− 99.4	17.4
H		cryst.	− 90.1	− 101.99	12.5
H	un-ionized	aq.	− 99.23		
1		aq.	− 93.939	− 97.302	27.6
CO₃		aq.	− 251.4		
ICO₃		aq.	− 202.89	− 222.5	37.1
ICO₃	un-ionized	aq.	− 202.56		
CO₃		aq.	− 190.54		
SO₄		aq.	− 240.91		
₂O, SiO₂		cryst.		− 360.4	27.2
a₂O, SiO₂		cryst.		− 490.3	46.8
₂O, Al₂O₃, 2SiO₂	nepheline	cryst.		− 939.3	29.7
₂O, Al₂O₃, 4SiO₂	jadeite	cryst.	− 1356.6		31.9
tassium					
+	metal	cryst.	0	0	15.2
		aq.	− 67.466	− 60.04	25.5
₂O		cryst.	− 76.238	− 86.4	20.8
OH		cryst.	− 89.5	− 101.78	14.2

Table 6.5 cont'd

Formula	Description	State	$\Delta G°$ (kcal)	$\Delta H°$ (kcal)	$S°$ (caldeg^{-1})
KCl		cryst.	−97.592	−104.175	19.76
K_2CO_3		cryst.	−255.5	−273.93	33.6
KOH		aq.	−105.061	−115.00	22.0
$K_2O, 2SiO_2$		cryst.		−574.7	43.5
K_2O, SiO_2		cryst.		−357.1	34.9
$K_2O, 4SiO_2$		cryst.		−981.9	63.5
$K_2O, Al_2O_3, 2SiO_2$	kaliophilite	cryst.		−981.5	63.6
$K_2O, Al_2O_3, 4SiO_2$	leucite	cryst.		−1399.5	88.0
$K_2O, Al_2O_3, 6SiO_2$	orthoclase	cryst.		−1819.89	
$K_2O, Al_2O_3, 6SiO_2$	microcline	cryst.		−1815.9	105
$K_2O, Al_2O_3, 6SiO_2$	andularia	cryst.		−1814.5	112
$K_2O, Al_2O_3, 6SiO_2$	sanidine	cryst.		−1813.3	113.8
$K_2O, 3Al_2O_3, 6SiO_2, 2H_2O$	muscovite	cryst.		−1965.6	138
Rubidium					
Rb	metal	cryst.	0	0	16.6
Rb$^+$		aq.	−67.45	−58.9	29.7
Rb_2O		cryst.	−69.5	−78.9	26.2
RbOH		cryst.	−87.1	−98.9	16.9
RbOH		aq.	−105.05	−113.9	27.2
Rb_2CO_3		cryst.	−249.3	−269	23.3
$Rb_2O, 4SiO_2$		cryst.		−980.8	66.5
$Rb_2O, 2SiO_2$		cryst.		−564.7	46.5
Rb_2O, SiO_2		cryst.		−352.8	38.5
Cesium					
Cs	metal	cryst.	0	0	19.8
Cs$^+$		aq.	−67.41	−59.2	31.8
Cs_2O		cryst.	−65.6	−75.9	29.6
CsOH		cryst.	−84.9	−97.2	18.6
$Cs_2O, 4SiO_2$		cryst.		−977.8	70.0
$Cs_2O, 2SiO_2$		cryst.		−564.2	50
Cs_2O, SiO_2		cryst.		−353.3	42
Magnesium					
Mg	metal	cryst.	0	0	7.77
Mg^{2+}		aq.	−108.99	−110.41	−28.2
MgO		cryst.	−136.13	−143.84	6.4
$Mg(OH)_2$		cryst.	−199.27	−221.00	15.09
MgOH$^+$		aq.	−150.10		
$MgCl_2$		cryst.	−141.57	−153.40	21.4
$MgCO_3$		cryst.	−246	−266	15.7
$MgHCO_3^+$		aq.	−250.88		
$MgSO_4$		cryst.	−280.5	−305.5	21.9
$MgSO_4$	un-ionized	aq.	−289.55		
$MgCO_3$	un-ionized	aq.	−239.85		
MgO, SiO_2	clinoenstatite	cryst.		−357.9	16.2
$2MgO, SiO_2$	forsterite	cryst.		−508.1	22.75
$3MgO, 2SiO_2, 2H_2O$	serpentine	cryst.		−1018.8	53.1

	Description	State	$\Delta G°$ (kcal)	$\Delta H°$ (kcal)	$S°$ (cal deg^{-1})
O$_2$, H$_2$O	talc	cryst.		−1365.0	62.3
O$_2$, H$_2$O	anthophyllite	cryst.		−2791.3	133.6
l$_2$O$_3$, 5SiO$_2$	cordierite	cryst.		−2114.0	97.3
	metal	cryst.	0	0	9.95
		aq.	−132.18	−129.77	−13.2
	cubic	cryst.	−144.4	−151.9	9.5
	rhombic	cryst.	−214.22	−235.8	18.2
		aq.	−171.55		
	calcite	cryst.	−269.78	−288.45	22.2
	anhydrite	cryst.	−315.56	−342.42	25.5
	soluble	cryst.	−313.52	−340.27	25.9
	soluble	cryst.	−312.46	−339.21	25.9
	un-ionized	aq.	−312.67		
	un-ionized	aq.	−262.76		
$_3^+$		aq.	−273.67		
O$_2$	wollastonite	cryst.	−358.2	−378.6	19.6
O$_2$	pseudo-wollastonite	cryst.	−357.4	−377.4	20.9
SiO$_2$	B	cryst.	−512.7	−538.0	37.6
SiO$_2$	r	cryst.	−513.7	−539.0	37.6
SiO$_2$		cryst.		−688.1	40.3
gO, 2SiO$_2$	diopside	cryst.		−741.2	34.2
gO, SiO$_2$	monticellite	cryst.		−528.7	26.4
MgO, 2SiO$_2$	akermanite	cryst.		−902.3	50.0
5MgO, 8SiO$_2$,	tremolite	cryst.		−1144.5	131.2
, MgO, 2SiO$_2$	merwinite	cryst.		−1067.3	60.5
Al$_2$O$_3$, 2SiO$_2$	anorthite	cryst.		−983.6	48.4
ium	metal	cryst.	0	0	13.0
		aq.	−133.2	−130.38	−9.4
		cryst.	−133.8	−141.1	13.0
H)$_2$		cryst.	−207.8	−229.3	21
O$_3$	strontianite	cryst.	−271.9	−291.2	23.2
O$_4$		cryst.	−318.9	−345.3	29.1
, SiO$_2$		cryst.	−350.8	−371.2	22.586
O, SiO$_2$		cryst.	−495.7	−520.6	43
O, Al$_2$O$_3$, SiO$_2$		cryst.		−928.09	
ium	metal	cryst.	0	0	16
$_2{}^+$		aq.	−134.0	−128.67	3
O		cryst.	−126.3	−133.4	16.8
(OH)$_2$		cryst.	−204.7	−226.2	22.7
CO$_3$	witherite	cryst.	−272.2	−291.3	26.8
SO$_4$		cryst.	−323.4	−350.2	31.6
O, SiO$_2$		cryst.	−338.7	−359.5	24.2

Table 6.5 cont'd

Formula	Description	State	$\Delta G°$ (k cal)	$\Delta H°$ (k cal)	$S°$ (cal deg^{-1})
$2BaO, SiO_2$		cryst.	− 470.6	− 496.8	**46.4**
$2BaO, SiO_2$		cryst.	− 470.6	− 496.8	46.4
$BaO, 2SiO_2$		cryst.		− 585.5	
$2BaO, 3SiO_2$		cryst.		− 965.4	
Zinc					
Zn	metal	cryst.	0		0.95
Zn^{2+}		aq.	− 35.184	− 36.43	− 25.45
ZnO	orthorhombic	cryst.	− 76.14	− 83.3	10.4
$Zn(OH)_2$	-white	cryst.	− 133.31		
$Zn(OH)_2$	amorphous		− 131.85		
$HZnO_2^-$		aq.	− 110.9		
$ZnCO_3$		cryst.	− 174.8	− 194.2	19.7˙
$ZnSO_4$		cryst.	− 208.31	− 233.88	29.8
$ZnCl_2$		cryst.	− 88.255	− 99.40	25.9
ZnO, SiO_2		cryst.	− 274.8	− 294.6	21.4
$2 ZnO, SiO_2$	willemite	cryst.	− 352.7	− 379.8	31.4
Cadmium					
Cd	metal	cryst.	0	0	12.3
Cd^{2+}		aq.	− 18.58	− 17.30	− 14.6
CdO	cubic	cryst.	− 53.79	− 60.86	13.1
$Cd(OH)_2$	active	cryst.	− 112.46	− 133.26	22.8
$HCdO_2^-$		aq.	− 86.5		
$CdCl_2$		cryst.	− 81.88	− 93.00	28.3
$CdCl^+$		aq.	− 51.8		5.6
$CdCl_2$	un-ionized	aq.	− 84.3		17
$CdCO_3$		cryst.	− 160.2	− 178.7	25.2
$CdSO_4$		cryst.	− 195.99	− 221.36	32.8
CdO, SiO_2		cryst.		− 271.06	23.3
Lead					
Pb	metal	cryst.	0	0	15.51
Pb^{2+}		aq.	− 5.81	0.39	5.1
PbO	yellow	cryst.	− 45.05	− 52.07	16.6
PbO	red	cryst.	− 45.25	− 52.40	16.2
$Pb(OH)_2$		cryst.	− 100.6	− 123.0	21
$HPbO_2^-$		aq.	− 81.0		
$PbSO_4$		cryst.	− 193.89	− 219.50	35.2
$PbCl_2$		cryst.	− 75.04	− 85.85	32.6
$PbCO_3$		cryst.	− 149.7	− 167.3	31.3
$PbO, PbCO_3$		cryst.	− 195.6	− 220.0	48.5
$2PbO, PbCO_3$		cryst.	− 242	− 273	65
$Pb_3(OH)_2(CO_3)_2$		cryst.	− 406.0		
PbO, SiO_2		cryst.	− 239.0	− 258.9	27
$2PbO, SiO_2$		cryst.	− 285.7	− 312.7	43
$4 PbO, SiO_2$		cryst.		− 417.68	
Boron					
B		cryst.	0	0	1.56
B_2O_3		glass	− 280.4	− 297.6	18.8

	Description	State	ΔG° (kcal)	ΔH° (kcal)	S° (cal deg^{-1})
		cryst.	-230.2	-260.2	21.41
		aq.	-230.16		
		aq.	-616		
		aq.	-217.63	-251.8	7.3
		aq.	-200.29		
		aq.	-181.48		
		gas	-261.3	-265.4	60.70
		aq.	-343	-365	40
n	metal	cryst.	0	0	6.769
		aq.	-115.0	-125.4	-74.9
	corundum	cryst.	-376.77	-399.09	12.186
$_2O$	boehmite	cryst.	-435.0	-471.0	23.15
	amorph.	amorph.	-271.9	-307.9	-305
		aq.	-257.4		
		aq.	-200.7		
$(OH)_4$	kaolinite	cryst.	-884.5		
	metal	cryst.	0	0	4.47
	quartz II	cryst.	-192.4	-205.4	10.00
	cristobalite II	cryst.	-192.1	-205.0	10.19
	tridymite IV	cryst.	-191.9	-204.8	10.36
		vitreous	-200.9	-190	
$_4$		aq.	-511	-558.5	-12
$_4$		aq.	-300.3		
		aq.	-286.8		
		gas	-360	-370	68.0
		gas	-136.2	-145.7	79.2
$_3$		aq.	-242		
$_3^-$		aq.	-228.36		
		aq.	-212.0		
$anium$	metal	cryst.	0	0	10.14
$_2$		cryst.	-136.1		
eO_3		aq.	-186.8		
eO_3^-		aq.	-175.2		
$_3$		aq.	-157.9		
$nium$	metal	cryst.	0	0	7.24
$_2$	rutile	cryst.	-212.3		
$^{2+}$		aq.	-138.0		
$(OH)_2$		cryst.	-253.0		
iO_3^-		aq.	-111.7		
$rconium$	metal	cryst. II	0	0	9.18

Table 6.5 cont'd

Formula	Description	State	$\Delta G°$ (k cal)	$\Delta H°$ (k cal)	$S°$ (cal deg^{-1})
Zr^{4+}		aq.	$-.142.0$		
ZrO^{2+}		aq.	-201.5		
ZrO_2		cryst.	-247.7		
$ZrO(OH)_2$		cryst.	-311.5	-338.0	22
$Zr(OH)_4$		cryst.	-370.0	-411.2	31
$HZrO_3^-$		aq.	-287.7		
Phosphorous					
P	metal	white cryst.			
		III	0	0	10.6
		red cryst. II	-3.3	-4.4	7.0
		black cryst. I		-10.3	
H_3PO_4		aq.	-274.2	-308.2	42.1
$H_2PO_4^-$		aq.	-271.3	-311.3	21.3
HPO_4		aq.	-261.5	-310.4	-8.6
PO_4^{3-}		aq.	-245.1	-306.9	-52.0
P_2O_5		cryst.		-356.6	27.35

6.5.1 Effect of pH of the solution on the solubility of SiO_2 (quartz)

The reaction between water and silica may be represented as

$$SiO_2 \text{ (quartz)} + H_2O \rightleftharpoons H_2SiO_3 \qquad (6.25)$$

$$K = \frac{[a_{H_2SiO_3}]}{[a_{SiO_2}][a_{H_2O}]} \qquad (6.26)$$

$\Delta G°$ of reaction (6.25) is $+7090$ cal. Therefore $\log K = -5.198$. In this particular case, pure quartz is reacting with pure water, with very little formation of H_2SiO_3, thus

$$a_{SiO_2} = a_{H_2O} = 1$$

Therefore

$$a_{H_2SiO_3} = K = 10^{-5.198} \sim C_{H_2SiO_3} \qquad (6.27)$$

From equation (6.27) it will appear that the solubility of SiO_2, in terms of H_2SiO_3 in solution, is independent of pH, but in the presence of alkali, additional silica passes into solution as silicate ions. The dissociation of silicic acid can be written as

$$H_2SiO_3 \overset{K_1}{\rightleftharpoons} H^+ + HSiO_3^- \qquad (6.28)$$

$$HSiO_3^- \overset{K_2}{\rightleftharpoons} H^+ + SiO_3^{2-} \qquad (6.29)$$

$\Delta G^{\circ} = +13\,640$ cal, and $\log K_1 = -10$.

$$K_1 = \frac{[a_{H^+}]\, a_{HSiO_3}]}{[a_{HSiO_3}]}$$

$$\log K_1 = \log a_{HSiO_3^-} - \log a_{H_2SiO_3} - pH$$

$$\log a_{HSiO_3^-} = -15.198 + pH \tag{6.30}$$

...vs that although the solubility of silica near the neutral point (pH ~ 7) is ...ly affected by pH, the solubility increases rapidly with alkalinity at pH ...ilarly for the reaction (6.29), $\Delta G^{\circ} = +16360$ cals and log $K_2 =$

Thus $\log a_{SiO_3^=} = -11.994 + \log a_{HSiO_3^-} + pH$

$$= -27.192 + 2pH \tag{6.31}$$

...7), (6.30) and (6.31) the equilibrium activity (or, loosely, concentration in ...r litre) of different species of silica in aqueous solution has been ...ed at various pH values, and these are shown in Table 6.6, along with the ...onding solubility values for vitreous silica. The results of Table 6.6 are ...in Figure 6.6.

Table 6.6.

...quilibrium activity of different silicate species in aqueous solution at 25° C

	Quartz				Vitreous silica		
$\log$ $a_{H_2SiO_3}$	$\log$ $a_{HSiO_3^-}$	$\log$ $a_{SiO_3^{2-}}$	$\log$ (total silica)	$\log$ $a_{H_2SiO_3}$	$\log$ $a_{HSiO_3^-}$	$\log$ $a_{SiO_3^{2-}}$	$\log$ (total silica)
-5.198	-7.198	-11.192	-5.194	-4.098	-6.098	-10.092	-4.093
-5.198	-6.198	-9.192	-5.157	-4.098	-5.098	-8.092	-4.049
-5.198	-5.198	-7.192	-4.895	-4.098	-4.098	-6.092	-3.754
-5.198	-4.198	-5.192	-4.118	-4.098	-3.098	-4.092	-2.952
-5.198	-3.198	-3.192	-2.892	-4.098	-2.098	-2.092	-1.751
-5.198	-2.198	-1.192	-1.151	-4.098	-1.098	-0.092	-0.043
-5.198	-1.198	$+0.808$	$+0.812$	-4.098	-0.098	$+1.908$	$+2.097$

...ne can divide Figure 6.6 into three distinctly different pH zones based on the ...dominance of one particular silicate species. For example, in the first zone ...($\leqslant 10$), the minimum solubility is represented by the undissociated but ...ble portion of H_2SiO_3, reaching about $10^{-5.2}$ M, this species predominates ...to pH = 9 (independent of pH). In the second zone (pH = 10 to 12) most of ...silica which passes into the solution is due to the formation of $HSiO_3^-$ species. ...the third zone (pH $\geqslant 12$) SiO_3^{2-} predominates. ...From Figure 6.6 it is evident that the quantity of silica extracted from both ...artz and vitreous silica follows the same pattern but the solubility of silica from ...e glassy form is more than that from quartz. This is because the free

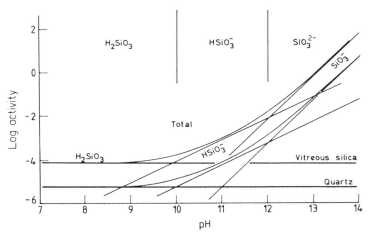

Fig. 6.6 Stability diagram of quartz and vitreous silica in aqueous solution at 25°C.

energy of quartz is more negative than that of vitreous silica; or in other words quartz is thermodynamically more stable than vitreous silica.

From Figure 6.6 and the preceding discussions it is now obvious that all silica-based glasses will follow the same trend with respect to pH changes of the solution, the absolute magnitude on the ordinate of Figure 6.6, however, will be altered by the amount $\log a_{SiO_2}$, where a_{SiO_2} denotes activity of silica in the respective glass with fused silica as the standard state.

Let us now consider the effect of pH of the solution on the stability of binary sodium silicate glasses. The reaction between a sodium silicate glass, for simplicity say Na_2O, SiO_2 (crystal), and water may be typically written as:

$$Na_2SiO_3 \text{ (crystal)} + 2H^+ \text{ (aq)} \rightleftharpoons H_2SiO_3 + 2Na^+ \text{ (aq)} \qquad (6.32)$$

$$K = \frac{a_{H_2SiO_3} \cdot a^2_{Na^+ (aq)}}{a_{Na_2SiO_3} \cdot a^2_{H^+ (aq)}}$$

$\Delta G°$ for the reaction (6.32) is -28.880 cal, and $\log K = +21.44$.

Therefore

$$\log H_2SiO_3 = 21.44 - 2 \log a_{Na^+ (aq)} - 2pH \qquad (6.33)$$

From (6.33) it is clear that, unlike pure silica, in the case of sodium silicate glass, the activity of H_2SiO_3 or the extent of ion exchange represented by (6.32) will depend on the pH as well as on the activity of Na^+ (aq) in the leach solution. According to this equation, when a sodium silicate glass is brought into contact with aqueous solution in which the activity of Na^+ is very small (say 1 p.p.m.) e.g. distilled water, then:

$$a_{Na^+ (aq)} = 10^{-6} \text{ and } 2 \log a_{Na^+ (aq)} = -12 \qquad (6.34)$$

Substituting (6.34) in (6.33) we have

$$\log a_{H_2SiO_3} = 33.44 - 2pH \qquad (6.35)$$

204

s that if the Na^+ (aq) ions are not allowed to build up in the leach
in almost complete removal of Na^+ (glass) is possible even up to
. As we have seen before, H_2SiO_3 becomes unstable due to ionization
quent solution at pH $\sim$ 9–10, which is why binary sodium–silicate
not durable in water. In fact, as can be seen from Table 6.4 none of the
ali–silicate glasses should be durable in water, and from the thermo-
point of view absolute stability should increase in the order: K_2SiO_3
$O_3 < Li_2SiO_3$.

the better durability of lithium–silicate glass over sodium– or
n–silicate glasses of the same molar composition is well known [29].
aring the standard free energy of formation of Na_2SiO_3 ($-338\ 370$ cal)
of H_2SiO_3 ($-242\ 000$ cal), sodium silicate appears to be more stable,
reality is not the case. The reason for this is that Na_2SiO_3 in contact with
solutions releases Na^+ ions to solution which become hydrated, and in
ration of Na^+ ions causes a considerable free energy difference. The effect
tion of Na^+ ion on the decomposition of sodium–silicate glass can be
d if Na_2SiO_3 is brought into contact with a solvent such as benzene or
doped with anhydrous protons (this can be achieved by dissolving organic
lic acids like adipic acid in dioxan in which the acid ionizes, producing
which Na^+ cannot be hydrated, and the glass is expected to be durable.
such anhydrous conditions, binary alkali–silicate glasses have been found
se very little alkali into the solution phase [30]. Some typical results are
in Figure 6.7.

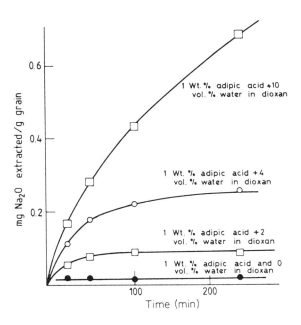

Fig. 6.7 Rate of extraction of Na_2O from $25\ Na_2O$, $75\ SiO_2$ glass grains in
water–dioxan–adipic acid mixtures at $65°C$.

6.5.2 Effect of CaO on the durability of silicate glasses at different pH values

When $CaSiO_3$ glass is brought into contact with an aqueous solution the ion exchange reaction may be written as

$$CaSiO_3 \text{ (glass)} + 2H^+ \text{ (aq)} \rightleftharpoons H_2SiO_3 + Ca^{2+} \text{ (aq)} \qquad (6.36)$$

$\Delta G°$ of (6.36) is $-16\,780$ cal/mol*, thus log $K = 12.30$

$$K = \frac{[a_{H_2SiO_3}] \, [a_{Ca^{2+} \text{(aq)}}]}{[a_{CaSiO_3 \text{(glass)}}] \, [a^2_{H^+}]}$$

and

$$\log K = \log a_{H_2SiO_3} + \log a_{Ca^{2+} \text{(aq)}} + 2pH$$

or

$$\log a_{H_2SiO_3} = 12.3 - \log a_{Ca^{2+} \text{(aq)}} - 2pH \qquad (6.37)$$

From (6.37) it is clear that the exchange of Ca^{2+} (glass) for H^+ (aq) or 'hydration' of calcium silicate glass is dependent on the pH and on the activity of Ca^{2+} ions of the leach solution. It may be pointed out that, unlike NaOH, the solubility of $Ca(OH)_2$ in water is low. Thus even a small concentration of Ca^{2+} (aq) ion will produce a significant activity in the solution.

From similar calculations as those for sodium–silicate it can be shown that even for $a_{Ca^{2+} \text{(aq)}}$ as low as 10^{-6} in the solution,

$$a_{Ca^{2+} \text{(soln)}} = a_{SiO_3^-} \sim 0.0005 \text{ at pH} = 10.8$$

This corresponds to about 0.05 per cent of calcium exchange from the $CaSiO_3$ (glass). At lower pH values, although exchange of calcium from the glass is favoured (about 2 per cent calcium exchange at pH $= 10$) in acidic solutions, the silicic acid does not ionize and probably offers a very high activation barrier for the diffusion of calcium ions through it. Thus calcium-containing glasses should appear durable up to pH ~ 10.

6.5.3 Durability of glasses containing ZnO

ZnO, though costly relative to CaO, has been used in the past in making chemically resistant scientific glasses; a typical example is JENA glass, containing about 7 wt% ZnO. The stability of ZnO in aqueous solutions of different pH values is shown in Figure 6.8. From this diagram it is clear that hydration of ZnO is easier than that of vitreous silica. Below the pH range 6.1 to 5.5 (depending on the type of hydrated ZnO formed) the activity of Zn^{2+} (aq) will exceed that of hydrated ZnO, and zinc will be extracted into the solution as Zn^{2+} (aq). Thus zinc-containing glasses will be susceptible to acid attack up to pH $\leqslant 5.5$. In the alkaline range, ZnO forms $HZnO_2^-$ and ZnO_2^{2-} ions, the activity of either of these species being smaller than that of hydrated ZnO. The iso-activity point corresponding to $HZnO_2^-$ and ZnO_2^{2-} species occurs at pH $= 13.1$. This means that zinc-containing glasses will be susceptible to vigorous alkaline attack above

---

* This value of $-16\,780$ cal mol^{-1} is for crystaline $CaSiO_3$; the corresponding figure for glassy $CaSiO_3$ is not known.

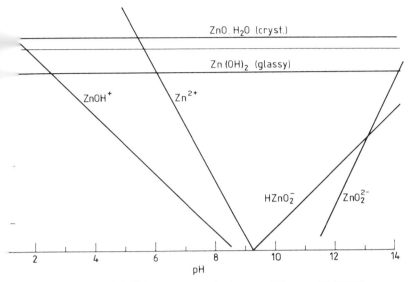

Fig. 6.8 Stability of ZnO in aqueous solutions at different pH (25°C).

13. It may be recalled that the corresponding critical pH for the
...minance of SiO_3^{2-} is ~ 12. Thus addition of ZnO to a silicate glass is
...ed to increase its chemical durability in the alkaline range up to a pH of
13. That this expectation is nicely borne out can be seen from Figure 6.9,
... the leaching behaviour of a 15 Na_2O, 10 ZnO, 75 SiO_2 glass is shown at
...ent pH from 0 to 14.

Durability of glasses containing PbO

is used in many commercial glasses, the most important being crystal glass.
... is a poison – small amounts posing a health hazard. Thus, recently, much
...ation has been focused on the leaching behaviour of lead from glass surfaces,
...icularly those coated with low-melting lead borosilicate enamels. The
...ility diagram of PbO in aqueous solutions of different pH value is shown in
...re 6.10. Hydration of PbO is small. In the acidic range (pH ⩽ 6.8) lead
...olves as Pb^{2+} (aq) and $PbOH^+$ (aq), the activity of the former being always
...ch greater than that of $PbOH^+$ (aq). In the alkaline range lead forms
...bO_2 (aq) and the activity of $HPbO_2^-$ (aq) becomes greater than that of
...$(OH)_2$ only above pH ~ 14.5. Thus PbO in a silicate glass is expected to
...rease the alkaline durability, whereas the acidic durability should decrease.
...is is indeed the fact as can be seen from Figure 6.11 where the leaching
...haviour of a typical lead crystal glass is shown. Recently Shamy [31] has
...ported the durability of binary PbO–SiO_2 glasses in different acidic media and
...served a sharp increase in lead extraction when the PbO content was increased

207

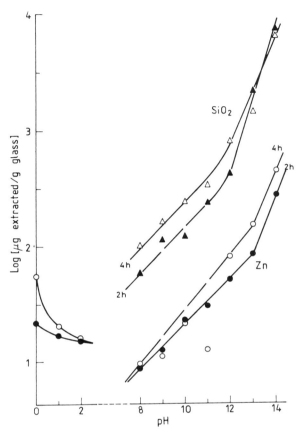

Fig. 6.9. Extraction of zinc and silica from 15 Na$_2$O, 10 ZnO, 75 SiO$_2$ glass grains at different pH (70°C).

from 35 to 40 mol %. In the PbO–SiO$_2$ system silica saturation occurs around 66 mol % SiO$_2$. The activity of PbO in binary lead–silicates, containing less than 34 mol % PbO, is very low and increases sharply above this critical concentration, and this is probably related to the enhanced lead extraction from glasses containing more than 35 mol % PbO.

6.5.5 Durability of glasses containing Al$_2$O$_3$

Al$_2$O$_3$ is addɘd to many commercial glasses. Small amounts of Al$_2$O$_3$ are known to accelerate glass batch reaction [32 (a)] and are also known to improve the durability of silicate glass [32b]. The stability diagram of Al$_2$O$_3$ in aqueous solution of different pH values is shown in Figure 6.12. In the acid range at pH ⩽ 3.2 the predominant species is Al^{3+} (aq), which is expected to be leached out from the glass surface. In the alkaline range, AlO$_2^-$ (aq) is formed and becomes of

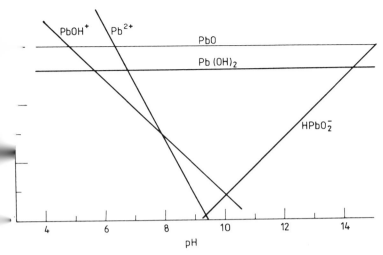

Fig. 6.10 Stability of PbO in aqueous solutions at different pH (25°C).

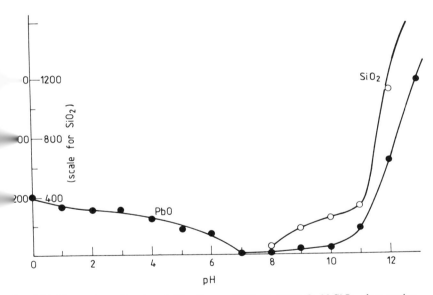

Fig. 6.11 Extraction of lead and silica from a $10\,K_2O$, $10\,PbO$, $80\,SiO_2$ glass grains at different pH (50°C).

gnificance above pH $\sim$ 14. Thus Al_2O_3 in glass is expected to increase the kaline durability of glass. As will be shown in the section on glass electrodes, ation-sensitive glasses containing Al_2O_3 become H^+ sensitive below pH $\sim$ 4. his is presumably due to the fact that the alkali cations are adsorbed on the

209

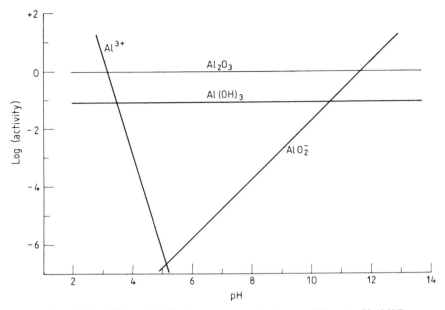

Fig. 6.12 Stability of Al_2O_3 in aqueous solutions at different pH (25°C).

hydrated alumina sites on the glass surface; below a pH $\sim$ 4, these sites are leached into the solution as Al^{3+} (aq), and the cation-sensitivity of the surface is lost.

6.5.6 Durability of glasses containing ZrO_2

Of all the oxides, addition of ZrO_2 is known to increase the durability of silicate glasses most; even a small amount of ZrO_2 (about 2 wt %) increases acid and alkaline durability of a glass significantly [33]. This is the reason why commercial glass fibres for cement reinforcement, where a very high alkaline durability is desired, contain about 16 wt% ZrO_2 [34a]. The extreme durability of ZrO_2 containing glasses is apparent from the stability diagram of ZrO_2 in aqueous solution shown in Figure 6.13. Although hydration of ZrO_2 is energetically very favourable, the predominance of ionic species like ZrO^{2+}, Zr^{4+}, and $HZrO_3^-$ will only occur below pH $\sim$ O and pH $\sim$ 17 respectively. Thus a hydrated ZrO_2 surface, unlike any other known hydrated oxide surface, is stable at all conceivable pH values of the solution [34b].

6.6 GLASS ELECTRODES

In 1906 Cremer [35] first noticed that a bulb-type glass electrode changed its potential when immersed in solutions of different hydrogen ion content. Haber and Klemensiewiez [36] in 1909 proved that glass electrode potentials varied in

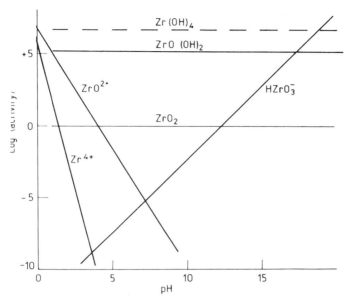

3. 6.13 Stability of ZrO_2 in aqueous solutions at different pH (25°C).

ct hydrogen electrode manner according to the equation:

$$E = E^\circ + \frac{RT}{F} \ln H^+ \qquad (6.38)$$

R = gas constant
T = absolute temperature
F = Faraday constant

H^+ = activity of hydrogen ion

:rode potential given by a glass electrode when dipped into a solution can ated into two parts:

ioundary potential V_S which results from the exchange equilibrium at the idary between the glass electrode and the solution, and
liffusion potential V_D which results from the inter-diffusion of ions in the ;.

2 total membrane potential $V = V_S + V_D$.

ioundary potentials

:r two different phases (two immiscible liquids) at constant temperature ;sure containing a non-ionic component that can exchange between them.

If the two phases are at equilibrium the chemical potential of the solute must be equal in both phases

$$\mu_1 = \mu_2$$

Now

$$\mu_1 = \mu_1^\circ + RT \ln a' \qquad (6.39)$$

$$= \theta'(T) + RT \ln a'$$

where $\theta(T)$ is the standard chemical potential which is a function of temperature only. The value of θ is fixed by the equation for activity

$$\varepsilon \rightarrow \varepsilon^* \, \varepsilon = 1$$

Here, ε is some convenient variable, such as mol fraction or concentration, and ε^* is the reference state of that variable.
Now

$$\theta'' - \theta' = RT \ln \frac{a'}{a''} \qquad (6.40)$$

Thus the relative values of the ratio a'/a'' depend only on temperature.

If the solute is ionic and only ions of one sign can exchange (Na^+ in the case of sodium silicate glass and not the silicate group), electrical fields can arise. At equilibrium between the two phases, the total of the electrochemical potentials of the exchanging cations in them must be equal. However, this equality cannot be achieved by gross changes of the cation concentration in glass, for the anionic concentration is constant. As the cations leave or enter the glass, giving an excess or deficient total charge, an electrical potential difference between the two phases builds up. At equilibrium this electrical potential difference $(X' - X'')$ just balances the differences in thermodynamic activity of the cations between the two phases. Thus

$$0 = \mu' - \mu'' = \theta' - \theta'' + RT \ln \frac{a'}{a''} + Z(X' - X'')F \qquad (6.41)$$

where F is the Faraday constant, and
        Z is the valency ($+1$ for Na^+) of the exchanging ions.
One prime represents the solutions and two primes the glass.
In an experiment of the following type:

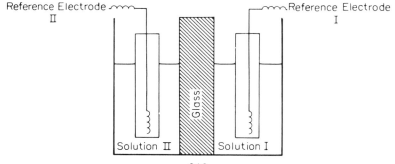

$$V_s = (X'_2 - X''_2) - (X'_1 - X''_2)$$

$$= (\theta'_2 - \theta''_1) - (\theta'_2 - \theta''_2)/F + \frac{RT}{F} \ln \frac{a'_1 a''_2}{a'_2 a''_1} \qquad (6.42)$$

⚫n II is considered positive. If the glass is uniform and homogeneous ⚫e definition of activity is used throughout, then

$$\theta''_1 = \theta''_2 \text{ and } a''_1 = a''_2$$

⚫ents are the same and the same definition of activity of the ion is used ⚫hen $\theta'_1 = \theta'_2$

$$V_s = \frac{RT}{F} \ln \frac{a'_1}{a'_2} \qquad (6.43)$$

⚫rmal use of glass electrodes, the conditions in solution II are held ⚫and the relative variation of V_s then measures the activity of the ion in ⚫. If only one ion is present, there is no diffusion potential in the glass. ⚫case of exchange of two cationic species A and B at the glass surface

$$A'' + B' \underset{\rightleftharpoons}{\overset{K}{}} A' + B'' \qquad (6.44)$$

⚫ prime represents solution and two represent glass.

$$K = \frac{a' . b''}{a'' . b'} \qquad (6.45)$$

n [37] has found experimentally that, for many ion exchanges

$$K = \left[\frac{a'}{b'} \right] . \left[\frac{N_B}{N_A} \right]^n \qquad (6.46)$$

⚫$_i$ is the mol fraction of i in the glass and n is an integer.

$$V_s = \frac{RT}{F} \left[\ln \frac{a' + Kb'}{a'_2 + Kb'_2} - \ln \frac{a'_1 + b''_1}{a''_2 + b''_2} \right] \qquad (6.47)$$

⚫sum of the activities at both the surfaces of the glass are equal, then

$$V_s = \frac{RT}{F} \ln \frac{a'_1 + kb'_1}{a'_2 + Kb'_2} \qquad (6.48)$$

Diffusion potential

a glass is dipped in an aqueous solution, generally the intruding and ⚫al ions will have different mobilities, thus as diffusion proceeds one ion to out-run the other, leading to a build up of electrical charge. Accompany-⚫is charge is a gradient in electrical potential that slows down the fast ion and ⚫s up the slow one. To preserve electroneutrality the fluxes of the two ions be equal and opposite.

213

The driving force for the transport of an ion is assumed to be the gradient of its electrochemical potential μ^*, where

$$\mu^* = \theta(T) + RT \ln a + ZFX \tag{6.49}$$

X is the electrical potential F is the Faraday constant and Z is the charge of the ion.

Then the gradient of μ^* in the x-direction of constant temperature and pressure

$$\frac{\partial \mu^*}{\partial x} = \left[\frac{RT}{c} \right] \cdot \left[\frac{\partial c}{\partial x} \right] \left[\frac{\partial \ln a}{\partial \ln C} \right] + ZF \frac{\partial X}{\partial x} \tag{6.50}$$

Here a is the thermodynamic activity and C is the concentration of the ion in the glass. Now the flux, J per unit time and area is equivalent to Cv, where v is the average molecular velocity of the ion. Thus J is given by

$$J = -u \left[RT \left(\frac{\partial C}{\partial x} \right) \left(\frac{\partial \ln a}{\partial \ln C} \right) + CFE \right] \tag{6.51}$$

Where

$$E = \frac{X}{x}$$

If two monovalent cations A and B interdiffuse, each has a flux equation as above the conditions of electroneutrality requires

$$J_A = -J_B$$

and

$$\frac{\partial C_A}{\partial x} = \frac{\partial C_B}{\partial x}$$

From all these relations, the electrical potential gradient can be deduced as follows:

$$\frac{\partial X}{\partial x} = E = \frac{RT}{F} \cdot \frac{U_B - U_A}{C_A U_A + C_B U_B} \cdot \frac{\partial C_A}{\partial x} \frac{\partial \ln a}{\partial \ln C_A} \tag{6.52}$$

And the diffusion potential, V_D is, considering solution (2) as positive

$$V_D = \int_1^2 E \, dx = \frac{RT}{F} \int_{C_A(1)}^{C_A(2)} \frac{(U_B/U_A) - 1}{C_B(U_B/U_A) - C_A} \frac{\partial \ln a_A}{\partial \ln C_A} \cdot dC_A \tag{6.53}$$

Thus the diffusion potential depends upon the ionic concentrations of the two surfaces $C_A(2)$ and $C_A(1)$, the mobility ratio U_B/U_A and the thermodynamic factor $\partial \ln a_A / \partial \ln C_A$.

To measure the e.m.f. of a glass membrane, it is necessary for the resistance of the measuring circuit to be very much greater than that of the membrane. Hence the resistance through the glass bulb which forms the glass electrode must be less than that of the measuring instrument. Even with an instrument of infinite resistance, the resistance through the glass must be small enough for the poles of the cell to attain their equilibrium potentials in the time of experiment, i.e. the CR (capacitance × resistance) time constant of the circuit must not be greater than a few seconds. In practice, this means that the specific resistance of the glass may

lly be larger than about 10^{12} ohm cm. Since the resistivity of glass
ly with composition and may reach extremely high values, especially at
erature, it limits the range of glass compositions which can usefully be
as electrodes.

stivity limitation for electrode glasses excludes the practical use at
mperatures of alkali-free glasses such as fused silica, lead-silicate, or the
irth–boroaluminate glasses.

lrogen-ion response of a glass electrode was first proved by Haber and
wiez [36], but the effect was not studied to any great extent until 1920.
0 to 1930, many investigations were carried out with the resultant
ent of Corning 015 glass [38], this glass contains (wt %) 22 % Na_2O,
ind 72 % SiO_2. During the next decade the superiority of lithium glasses
easurement in many applications was discovered.

) Hughes [39], while studying H^+ ion response of a typical pH glass
observed that relation (6.38) is obeyed satisfactorily upto a pH $\sim$ 10
after the relationship between pH and electrode potential is non-linear
ecise, or in his words 'some alkaline error' was involved. The alkaline
Corning 015 glass in 0.1 N solutions of Na^+, Li^+ and K^+ are shown in
: position of Figure 6.14. It was also noted that small amounts of Al_2O_3
in the glass increases the 'alkaline error'. Horovitz [40] and Schiller [41]
at the introduction Al_2O_3 or B_2O_3 causes glass electrodes to become
sensitive to Na^+ as to H^+, and the less Al_2O_3 the glass contained, the
was for measuring H^+. These observations were extended by Lengyel
m [42] and subsequently by Eisenman [43] who conclusively dem-

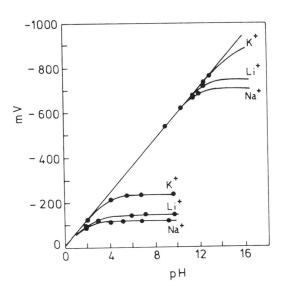

6.14 Effect of Al_2O_3 on glass electrode potentials. The upper data points are
observed in a sodium silicate glass containing no Al_2O_3, while the lower
data are from a sodium silicate glass containing 1.7 % Al_2O_3.

onstrated that the introduction of Al_2O_3 or B_2O_3 (or both) into the glass causes its potential to follow the classical Nernst equation with respect to Na^+ ion concentration in solution. The lower portion of Figure 6.14 illustrates the effect of adding a small amount of Al_2O_3 to a typical pH glass.

In 1937 Nicolskii[44] deduced the following equation which is capable of describing not only the H^+ response and the initial portions of the cation errors but also the Na^+ response of glass electrodes in the region when the electrodes respond essentially to Na^+ alone.

$$V = E_0 + \frac{RT}{F} \ln [a_{H^+} + K^{pot} \cdot a_{Na^+}] \tag{6.55}$$

where K^{pot} corresponds to an ion exchange equilibrium constant of the type:

$$Na^+ \text{ (solution)} + H^+ \text{ (glass)} \rightleftharpoons H^+ \text{ (solution)} + Na^+ \text{ (glass)} \tag{6.56}$$

In deducing (6.55) Nicolskii assumed that the same number of sites are available to each exchanging cation, that the activities of Na^+ and H^+ in the glass are proportional to their mole fractions in the glass, and that the observed glass electrode potential represents the sum of the phase-boundary potentials (neglecting the possibility of a diffusion potential within the membrane). Eisenman et al.[43] found that a slight modification of Nicolskii equation as (6.57) gives a better description of a wider range of data:

$$V = \text{constant} + \frac{nRT}{F} \ln [a_i^{1/n} + (K_{ij}^{pot} \cdot a_J)^{1/n}] \tag{6.57}$$

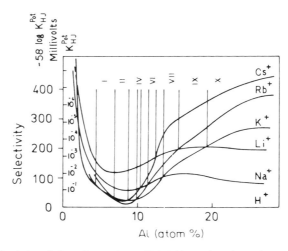

Fig. 6.15 Selectivity of glass electrodes to H^+, Li^+, Na^+, K^+, Rb^+, and Cs^+ ions as a function of atom % aluminium. The abscissa represents x, the atom % Al, in glasses, of composition $(Na_2O)_{50-x}$, $(Al_2O_3)_x$, $(SiO_2)_{50}$. The ordinate gives the value of the sensitivity of the cation J^+ relative to H^+, K_{HJ}^{pot} [43].

additional parameter, n (which is a constant specific for a given glass
cations) has been introduced.

.ve sensitivity of glass to various cations is defined by K_{ij}^{pot} of equation
effect of varying glass composition on relative sensitivities to various
trated in Figure 6.15, where the selectivities of the indicated cations
H^+ are plotted as a function of the aluminium content of sodium
sses $(50 - x) Na_2O$, $x Al_2O_3$, $50 SiO_2$. At the extreme left, where the
content is zero, a preference in excess of 10^{10} in favour of H^+ is seen,
ne addition of aluminium, an abrupt enhancement of the sensitivity to
ations occurs. Not only do the sensitivities relative to H^+ change with
on, but in addition the relative sensitivities among the alkali cations can
vary systematically.

erstand the mechanism of glass electrodes, three important character-
lass structure were implicitly presumed:

oxide glasses are 'solid electrolytes' consisting of atomic network with
immobile, anionic groups, or sites, and containing charge-balancing
ns with finite mobilities. The sum of the transport numbers of all mobile
ns is unity.

mobile cations respond to concentration (activity), potential, and
erature gradients and may leave the network if charge-neutrality is
tained, e.g., by replacement of other cations.

nic groups of the network positioned at the immediate glass surface,
ough bound to the network, are subject to reactions with contacting
ses, e.g. to the mass action law governing ionic equilibria in a neigh-
ring solution.

ntly Baucke et al. [45] have pointed out that inorganic oxide glasses are
n exchangers' although they may take up some water in surface-near
s. They proposed a Dissociation Mechanism, which is based on the
ria of functional groups at the glass surface and ions in contacting
ns. They also identified some of the proposed functional groups by
s surface analytical techniques.

REFERENCES

Pantano, C. G., Jr., Dove, D. B. and Onoda, G. Y., Jr., (1975), J. Non-Cryst. Solids,
9, 41.
Von Hickson, K. (1971), Glastech. Ber., 44, 537.
Heuch, L. L. (1975), in Characterization of Materials in Research Ceramics and
Polymers, (ed. J. J. Burke and V. Weiss) Syracuse University Press, 211.
Mass Spectrum No. 73–185 (1973) Commonwealth Scientific Corporation.
Bach, H. and Baucke, F. G. K. (1974), Phys. Chem. Glasses, 15, 123.
Sanders, D. M., Person, W. D. and Hench, L. L. (1972), Appl. Spectrosc., 26, 530.
Sykes, R. F. R. (1965), Glass Technol., 6, 178.
Williams, H. S. and Weyl, W. A. (1945), Glass Ind., 35, 347.
Persson, H. R. (1962), Glass Technol., 3, 17.
Gokularathnam, C. V., Gould, R. W. and Hench, L. L. (1975), Phys. Chem. Glasses,
16, 13.
] Charles, R. J. (1958), J. Appl. Phys., 11, 1549.

[12] Scholze, H., Helmreich, D. and Bekardjiev, I. (1975), *Glass Tech. Ber.*, **48**, 237.
[13] Iler, R. K. (1955), *Colloid Chemistry of Silica and Silicates*, Cornell University Press, Ithaca, New York.
[14] Douglas, R. W. and Isard, J. O. (1949), *J. Soc. Glass Tech.*, **33**, 289.
[15] Rana, M. A. and Douglas, R. W. (1961), *Phys. Chem. Glasses*, **2**, 179.
[16] (a) Hlavac, J. and Matej, J. (1963), *Silikaty*, **1**, 261.
(b) Matej, J. and Hlavac, J. (1967), *ibid.*, **11**, 3.
[17] (a) Boksay, Z., Bouquet, G. and Dobbs, S. (1970), *Phys. Chem. Glasses*, **11**, 140.
(b) Basu, S., Das, D. and Chakraborty, M. (1979), *J. Mater. Sci.*, **14**, 2303.
[18] Lyle, A. K. (1943), *J. Am. Ceram. Soc.*, **26**, 201.
[19] Dimbleby, V. and Turner, W. E. S. (1926), *J. Soc. Glass Technol.*, **10**, 314.
[20] El-Shamy, T. M. (1966), Ph.D. Thesis, University of Sheffield.
[21] Budd, S. M. and Frackiewiez, J. (1962), *Phys. Chem. Glass*, **3**, 116.
[22] Dubravo, S. K. and Schmidt, A. (1955), *Bull. Acad. Sci., U.S.S.R.*, Div. Chem. Sc., 403 (Consultants Bu. Translation).
[23] Budd, S. M. and Frackiewiez, J. (1962), *Phys. Chem. Glasses*, **3**, 116.
[24] Gastev, Yu. A. (1958), *The Structure of Glass*, p. 144, (Consultants Bu. Translation).
[25] Das, C. R. (1963), Ph.D. Thesis, University of Sheffield.
[26] Lewins, J. (1965), Ph.D. Thesis, Sheffield.
[27] El-Shamy, T. M., Lewins, J. and Douglas, R. W. (1972), *Glass Technol.*, **13**, 81.
[28] Morey, G. W., Fournier, R. O. and Rowe, J. J. (1962), *Geschim. Cosmochim. Acta*, **26**, 1029.
[29] Rana, M. A. (1961), Ph.D. Thesis, University of Sheffield.
[30] Nagai, N. (1973), M.Sc. Tech. Thesis, University of Sheffield.
[31] El-Shamy, T. M. and Taki-Eldin, H. D. (1974), *Glass Technol.*, **15**, 48
[32] (a) Warburton, R. S. and Wilburn, F. W. (1963), *Phys. Chem. Glasses*, **4**, 91.
(b) Paul, A. and Zaman, M. S. (1978), *J. Mater. Sci.*, **13**, 1399.
[33] Dimbleby, V. and Turner, W. E. S. (1926), *J. Soc. Glass-Technol.*, **10**, 304.
[34] (a) Majumdar, A. J. and Ryder, J. F. (1968), *Glass Technol.*, **9**, 78.
(b) Paul, A., Chakraborty, M., Das, D. and Basu, S. (1979), *Int. J. Cement Composites*, **1**, 103.
[35] Cremer, M. (1906), *Z. Biol.*, **47**, 562.
[36] Haber, F. and Klemensiewiez, Z. (1909), *Z. Phys. Chem.*, **67**, 385.
[37] Eisenman, G. (1962), *The Biophys. Journal*, **2**, pt. 2, 259.
[38] Dole, M. (1941), *The Glass Electrode*, Wiley, New York.
[39] Hughes, W. S. (1922), *J. Am. Chem. Soc.*, **44**, 2860.
[40] Horovitz, K. Z. (1923), *Z. Physik*, **15**, 369.
[41] Schiller, H. (1924), *Am. Physik*, **74**, 105.
[42] Lengyel, B. and Blum, E. (1934), *Trans. Farad. Soc.*, **30**, 461.
[43] Eisenman, G., Rudin, O. O. and Casby, J. V. (1957), *Science*, **126**, 3278, 931.
[44] Nicolskii, B. P. (1937), *Acta Physicochim, USSR.*, **7**, 597.
[45] Baucke, F. G. K. (1985), *J. Non-Cryst. Solids*, **73**, 215.

)xidation–reduction equilibrium in glass

7.1 GENERAL

is the loss of an electron′and reduction is the reverse:

$$Fe^{3+} + \text{electron} \xrightleftharpoons[\text{oxidation}]{\text{reduction}} Fe^{2+}$$

and reduction are complementary; without one the other cannot take
ᴇ simultaneous occurrence of reduction and oxidation is commonly
a 'redox process'. When a metal, M forms an oxide MO_x, it can be
that the metal electrons are transferred to the empty 2p orbitals of the
om and thus the elementary steps of the reaction can be hypothetically
ᴊ:

$$M \longrightarrow M^{2x+} + \text{electrons} \qquad \text{(oxidation)}$$
$$xO + 2x \text{ electrons} \longrightarrow xO^{2-} \qquad \text{(reduction)}$$

, in reality some oxides may be quite covalent involving no actual
transfer from the metal to the oxygen.
an element is capable of existing in more than one oxidation state, the
ʼm is termed a redox oxide. Thus cerium having two oxides: CeO_2 and
nd arsenic having two oxides: As_2O_5 and As_2O_3 can provide redox
ᴇt us consider a simple oxidation–reduction reaction between a metal, M
ɔxide, M_xO_{2y}. Writing the chemical reaction in terms of one mol of

$$\frac{x}{y}M + O_2 \overset{K}{\rightleftharpoons} \frac{1}{y} M_xO_{2y} \qquad (7.1)$$

$$K = \frac{[a_{M_xO_{2y}}]^{\frac{1}{y}}}{[a_M]^{\frac{x}{y}}.[a_{O_2}]}$$

ᴀse of negligible solid solution and oxygen pressure near one atmosphere,
al and its oxide may be considered at their standard states, and thus:

$$a_{M_xO_{2y}} = a_M = 1 \quad \text{and} \quad a_{O_2} = pO_2$$

ɔre,

$$K = \frac{1}{pO_2}$$

219

Now, the change of standard Gibbs free energy, $\Delta G°$ is related to the equilibrium constant of any chemical reaction by the equation:

$$\Delta G°_T = -RT \ln K = RT \ln pO_2 \qquad (7.2)$$

Thus by knowing $\Delta G°_T$ of any oxidation reaction, the equilibrium oxygen pressure of the system can be calculated.

The connection between heat (enthalpy) and entropy of reaction and the Gibbs free energy is

$$\Delta G = \Delta H - T \Delta S \qquad (7.3)$$

Equation (7.3) is valid for values of ΔG, ΔH and ΔS at the one temperature T. Since values for ΔH and ΔS are usually reported in the literature for room temperature (298 K), the values of ΔG may be calculated directly for this temperature only. However, ΔG values at higher temperatures can be calculated if heat capacities are known, for

$$\Delta H_T = \Delta H_{298} + \int_{298}^{T} \Delta C_p \, dT \qquad (7.4)$$

and

$$\Delta S_T = \Delta S_{298} + \int_{298}^{T} \frac{\Delta C_p}{T} \, dT \qquad (7.5)$$

Putting these in equation (7.3) leads to

$$\Delta G_T = \Delta H_{298} - T \Delta S_{298} + \int_{298}^{T} \Delta C_p \, dT - T \int_{298}^{T} \frac{\Delta C_p}{T} \, dT \qquad (7.6)$$

An example will make the use of equation (7.6) clear.

Let us consider the free energy of the reaction

$$2\,Al\,(solid) + \tfrac{3}{2}O_2\,(gas) \rightleftharpoons Al_2O_3\,(solid) \qquad (7.7)$$

The heat of formation of Al_2O_3 is $\Delta H°_{298} = -400\,000$ cal mol^{-1} and its entropy of formation is

$$\Delta S°_{298} = S°_{Al_2O_3} - 2S°_{Al} - \tfrac{3}{2}S°_{O_2}$$
$$= 12.20 - 13.54 - 73.53$$
$$= -74.87 \text{ cal deg}^{-1}$$

The sum of the heat capacities of the reactants is

$$C°_p(\tfrac{3}{2}O_2) = 10.74 + 1.50 \times 10^{-3}T - 0.60 \times 10^5 T^{-2}$$
$$\frac{C°_p(2Al) = 9.88 + 5.92 \times 10^{-3}T}{\Sigma C°_p = 20.62 + 7.42 \times 10^{-3}T - 0.60 \times 10^5 T^{-2}} \qquad (7.8)$$

The heat capacity of the product is

$$C°_p(Al_2O_3) = 27.43 + 3.06 \times 10^{-3}T - 8.47 \times 10^5 T^{-2} \qquad (7.9)$$

(7.8) from (7.9)

$$\Delta C_p^\circ = 6.81 - 4.36 \times 10^{-3}T - 7.87 \times 10^5 T^{-2}$$

on

$$\int \Delta C_p^\circ \cdot dT = 6.81T - 2.18 \times 10^{-3}T^2 + 7.87 \times 10^5 T^{-1} + K_1$$

8 K

$$\int \Delta C_p^\circ \cdot dT = 4477 + K_1$$

$$\int_{298}^{T} \Delta C_p^\circ \cdot dT = 6.81T - 2.18 \times 10^{-3}T^2 + 7.87 \times 10^5 T^{-1} - 4477 \quad (7.10)$$

equation divided by T and integrated gives in the same manner

$$\frac{\Delta C_p^\circ}{T} \cdot dT = 6.81 \ln T - 4.36 \times 10^{-3}T + 3.94 \times 10^5 T^{-2} - 41.94 \quad (7.11)$$

8

the values of equations (7.10) and (7.11) into (7.6) gives

$$- 400\,000)$$
$$- 4477 + 6.81\,T - 2.18 \times 10^{-3}\,T^2 + 7.87 \times 10^5\,T^{-1})$$
$$- 74.87T)$$
$$- 41.94T - 4.36 \times 10^{-3}T^2 + 3.94 \times 10^5 T^{-1} + 6.81T \ln T)$$

$$- 404\,477 + 123.62T + 2.18 \times 10^{-3}T^2 + 3.93 \times 10^5 T^{-1} + 6.81T \ln T$$
$$(7.12)$$

on (7.12) is typical of the formulae obtained for change in free energy with ature. In oxide systems most of the free energy changes can be represented eneral equation of the form:

$$\Delta G^\circ_T = A + BT \log T + CT$$

A, B and C are three experimentally determined constants for a particular 1. The values of these constants for a number of oxide system of interest in echnology is given in Table 7.1.
valuation of the free energy of formation of Al_2O_3 it is to be noted that the apacity values used were valid in the following ranges:

Al_2O_3 (solid)   273–3000 K
O_2 (gas)   298–1500 K
Al(solid)   273–932 K (melting point of aluminium)

nula (7.12) is therefore applicable at temperatures between 298 and 932 K. To uate the free energy of reaction above this temperature range, we must oduce the free energy of melting of aluminium. This can be done by adding the

Table 7.1
Standard Free Energy of some Oxidation–reduction reactions
⟨ ⟩-solid　　　　　　　{ } -liquid　　　　　　　( ) -gas

$$\Delta G_T^\circ = A + BT \log T + CT \text{ (calories)}$$

Reaction	A	$\Delta G_T^\circ (cals)$ B	C	Temperature range (K)
$4\langle Ag\rangle + (O_2) = 2\langle Ag_2O\rangle$	$-14\,600$	–	31.60	298–?
$4/3\langle Al\rangle + (O_2) = 2/3\langle Al_2O_3\rangle$	$-267\,207$	-2.653	58.427	298–923
$4/3\{Al\} + (O_2) = 2/3\langle Al_2O_3\rangle$	$-270\,507$	-2.500	61.480	923–1800
$4/3\langle B\rangle + (O_2) = 2/3\langle B_2O_3\rangle$	$-204\,070$	–	42.3	298–?
$2\langle Ba\rangle + (O_2) = 2\langle BaO\rangle$	$-271\,600$	–	46.400	298–983
$2\{Ba\} + (O_2) = 2\langle BaO\rangle$	$-278\,000$	–	53.200	983–1600
$2\langle Be\rangle + (O_2) = 2\langle BeO\rangle$	$-286\,900$	-3.32	56.100	298–1557
$2\{Be\} + (O_2) = 2\langle BeO\rangle$	$-291\,900$	-3.32	59.320	1557–2000
$4/3\langle Bi\rangle + (O_2) = 2/3\langle Bi_2O_3\rangle$	$-91\,450$	–	42.95	298–544
$2\langle C\rangle + (O_2) = 2(CO)$	$-53\,400$	–	41.900	298–2500
$\langle C\rangle + (O_2) = (CO_2)$	$-94\,200$	–	-0.20	298–2000
$2\langle Ca\rangle + (O_2) = 2\langle CaO\rangle$	$-302\,650$	–	47.32	298–1124
$2\{Ca\} + (O_2) = 2\langle CaO\rangle$	$-307\,100$	–	51.28	1124–1760
$2(Ca) + (O_2) = 2\,CaO$	$-380\,200$	–	93.24	1760–2500
$2(Cd) + (O_2) = 2\langle CdO\rangle$	$-174\,500$	–	99.9	1241–1379
$\langle Ce\rangle + (O_2) = \langle CeO_2\rangle$	$-260\,900$	-5.00	65.25	298–1000
$4/3\langle Ce\rangle + (O_2) = 2/3\langle Ce_2O_3\rangle$	$-290\,000$	–	47.13	298–?
$2\langle Co\rangle + (O_2) = \langle CoO\rangle$	$-111\,800$	–	33.8	298–1400
$6\langle CoO\rangle + (O_2) = 2\langle Co_3O_4\rangle$	$-87\,600$	–	33.8	298–1400
$4/3\langle Cr\rangle + (O_2) = 2/3\langle Cr_2O_3\rangle$	$-178\,500$	–	41.1	298–2100
$4\langle Cu\rangle + (O_2) = 2\langle Cu_2O\rangle$	$-81\,000$	-7.84	59.0	298–1356
$4\{Cu\} + (O_2) = 2\langle Cu_2O\rangle$	$-93\,400$	-7.84	68.2	1356–1503
$2\langle Cu_2O\rangle + (O_2) = 4\langle CuO\rangle$	$-69\,900$	-12.20	88.6	298–1300
$2\langle Fe\rangle + (O_2) = 2\langle FeO\rangle$	$-124\,100$	–	29.9	298–1642
$2\{Fe\} + (O_2) = 2\{FeO\}$	$-111\,240$	–	21.66	1808–2000
$6\langle FeO\rangle + (O_2) = 2\langle Fe_3O_4\rangle$	$-149\,240$	–	59.80	298–1642
$4\langle Fe_3O_4\rangle + (O_2) = 6\langle Fe_2O_3\rangle$	$-119\,240$	–	67.24	298–1460
$2(Geo) = \langle Ge\rangle + \langle GeO_2\rangle$	$-109\,200$	-13.8	124	
$2(H_2) + (O_2) = 2(H_2O)$	$-114\,500$	8.96	-4.42	298–2500
$4\langle K\rangle + (O_2) = 2\langle K_2O\rangle$	$-173\,200$	-9.2	89.2	298–?
$4\langle Li\rangle + (O_2) = 2\langle Li_2O\rangle$	$-285\,200$	–	58.68	298–454
$2\langle Mg\rangle + (O_2) = 2\langle MgO\rangle$	$-288\,700$	-5.9	67.9	298–923
$2\{Mg\} + (O_2) = 2\langle MgO\rangle$	$-290\,700$	-0.48	53.9	923–1380
$2(Mg) + (O_2) = 2\langle MgO\rangle$	$-363\,200$	-14.74	151.4	1380–2500
$2\langle Mn\rangle + (O_2) = 2\langle MnO\rangle$	$-183\,900$	–	34.8	298–1500
$2\{Mn\} + (O_2) = 2\langle MnO\rangle$	$-190\,800$	–	39.4	1500–2050
$4\langle Na\rangle + (O_2) = 2\langle Na_2O\rangle$	$-200\,000$	–	64.0	298–1193
$2\langle Ni\rangle + (O_2) = 2\langle NiO\rangle$	$-116\,900$	–	47.10	298–1725
$\langle P\rangle_{white} + (O_2) = 2/5\langle P_2O_5\rangle$	$-142\,600$	–	46.50	298–317
$2\{Ni\} + (O_2) = 2\langle NiO\rangle$	$-125\,300$	–	51.96	1725–2200
$2\langle Pb\rangle + (O_2) = 2\langle PbO\rangle$	$-105\,700$	-6.9	68.86	298–600
$2\{Pb\} + (O_2) = 2\langle PbO\rangle$	$-107\,500$	–	52.6	600–760
	$-106\,600$	–	51.4	760–1150
$2\langle Pd\rangle + (O_2) = 2\langle PdO\rangle$	$-44\,800$	-11.50	73.8	298–1133

	A	$\Delta G_T^\circ (cals)$ B	C	Temperature range (K)
$O_2) = (PtO_2)$	$-39\,270$	$-$	0.93	$1373-1823$
$O_2) = (RhO_2)$	$-45\,140$	$-$	4.94	$1473-1773$
$(O_2) = (RuO_2)$	$-73\,600$	-6.9	62.4	$298-1850$
$(O_2) = (SO_2)$	$-86\,620$	$-$	17.31	$298-2000$
$(O_2) = 2/3(SO_3)$	$-72\,813$	$-$	25.78	$318-1800$
$(O_2) = \langle SiO_2 \rangle$	$-215\,600$	$-$	41.5	$700-1700$
$(O_2) = \langle SiO_2 \rangle$	$-227\,700$	$-$	48.7	$1700-2000$
$SiO_2 \rangle + \langle Si \rangle$	$-169\,600$	-12.9	124.3	$298-1700$
$(O_2) = \langle SnO_2 \rangle$	$-140\,180$	$-$	51.52	$770-980$
$(O_2) = 2\langle TiO \rangle$	$-244\,600$	$-$	42.6	$600-2000$
$(O_2) = 2\langle Ti_2O_3 \rangle$	$-228\,300$	$-$	38.1	$298-2000$
$+ (O_2) = 4\langle Ti_3O_5 \rangle$	$-177\,000$	$-$	39.4	$700-2000$
$+ (O_2) = 6\langle TiO_2 \rangle$	$-146\,000$	$-$	46.0	$298-2123$
$+ (O_2) = 2\langle VO \rangle$	$-411\,800$	$-$	71.8	$900-1800$
$+ (O_2) = 2\langle V_2O_3 \rangle$	$-175\,300$	$-$	40.5	$823-1385$
$+ (O_2) = 4\langle VO_2 \rangle$	$-102\,800$	$-$	33.5	$1020-1180$
$+ (O_2) = 2\langle V_2O_5 \rangle$	$-64\,480$	$-$	45.6	$298-943$
$+ (O_2) = 2\{V_2O_5\}$	$-84\,300$	$-$	40.3	$943-1800$
$+ (O_2) = 2\langle ZnO \rangle$	$-168\,200$	-13.8	88.28	$298-693$
$+ (O_2) = 2\langle ZnO \rangle$	$-230\,840$	-20.7	164.76	$1170-2000$
$+ (O_2) = \langle ZrO_2 \rangle_\alpha$	$-259\,940$	-4.33	59.12	$298-1143$
$+ (O_2) = \langle ZrO_2 \rangle_\beta$	$-260\,200$	-6.44	65.99	$1478-2138$

ergy of reaction (7.13) to the formula (7.12)

$$2Al\,(solid) + \tfrac{3}{2}O_2\,(gas) \rightleftharpoons Al_2O_3\,(solid) \qquad (7.7)$$

$$\underline{2Al\,(liq) \rightleftharpoons 2Al\,(solid)} \qquad (7.13)$$

$$2Al\,(liq) + \tfrac{3}{2}O_2\,(gas) \rightleftharpoons Al_2O_3\,(solid) \qquad (7.14)$$

for the reaction (7.13)

$$\Delta H_{932}^\circ = -2 \times 2500 = -5000\,cal$$

$$\Delta S_{932}^\circ = \frac{-5000}{932} = -5.35\,cal\,deg^{-1}$$

$$2C_p^\circ(Al,\,solid) = 9.88 + 5.92 \times 10^{-3}T$$
$$\underline{2C_p^\circ(Al,\,liq) = 14.0\,(932 - 1273\,K)}$$

$$\Delta C_p^\circ = -4.12 + 5.92 \times 10^{-3}T$$

$$\int_{932}^{T} \Delta C_p^\circ \cdot dT = -4.12 + 2.96 \times 10^{-3}T^2 + 1269$$

223

and

$$\int_{932}^{T} \frac{\Delta C_p^\circ}{T} \cdot dT = -4.12 \ln T + 5.92 \times 10^{-3} T + 22.66$$

Thus ΔG_T° of the reaction (7.13) is

$$
\begin{aligned}
&= (-5000) \\
&+ (+1269 - 4.12T + 2.96 \times 10^{-3} T^2) \\
&- (-5.35T) \\
&- (+22.66T + 5.92 \times 10^{-3} T^2 - 4.12T \ln T)
\end{aligned}
$$

$$\Delta G^\circ{}_T = -3731 + 4.12T \ln T - 2.96 \times 10^{-3} T^2 \qquad (7.15)$$

Adding (7.15) to (7.12) results ΔG_T° of the reaction (7.14) and is ΔG_T° (reaction 5.14) $= -408\,208 - 2.68\,T \ln T - 0.78 \times 10^{-3}\,T^2$

$$+ 3.93 \times 10^{+5}\,T^{-1} + 102.19\,T \qquad (7.16)$$

Since C_p formula for liquid aluminium is valid up to 1273 K, equation (7.16) is applicable in the temperature range 932 to 1273 K only.

In 1944 Ellingham compiled diagrams of standard free energy changes against temperature for oxides and sulphides [1]. For the present purpose, only the oxide diagram will be described. In this diagram Ellingham plotted $-\Delta G^\circ$ in kilocalories, against temperature, T in degrees Celsius for a number of oxides. A scale of reversible electrochemical potential, E° was also included in this diagram, since $-\Delta G^\circ = nFE^\circ$ where n is the number of electrons involved in the appropriate chemical equation, and F is the Faraday constant (96 460 coulombs per gram equivalent).

Ellingham's diagrams allow immediate visual comparisons to be made of the standard free energies of a number of reactions of metallurgical importance over a wide range of temperatures without the necessity for calculation from standard tables as discussed before. In particular, the magnitude of the negative free energy of formation of a compound is a measure of its stability, hence the order of increasing stability of, for example, oxides (and hence of decreasing ease of reduction) is immediately apparent on looking down the appropriate oxide diagram at any chosen temperature.

Ellingham's diagrams were extended by Richardson [2], who added the nomographic scales (Figure 7.1) which allow equilibrium gas compositions to be read directly at any temperature for a number of reactions. To illustrate this, the oxygen partial pressure in equilibrium with manganese and manganous oxide (MnO) at 1000° C may be obtained by laying a straight-edge through the point representing zero free energy at the absolute zero of temperature and the point on the free energy curve for the formation of 2MnO at 1000° C; the equilibrium oxygen partial pressure is then found at the intersection of the straight-edge and the outer '$p_{\cdot}O_2$' scale, i.e. $pO_2 = $ approx. 10^{-24} atmosphere. Any oxygen pressure higher than this will cause the formation of MnO, while lower oxygen pressures will cause the reduction of MnO.

On the oxide diagram, nomographic scales are also given for direct reading of the equilibrium H_2/H_2O and CO/CO_2 ratios. For these scales the nomographic

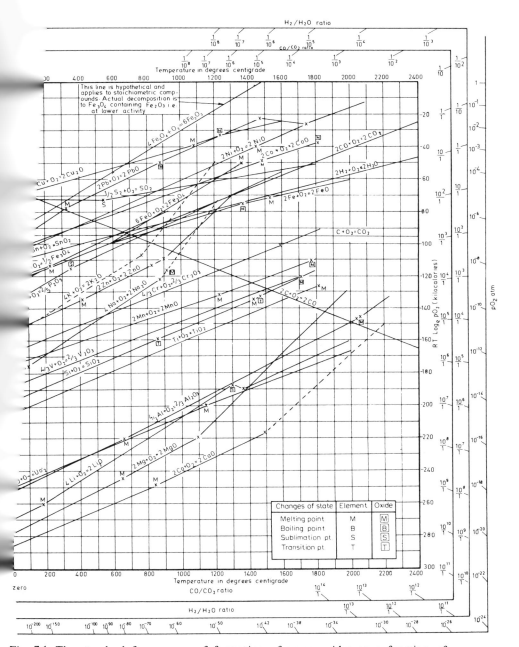

Fig. 7.1 The standard free energy of formation of some oxides as a function of temperature.

pivots are the points H and C respectively on the ordinate at the absolute zero of temperature, which are the zero temperature free energies for the reactions:

$$2CO + O_2 \rightleftharpoons 2CO_2 \tag{7.17}$$

$$2H_2 + O_2 \rightleftharpoons 2H_2O \tag{7.18}$$

On the Ellingham diagram it may be noticed that the lines for the formation of condensed metallic oxides all slope upwards. The slopes correspond to an entropy change of $\Delta S^\circ \sim -45\,cal/^\circ C$. This negative entropy change indicates a considerable decrease in randomness of the system and arises almost entirely from the disappearance of one gram mol of gaseous oxygen during reaction, which far outweighs the small entropy differences between the ordered metal and metal oxide phases that account for the small differences in slope between the oxides. The reaction for the formation of CO_2:

$$C_{(solid)} + O_{2(gas)} \rightleftharpoons CO_{2(gas)}$$

clearly involves only a small entropy change as the gas content is same on both sides of the equation and the free energy curve is consequently almost horizontal.

Changes in slope occur on the free energy plots at the temperatures of phase changes in the metal or oxide, e.g. at the melting point of the metal the solid and liquid metal are in equilibrium, so that the free energy of the reaction:

$$M_{(solid)} \rightleftharpoons M_{(liquid)} \text{ is zero, i.e.}$$

$$\Delta G_{fusion} = \Delta H_{fusion} - T\Delta S_{fusion} = 0$$

$$\text{or } \Delta S_{fusion} = \frac{\Delta H_{fusion}}{T}$$

This is positive, hence the entropy change for the reaction becomes more negative and the slope of the plot increases. Similarly the plots increase their slopes at the boiling points of the oxides.

Although there are many advantages of plotting ΔG° against T for various compounds, it always has to be remembered in using these diagrams that:

(a) The free energy changes shown refer to standard states only.
(b) The assumption is made that the oxides are compounds of definite composition, and although in practice this may not be so, at least for many oxides this assumption is true.
(c) No account is taken of the possibility of distribution of reactants and products between different phases (e.g. solid or liquid solutions).
(d) The possibility of the formation of intermetallic compounds is not taken into account.
(e) This diagram *only* indicates whether processes are thermodynamically possible, and do not give any information regarding the rate of the process under consideration.

7.2 ACTIVITY CORRECTIONS

The Ellingham diagrams give the standard free energies of formation of oxides, i.e. the free energy for the conversion of the reactants on the left-hand side of the

ιto the product on the right-hand side, all components being in the
standard state of pure condensed phases and gases at one atmosphere
ssure. In practice components frequently occur with activities not equal
ιd the free energy curves must be suitably adjusted.

: energy under non-standard conditions, ΔG, is obtained from the
ree energy, $\Delta G°$, by the Van't Hoff Isotherm:

$$\Delta G = \Delta G° + RT \ \ln \frac{\Pi a_{products}}{\Pi a_{reactants}}$$

s a convenient abbreviation for 'the product of'. The final correction
arly a linear function of temperature (as of course is $\Delta G°$ itself) being
: absolute zero of temperature, so the correction term effectively rotates
ιergy plot about its intercept on the ordinate. This may be illustrated by
ιg the following equation:

$$2\,Fe + O_2 \rightleftharpoons 2\,FeO$$

$$\Delta G = \Delta G° + RT \ \ln \frac{[a_{FeO}^2]}{[a_{Fe}^2] \cdot pO_2} \tag{7.19}$$

ι oxide activity is lowered below unity, e.g. due to solution in a glass, the
ι term is negative and the standard free energy plot is rotated clockwise.
a reduction in the activity of iron or pO_2 results in an anticlockwise
of the curve. These rotations of the free energy lines can seriously affect
·ent sequence of compound stability. For example, from a cursory glance
llingham diagram it will appear that SiO_2 will not be reduced by
:se dissolved in steel. However, due to very low activity of silicon in
·on, the line for the formation of SiO_2 is rotated strongly anticlockwise as
ι Figure 7.2 and manganese-steel cannot be made in a silica-brick furnace.

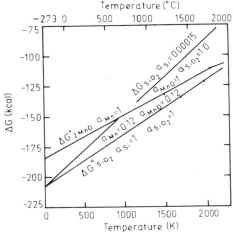

7.2 Illustration of the influence of changing activities on the relative stabilities
of SiO_2 and MnO, showing the possibility of reducing silica by manganese
in liquid steel.

227

7.3   OXIDATION–REDUCTION IN GLASS

When a redox oxide is introduced into a glass melt, it distributes itself into different states of oxidation depending upon the time and temperature of melting, the glass composition, the furnace atmosphere and the batch composition. In a given set of conditions, after sufficient time of melting, a melt comes to equilibrium with the partial pressure of oxygen in the ambient atmosphere and the relative concentrations of the different oxidation states reach equilibrium values. Figure 7.3 shows the attainment of cerous–ceric equilibrium in a 30 Na_2O, 70 SiO_2 glass at 1400° C after 60 hours of melting.

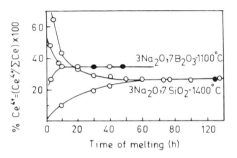

Fig. 7.3 Attainment of cerous–ceric equilibrium with time of melting of glasses.

The redox reaction in a glass melt may be written in different ways. Taking the redox reaction of ferrous and ferric iron as an example, the reaction may be written in terms of the pure oxides as:

$$4\,FeO(glass) + O_2(glass) \rightleftharpoons 2\,Fe_2O_3(glass) \qquad (7.20)$$

$$K = \frac{[a^2_{Fe_2O_3(glass)}]}{[a^4_{FeO(glass)}]} \cdot \frac{1}{pO_{2(glass)}} \qquad (7.21)$$

This method, which is used by metallurgical chemists, has the pure substances as standard states of the metal oxides, even though in the case of FeO this is a hypothetical state since FeO is non-stoichiometric. The activity of an oxide dissolved in glass, which is normally a complicated function of its concentration and the composition of the glass melt, approaches proportionality with the concentration, as the concentration tends to zero.

$$a_i = \gamma C_i$$

(where γ is a constant, C_i is the concentration of the species i, expressed in a convenient form, and a_i is the activity of the species i).

The redox reaction may also be written in terms of the ionic species present the system as:

$$4\,Fe^{2+}(glass) + O_2(glass) \rightleftharpoons 4\,Fe^{3+}(glass) + 2\,O^{2-}(glass) \qquad (7.$$

in which the ions are solvated by the medium in which they react. Equation (7. is formally equivalent to equation (7.20) but suffers from the disadvantage

228

ıg species whose activity has not yet been successfully measured.

bsence of reliable activity data the redox equilibrium is often expressed f concentration units instead of activities. The replacement of activities ntration is only justified thermodynamically for a substance which deal behaviour, and if this approximation is applied to a non-ideal the equilibrium constant for the reaction is not the true thermodynamic but is a function of both temperature and composition. It is sometimes at within the limits of sensitivity of a particular experimental investi-:he apparent equilibrium constant is seemingly independent of the tion. This may apply in a concentration range where the activities are not t vary almost linearly with the concentration or the activities of all the s and products vary at approximately the same rate with change in the ition. However, it is improbable that these approximations are valid over nan a limited concentration range and use of apparent constants at ;ition outside the range from which they are evaluated may give rise to error.

n equation (7.22) the factors determining redox equilibrium in glass are:

.e activity of the oxygen in the melt,

ıe activity of the oxygen ion in the melt,

ıe equilibrium constant which is a function of the free energy change and mperature of the reaction, and

he activity of the redox ions in the melt.

The activity of oxygen inside the glass melt

ıuilibrium, the activity of oxygen in the furnace atmosphere and that in the melt are identical. For simplicity the activity of oxygen in the furnace ısphere is generally replaced by the partial pressure of oxygen. This ıcement of activity by partial pressure is permissible only at low pressures. At ıer pressures deviations are quite frequent especially in the cases where the gas ts chemically with the melt.

ome workers [3] have attempted to study the effect of pO_2 on the redox ilibria in slag or glasses and have observed the theoretical depen-ıce – according to equation (7.22) – of the equilibrium redox ratio on the pO_2, gure 7.4). However, on a close examination, much unsatisfactory scatter may observed in some of the reported results. Platinum crucibles cannot be used der reducing conditions and hence some workers used alumina crucibles but in ıt case an appreciable amount of alumina was dissolved in the glass from the ıcible, probably yielding more acidic glasses, but no allowance was made for is inclusion of alumina.

.3.2 The activity of oxygen ion in glass

'he activity of the oxygen ion in glass can be calculated by studying the potential ıken up by a reversible oxygen electrode immersed in the molten glass, as well as ıy studying the equilibrium between an acidic gas such as H_2O, SO_2 or CO_2 and he melt at high temperature; some workers have also attempted to calculate the ıxygen ion activity from the molar refraction of oxygen in the glass, (for details of

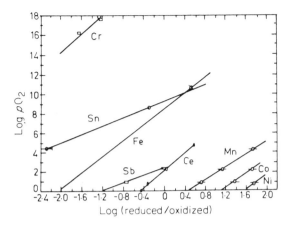

Fig. 7.4 Equilibrium dependence of $\log(M^{x+}/M^{(x+n)+})$ on $-\log pO_2$ in Na_2O, $2SiO_2$ melt at $1400°C$ [3b].

these methods see Chapter 8). Results from all these experiments generally indicate that the oxygen ion activity increases with increasing concentration of the modifier oxide (Na_2O, CaO etc.) in the glass.

7.3.3 Oxygen ion activity and redox equilibrium in glass

All experiments on redox equilibria reported in the literature, show that the proportion of the redox ion in the higher oxidation state increases with the basicity of the glass. Typical results are shown in Figure 7.5. Comparing these with equation (7.22) for the redox equilibrium in glass (in Figure 7.5 pO_2 is constant, K being assumed to be independent of composition) the conclusion

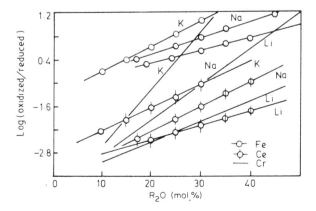

Fig. 7.5 Variation of redox ratio with alkali concentration in binary alkali–silicate melts at $1400°C$.

at $a_{O^{2-}}$, the oxygen ion activity decreases with increasing basicity. This
paradox is resolved by recognizing that the concentration equilibrium
K, which was measured in these cases, varies with composition. Further,
discussed in Chapter 9, transition metal ions occur as different
in glass, and not free ions as written in equation (7.22). These
can be spectroscopically identified and their concentrations can be
When oxidation–reduction equilibria are written in terms of these
s, instead of the free ions, a satisfactory qualitative correlation may be
between the oxygen ion activity and the redox equilibrium. A typical
vill clarify this point. From spectroscopic investigation it is known that,
glasses, chromium (VI) is present as a $[CrO_4]^{2-}$ group and not as the
ionically bonded with oxide ligands in glass (see Chapter 8). Thus the
relating the reaction between the components of a binary alkali silicate
nd chromium oxide may be written as follows:

$$\tfrac{1}{2} Cr_2O_3 + \tfrac{3}{4} O_2 \rightleftharpoons CrO_3 \tag{7.23}$$

$$2\, SiO_{4/2} + R_2O \rightleftharpoons 2\, SiO_{3/2}O + 2R^+ \tag{7.24}$$

$$2[SiO_{3/2}O]^- + CrO_3 \rightleftharpoons [CrO_4]^{2-} + SiO_{4/2} \tag{7.25}$$

$$\tfrac{1}{2} Cr_2O_3 + \tfrac{3}{4} O_2 + R_2O \rightleftharpoons [CrO_4]^{2-} + 2R^+ \tag{7.26}$$

re basic the alkali oxide and the greater its concentration, the more
(7.26) will move to the right and more $[CrO_4]^{2-}$ will be formed. Thus the
the ratio: [chromium(VI)/chromium(III)] is to increase with alkali
and the order: Li < Na < K is predicted, assuming that the basicity of the
xide increases with the atomic weight.

Effect of temperature on the redox equilibrium in glass

dation–reduction equilibrium in glass usually moves towards the reduced
th increasing temperature. This can easily be explained using standard
dynamic data. For the reaction:

Fe(solid) + O_2(gas) $\rightleftharpoons$ 2FeO(solid), $\Delta G_T^{\circ} = -124\,100 + 29.9T$ cal.

take two foils of solid metallic iron and equilibrate these in two furnaces
0° and 1200° C, both the furnaces having a constant $pO_2 = 10^{-16}$
phere.

0° C

$$T = 1273\ K$$
$$pO_2 = 10^{-16}\ \text{atmosphere}$$
$$\Delta G^{\circ}{}_{1273} = -86\,037\ \text{cal}$$
$$K = 10^{+14.7695}$$

the standard state condition, solid pure Fe and solid 'pure' FeO will exist
er, and $K = 1/pO_2$.

us the critical oxygen pressure, to keep both Fe and FeO at unit activity is
4.7695 atmosphere. Since the experimental oxygen pressure of the system

231

(10^{-16} atmosphere) is lower than this critical one, the system will have $a_{Fe} = 1$ and $a_{FeO} < 1$. Let us calculate the activity of FeO in the system

$$10^{14.7695} = K = \frac{[a_{FeO}^2]}{[a_{Fe}^2]} \cdot \frac{1}{pO_2}$$

$$= \frac{[a_{FeO}^2]}{10^{-16}}$$

$$a_{FeO} = 0.2435$$

At $1200°\,C$

$$T = 1473\,K$$

$$pO_2 = 10^{-16}\,\text{atmosphere}$$
$$\Delta G° = -80\,057\,\text{cal}$$
$$K = 10^{+11.87974}$$

As before in the system $a_{Fe} = 1$, and $a_{FeO} < 1$. Thus

$$10^{11.87974} = \frac{[a^2_{FeO}]}{10^{-16}}$$

$$a_{FeO} = 0.00871.$$

It is clear from the above calculations that, by increasing the temperature from 1000 to $1200°\,C$, under the same oxygen pressure of $pO_2 = 10^{-16}$ atmosphere, the activity of FeO (oxidized form) has decreased from 0.2435 to 0.0087; or in other words the system has moved towards the reduced side.

The redox equilibrium constant, K, is related to the temperature of reaction by the Van't Hoff isochore:

$$\frac{d \ln K}{dT} = \frac{\Delta H}{RT^2}$$

where ΔH is the enthalpy, R is the gas constant and T is the absolute temperature.

On integration

$$\ln K = -\frac{\Delta H}{RT} + I$$

where I is the integration constant.

From previous discussions it is clear that the ionic equilibrium constant K corresponding to equation (7.22) cannot be estimated due to uncertainty of the activities of the oxygen ion and redox ion. But if a glass of constant chemical composition is melted under an unchanged oxygen potential, and if the nature of redox ion complexes remain unchanged within the experimental temperature range, then a plot of the ratio: log [(concentration of the oxidized form)/ (concentration of the reduced form)] against reciprocal absolute temperature should give a straight line, with slope equivalent to $-\Delta H/(R \times 2.303)$. The same plot for different glasses, containing iron is shown in Figure 7.6. The slopes of the lines are all positive, indicating that these reactions are exothermic and that they

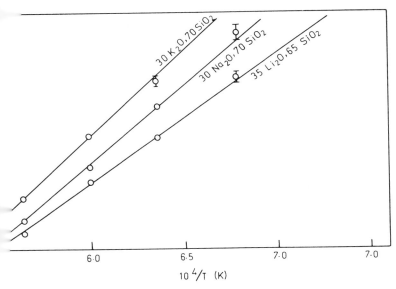

6 Variation of the ferrous–ferric equilibrium with temperature in some binary alkali–silicate melts.

...ceed towards the reduced side at higher temperatures. From these plots, ...culated heats of reaction are found to change with the nature and ...ration of alkali, which indicates the variation in the degrees of interaction ...1 the redox ion and the oxygen ligands of glass, when the composition of ...s changes.

The activity of the redox ions in the glass

ions of different oxidation states in the glass melt are not free ions as ...ented in equation (7.22). The interaction between a redox ion and the glass ...upon which the activity of the redox ion depends, is mainly determined by:

...e nature of the redox ion (size, charge, electronic arrangement, polariz-
...bility etc.),
...e coordination number and symmetry of the redox ion,
...e polarizability of the ligands associated with the redox ion (donor property
...nd the size effect), and
...ne temperature of the glass melt.

...ct, no systematic studies have yet been made to correlate the activity of redox ...in glass with the different factors mentioned above, but for simplicity all ...e factors are neglected and consequently the results obtained cannot be of ...ral applicability. In fact, as will be shown in Chapter 9 (Figure 9.35), the ...vity coefficient of the redox ion in glass changes critically with temperature ...concentration. Figure 7.7 shows the equilibrium solubility of NiO in different ...li–borate glasses at $1000°$ C. It should be noted that the activity of NiO in all

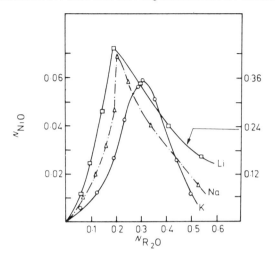

Fig. 7.7 Equilibrium solubility of NiO in different alkali–borate melts at 1000°C.

these glasses is the same (namely unity) but the concentration of dissolved NiO varies considerably, and goes through a maximum when plotted as a function of alkali oxide content of the glass. Although this type of solubility maximum can be qualitatively explained by considering the ligand field stabilization energies of the different nickel (II) complexes formed in these glasses and the acid–base equilibria involved, the quantitative understanding of the problem is still far from satisfactory [4].

The effect of the concentration of chromium oxide, iron oxide, and cerium oxide on chromium (VI)–chromium (III), iron (III)–iron (II), and cerium (IV)–cerium (III) equilibria respectively has been studied when glass composition, temperature of melting and furnace atmosphere were all kept constant. The redox ratios in terms of concentration were found to alter significantly especially when the concentrations of the redox oxides were very small in the melt; a typical result is shown in Figure 7.8.

The change of redox ion activities is also suggested from the fact that iron (III), cerium (IV) etc. may exist in glass in different coordination symmetries and these coordination equilibria are also found to be dependent on the concentration of the redox oxides. Since a coordination change causes a major change in the nature of bonding between the central ion and the ligands around it, the activity coefficient is also expected to change correspondingly.

It is known that an ordinary soda–lime–silica glass can dissolve some 30–40 wt % of iron at 1400° C. Since the solubility of FeO and Fe_2O_3 in ordinary glasses is very large, a saturated melt with pure iron oxide (standard states) will have a very different constitution from the parent glass. Previously it has been shown that if a solid iron crucible is oxidized at 1200°C with $pO_2 = 10^{-16}$ atmosphere, the activity of iron will be unity, and the activity of FeO will be 0.0087. If now a glass melt is brought in contact with the system, FeO will dissolve in it so that the activity of FeO in the solid phase and that in the glass become

234

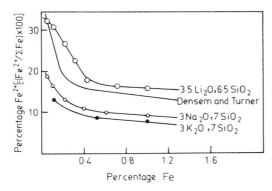

8 Effect of total iron concentration on the ferrous–ferric equilibrium in glass forming melts at 1400°C.

by estimating the FeO content in the glass the activity coefficient γ_{FeO} can ...lated from the relation

$$a_{FeO} = N_{FeO} \cdot \gamma_{FeO}$$

N_{FeO} is the mol fraction of FeO in the glass. Thus the activity coefficient for ...e species in glass can, in principle, be estimated. However, if any ordinary ...ime–silica glass of commercial interest is subjected to $pO_2 = 10^{-16}$ ...phere at 1200° C, the Na_2O in the glass is reduced and metallic sodium ...r comes out of the melt. In the system under discussion the effective ...on mechanism becomes

$$Na_2O(glass) + Fe(solid) \rightleftharpoons FeO(solid) + 2Na(gas)$$

$$\underline{FeO(solid) \rightleftharpoons FeO(glass)}$$

$$Na_2O(glass) + Fe(solid) \rightleftharpoons FeO(glass) + 2Na(gas)$$

...s the activity of FeO in the system is determined by the activity of Na_2O in ...s (which is not known) and not by the experimentally adjusted pO_2. Thus ...in the glass cannot be estimated.
... feasible way of overcoming this problem will be to use a Pt + Fe alloy of ...wn low iron activity (say $a_{Fe} = 10^{-3}$ in the alloy). If this alloy is oxidized at ...0° C and $pO_2 = 10^{-10}$ atmosphere (at this oxygen pressure the Na_2O in the ...ss is stable), then a_{FeO} on the alloy will be given by

$$K = 10^{11.87974} = \frac{[a_{FeO}^2]}{[10^{-3}]^2} \cdot \frac{1}{10^{-10}}$$

$a_{FeO}(alloy) = 0.0087.$

... a glass melt is now saturated with this alloy, since the activity of FeO in the

235

oxidized alloy is low, a small amount of FeO will dissolve into the melt. By estimating the dissolved FeO the activity coefficient of FeO into this melt at $1200°$ C can be estimated.

In most of the oxidation–reduction studies in glass the concentration of redox oxide used is very small (0.1 to 3 wt $\%$ in the case of iron). The activities of FeO and Fe_2O_3 in the equilibrated glasses were thus always less than unity.

When a glass is equilibrated at a temperature T and an oxygen pressure $pO_2 = x$, an invarient point is obtained on the free energy–temperature diagram (Ellingham diagram). In air $(pO_2 = 0.21$ atmosphere) at $1400°$ C, the free energy is given by

$$R \times (1400 + 273) \times \ln 0.21 = -5188 \, cal$$

If a glass containing a redox oxide is equilibrated under this condition, the redox equilibrium constant, K, will be adjusted in such a way that

$$-5188 = \Delta G° + RT \, \ln K$$

or

$$\log K = \frac{-5188 - \Delta G°}{4575 \times T}$$

where $\Delta G°$ is the standard free energy of the redox reaction as plotted in Ellingham's diagram at the temperature T.

For the reaction

$$4\,FeO(glass) + O_2(glass) \overset{K}{\rightleftharpoons} 2Fe_2O_3\,(glass) \tag{7.20}$$

$$K = \frac{[a^2_{Fe_2O_3\,(glass)}]}{[a^4_{FeO(glass)}] \cdot pO_2}$$

$$= \frac{[N^2_{Fe_2O_3(glass)}]}{[N^4_{FeO(glass)}]} \cdot \frac{[\gamma^2_{Fe_2O_3(glass)}]}{[\gamma^4_{FeO(glass)}]} \cdot \frac{1}{pO_2}$$

K can be calculated from $\Delta G°$ for reaction (7.20). Experimentally the FeO and Fe_2O_3 concentrations can be estimated, and thus the ratio of the activity coefficients can be calculated. Some calculated values of the activity coefficient ratios for FeO–Fe_2O_3 equilibrium in different soda-lime–silica glasses are shown in Figure 7.9. The ratio of the activity coefficients increase as the CaO concentration increases. The increase is most marked at lower temperatures and it may be due to an increase in the activity coefficient of the oxidized form or a reduction in the activity coefficient of the reduced form. In order to settle this question satisfactorily, experiments must be carried out where either the reactants or the products are of known activity, as discussed before.

7.3.6 Effect of the nature of the melting crucible on the redox equilibrium in glass

Most of the oxidation–reduction equilibria studied so far by different workers were made by melting glasses in platinum crucibles and in some isolated cases in alumina crucibles. There have been various controversial reports about the effect

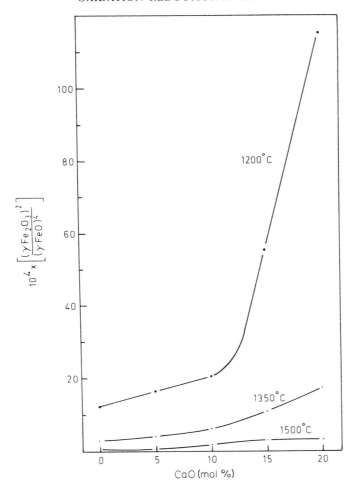

Fig. 7.9 Variation of the activity coefficient ratio for the ferrous–ferric equilibrium in $(30-x)\,Na_2O$, $x\,CaO$, $70\,SiO_2$ melts at different temperatures.

alumina and platinum on the iron (III)–iron (II) equilibrium in $Na_2O \cdot 2SiO_2$ ass. In order to study this aspect of the problem, the author melted several alkali icate glasses containing iron in platinum, alumina and silica crucibles. The asses melted in silica and alumina crucibles were chemically analysed and the lica and alumina enrichment of the glass from crucible corrosion was accounted or. From these studies the influence of platinum is distinctly evident particularly t low alkali concentrations.

Effect of alumina
The glasses when melted in alumina crucibles for 70 hours at 1400° C took up about 8–10 wt % Al_2O_3, and this changes the iron (III)–iron (II) equilibrium. To

study the effect of alumina in glass on redox equilibria some high alkali content glasses containing iron were melted in platinum (platinum uptake by these glasses did not disturb the iron (III)–iron (II) equilibrium) with varying amounts of Al_2O_3. When the equilibrium iron (II) content was plotted against wt % Al_2O_3 in the glass, a maximum was indicated at about 5 wt % Al_2O_3. It is interesting to note that Isard [5] while measuring the electrical conductivity of alkali–aluminosilicate glasses found a maximum in the activation energy at compositions having approximately 5 wt % Al_2O_3.

Comparision of melts in platinum and silica crucibles
It has been found that in equilibrium conditions the ratio [iron (III)/iron (II)] remains the same both for melts made in platinum as for those made in silica crucibles when the glasses contain thirty or more mol % alkali oxide (R_2O). When the alkali concentration was less than thirty mol %, platinum oxidized a part of the iron (II), the degree of oxidation increased with the decrease of alkali content in the glass. In glasses containing less than thirty mol % R_2O, flakes of metallic platinum were found to float on the surface of the molten glass; the amount of floating platinum appeared to increase in glasses containing less alkali oxide i.e. where the extent of oxidation of iron (II) was high. All these observations indicated a reaction between platinum, oxygen and the redox ions in the glass as follows:

$$Pt^0 + 2\,O_2 \rightleftharpoons PtO_2 \rightleftharpoons Pt^{4+} + 2\,O^{2-} \tag{7.27}$$

$$Pt^4 + \frac{4}{m}\,R^{n+} \rightleftharpoons Pt^0 + \frac{4}{m}\,R^{(m+n)+} \tag{7.28}$$

$$2\,O^{2-} + 4\,SiO_{4/2} \rightleftharpoons 4\,[SiO_{3/2}\,O]^- \tag{7.29}$$

Combining the above equations

$$\frac{4}{m}\,R^{n+} + 4\,SiO_{4/2} + O_2 \rightleftharpoons \frac{4}{m}\,R^{(m+n)+} + 4\,[SiO_{3/2}\,O]^- \tag{7.30}$$

The concentration of 'non-bridging oxygen' in a glass increases with alkali content. Thus at higher concentrations of alkali oxide (thirty mol % or more) the concentration of non-bridging oxygen is sufficient to stop the forward reaction of equation 7.30. At low concentration of alkali oxides, the reaction (7.30) shifts towards the right and consequently disturbs the iron (III)–iron (II) equilibrium in glass.

7.3.7 Mutual oxidation–reduction in glass

When a glass is melted with a single redox oxide, equilibrium between the different states of oxidation is established, depending upon the composition of the base glass and the conditions of melting, i.e. melting temperature, and the furnace atmosphere. This equilibrium may be represented by the general equation:

$$\text{Reductant} + \text{Oxygen} \; \overset{K}{\rightleftharpoons} \; \text{Oxidant} + \text{Oxygen ion}$$

y of the redox ions and the oxygen ion activity in glass change with
osition and unless measured quantitatively the exact value of K cannot
ed. When more than one redox oxide is used in the same glass, the
n becomes more complicated owing to the reciprocal action of the two
's.

al., [6] have tabulated some of the commonly used redox oxides in glass,
of increasing 'inner oxygen pressure'. Any oxide of this table will be
by any of the oxides placed above it. A similar table calculated from
thermodynamic data on pure oxides has been reported by Tress [7].

THEORY OF REDOX REACTIONS IN SOLUTIONS

x system containing both an oxidizing agent and its reduction product,
l be an equilibrium between them and their electrons. Thus if an inert
e like platinum is placed in a redox system: $R^{(m+n)+} - R^{(m)+}$, it will assume
e potential indicative of the position of equilibrium. If the system tends to
n oxidizing agent, then $R^{(m+n)+}$ will be converted to $R^{(m)+}$ and electrons
taken from the platinum electrode making it positively charged. The
de of the positive charge will be a measure of the oxidizing property of
em. The electrode potential developed in such a case is given by the Nernst
n:

$$E_1 = E_1^0 + \frac{RT}{nF} \ln \frac{[a_{R_1}^{(m+n)+}]}{[a_{R_1}^{(m)+}]}$$

$R_1^{(m+n)+} - R_1^{(m)+}$ system, and

$$E_2 = E_2^0 + \frac{RT}{yF} \ln \frac{[a_{R_2}^{(x+y)+}]}{[a_{R_2}^{(x)+}]}$$

nother $R_2^{(x+y)+} - R_2^{(x)+}$ system.
he two redox pairs are in equilibrium, then $E_1 = E_2$ and therefore

$$\Delta E^0 = E_1^0 - E_2^0 = \frac{RT}{F} \ln \left\{ \frac{[a_{R_1}^{(m)+}]}{[a_{R_1}^{(m+n)+}]} \right\}^n \left\{ \frac{[a_{R_2}^{(x+y)+}]}{[a_{R_2}^{(x)+}]} \right\}^y \qquad (7.31)$$

$$= \frac{RT}{F} \ln K$$

re K is the equilibrium constant of the overall reaction.
hus as ΔE^0 increases, K increases correspondingly. Or, in other words, the
nce of getting all the oxidation states of the redox ions in measurable amount
single solution decreases as the electromotive force difference between the
redox pairs increases.

Consider two redox equilibria in glass as follows:

$$R_1^{(m)+} + \frac{n}{4}O_2 \rightleftharpoons R_1^{(m+n)+} + \frac{n}{2}O^{2-} \qquad (7.32)$$

d

$$R_2^{(x)+} + \frac{y}{4}O_2 \rightleftharpoons R_2^{(x+y)+} + \frac{y}{2}O^{2-} \qquad (7.33)$$

239

If the redox potentials of R_1 and R_2 are different in the glass (actually how these potentials can be different in the same glass will be discussed later in this section) i.e. the free energy change of reactions (7.32) and (7.33) are different in the glass at a particular temperature, then the redox pairs will interact shifting each of the above equilibria, and the final equilibrium will be given by

$$y R_1^{(m)+} + n R_2^{(x+y)+} \overset{K}{\rightleftharpoons} y R_1^{(m+n)+} + n R_2^{(x)+} \tag{7.34}$$

which is similar to equation (7.31).

If in the molten glass $R_2^{(x+y)+} - R_2^{(x)+}$ pair has a higher oxidation potential or free energy than that of the pair $R_1^{(m+n)+} - R_1^{(m)+}$, the reaction (7.34) will proceed forward, and under the new equilibrium condition, some of the $R_1^{(m)+}$ ions will be oxidized at the expense of $R_2^{(x+y)+}$ ions. From equation (7.34)

$$\log \frac{[a_{R_1}^{(m+n)+}]}{[a_{R_1}^{(m)+}]} = \frac{\log K}{y} + \frac{n}{y} \log \frac{[a_{R_2}^{(x+y)+}]}{[a_{R_2}^{(x)+}]}$$

Thus a plot of log

$$\frac{[a_{R_2}^{(x+y)+}]}{[a_{R_2}^{(x)+}]} \text{ against } \log \frac{[a_{R_1}^{(m+n)+}]}{[a_{R_1}^{(m)+}]}$$

is expected to give a straight line of slope n/y and intercept $(\log K)/y$.

Mutual interaction of redox pairs such as chromium–cerium, manganese–cerium, manganese–arsenic etc. have been studied in simple binary alkali–borate and alkali–silicate glasses, and with the assumption that the activity of redox ion in glass is proportional to its concentration, the above relationships have been found to be obeyed satisfactorily [8].

The relationship in equation (7.31) gives the value of the overall equilibrium constant, K, but does not give any information as to the concentration of the different redox ions at equilibrium. The analytical techniques used are limited to certain concentration ranges, with lower concentrations the accuracy is poor. It has been found experimentally that to get the simultaneous presence of all the redox ions of the different oxidation states in measurable amounts in glass, introduction of stoichiometric amounts of oxidant and reductant at equilibrium gave the best result. However, it should be noted that this equivalent ratio between two redox oxides changes with glass composition and temperature of melting; it holds only over a limited concentration range of the redox ions [8].

7.4.1 Thermodynamics of mutual oxidation–reduction in glass

As discussed earlier, when a glass is equilibrated at a temperature, T, with an oxygen potential, $pO_2 = x$, an invariant point is obtained on the free energy–temperature diagram. Such a point for air at $1400°$ C is shown in Figure 7.10 as point A; the free energy corresponding to this point is -5188 cal.

Now if a redox oxide is added to this glass and brought to equilibrium, the equilibrium constant, K of the oxidation reaction will be adjusted in such a way so that

$$-5188 = \Delta G_f^{\circ} + RT \ln K \tag{7.35}$$

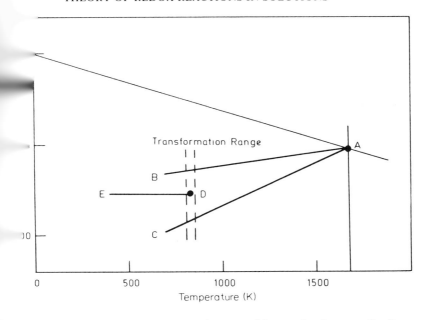

7.10 Schematic free energy diagram for mutual interaction between R_1–R_2 redox pairs in glass.

ΔG_f^0 is the standard free energy change of the reaction (this is plotted in ~ham's diagram) at temperature, T. If two redox oxides are added to the same ~together, each oxidation–reduction equilibrium will be separately adjusted ~mply with the relation written in equation (7.35). When this is cooled, the ~melt becomes virtually cut off from atmospheric oxygen, for the diffusion of ~cular oxygen inside the glass melt is a very slow process.

~)w, since the activity coefficients of different redox oxides dissolved in a glass ~change in different manners with temperature, the free energy of different ~x systems will follow different paths on cooling, as shown schematically by ~ AB and AC in Figure 7.10.

~he free energy difference between two redox systems increases with increased ~ling, and the two redox systems start interacting according to equation (7.31). ~wever, this interaction cannot continue for a long time because, eventually, the ~iperature of the glass falls below the transformation range and all long-range ~djustment in structure becomes restricted due to the high viscosity of the glass. ~us below the 'fictive temperature' the system cools down to a thermodynami-~ly metastable condition along some line DE shown in Figure 7.10.

Let us take an example to explain these points. Figure 7.11 shows a simplified ~lingham diagram for some of the commonly used redox oxides in glass ~dustry. In this diagram the MnO–Mn_2O_3 line lies well below the As_2O_3–As_2O_5 ~ie. But in a glass melt, arsenic oxide is well known to reduce to white the pink ~)lour due to Mn_2O_3. How is this possible?

241

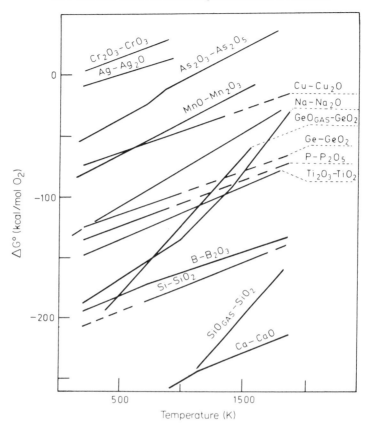

Fig. 7.11 Ellingham diagram for some oxides of interest to glass technologists.

The variation of activity coefficient of manganese and arsenic oxides with temperature is shown in Figure 7.12. Now if we equilibrate a glass at $T = 1673$ K and $pO_2 = 0.21$ atmosphere containing a certain amount of manganese oxide dissolved in it, the Mn_2O_3/MnO ratio will be adjusted in such a way that:

$$-5188 = \Delta G^{\circ}_{1763} (MnO–Mn_2O_3) + R \times 1673$$

$$\times \ln \left[\frac{(N^2_{Mn_2O_3})}{(N^4_{MnO})} \cdot \frac{(\gamma^2_{Mn_2O_3})}{(\gamma^4_{MnO})} \cdot \frac{1}{0.21} \right]$$

If we now isolate the melt from the atmosphere and do not allow any oxygen either to enter or leave the melt, then $N_{Mn_2O_3}$ and N_{MnO} cannot change, and since $\gamma_{Mn_2O_3}$ and γ_{MnO} will change with temperature on cooling to a particular temperature, the system will acquire a pseudo-equilibrium p^*O_2 which can be

242

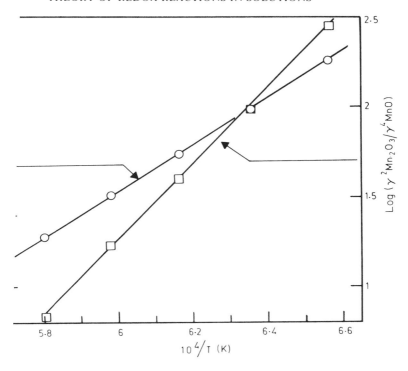

12 Variation of activity coefficients of manganese and arsenic oxides with temperature in $3\,Na_2O, 7\,SiO_2$ melt.

d from a knowledge of $\Delta G°\,(MnO–Mn_2O_3)$, and $\gamma^2_{Mn_2O_3}/\gamma^4_{MnO}$ at that ure. Such lines of equilibrium $p*O_2$ with $MnO–Mn_2O_3$ and s_2O_5 in a $30\,Na_2O, 70\,SiO_2$ glass melt are shown in Figure 7.13. It is n this figure that although the standard free energy line of $MnO–Mn_2O_3$ that of $As_2O_3–As_2O_5$ line, the equilibrium $p*O_2$ produced by arsenic glass on cooling is much lower than that due to manganese oxide. Thus, ig, As_2O_3 will reduce Mn_2O_3 in the melt and this process will continue igh the interaction zone until the high viscosity at lower temperatures is and stops any further mutual oxidation–reduction.

the important factors determining mutual oxidation–reduction in glass

initial conditions of equilibrium, this fixes the point 'A' on the free gy–temperature diagram.

variation of activity coefficient with temperature for the individual redox :m; the larger the difference between the two systems, the greater will be mutual interaction.

rate of cooling, particularly through the transformation range. If the rate ooling in the transformation range is faster than the time necessary for

243

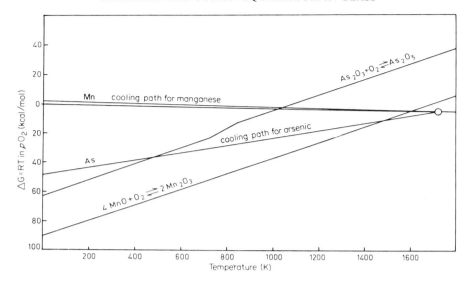

Fig. 7.13 Variation of effective pO_2 with temperature.

electron transfer between two redox ions and the consequent structural modifications involved with it, then the extent of mutual interaction will be sensitive to heat-treatment at the transformation range.

Recently Schaeffer *et al.* [9] have reported direct estimation of the valence ratios in glass-forming oxide melts by the measurement of Current-Potential curves, and observed: 'At equilibrium temperature a redox pair will not change its valence ratio in the presence of a second redox pair. During cooling, two redox pairs will react with each other until the reaction is frozen in; the extent of interaction depends on the temperature difference between the temperature of equilibration and the freezing-in temperature.'

7.4.2 Redox number of glass batch

The main ingredients used for the manufacture of soda-lime silicate glasses are sand, sodium carbonate, limestone, dolomite, and minerals containing alumina. In addition to melting aids such as borates and fluorspar there are two other groups of supplementary raw materials in use; those which have a reducing effect and those which are oxidizing. Manring and Hopkins [10] first suggested an empirical approach to the use and control of these reducing and oxidizing agents and a means of predetermining their effectiveness in the glass melting process. Based on shop floor experience they assigned numerical values to the commonly used oxidizing and reducing agents that are used for conventional glass melting. An example of these assigned values is given in Table 7.2; each factor of Table 7.2 is based on 1 kg of raw material per 2000 kg sand in a glass batch mixture.

Table 7.2
Redox values of various batch materials (after Ref. [11])

ents	Redox number	Oxidizing agents	Redox number
ɔm 'calumite'* slag	−9.00	SO_3 from calumite slag	+1.20
m calumite slag	−6.70	Arsenic	+0.93
0% C)	−6.70	Sodium bichromate	+0.77
C)	−5.70	Anhydrate ($CaSO_4$)	+0.70
65% C)	−4.36	Salt cake (Na_2SO_4)	+0.67
lphide	−1.60	Potassium bichromate	+0.65
ːs	−1.20	Gypsum	+0.56
	−0.10	Barytes	+0.40
iron and aluminium)		Sodium nitrate	+0.15
mite	−1.00		
	−0.93		

te is a processed blast furnace slag.

these factors a *redox number* can be assigned to a batch but, dependent ype of furnace, fuel used, etc., the redox number giving optimum results s of refining and colour) for any given furnace will vary. This ideal level determined after making the necessary adjustments to the batch. Once hed, it can be used to indicate the correct level necessary to return to, ιy other batch changes required for that particular melting unit are made.

REFERENCES

llingham, H. J. T. (1944), *J. Soc. Chem. Ind.*, **63**, 125.
ιichardson, F. D. and Jeffes, J. H. E. (1948), *IISI.*, **160**, 3261.
a) Irmann, F. (1952), *J. Amer. Chem. Soc.*, **74**, 4767.
b) Johnston, W. D. (1965), *J. Amer. Ceram. Soc.*, **48**, 184.
c) Richardson, F. D. (1955),*The Vitreous State*, The Glass Delegacy of the University of Sheffield p. 81.
d) Banerjee, S. and Paul, A. (1974), *J. Amer. Ceram. Soc.*, **57**, 286.
e) Johnston, W. D. and Chelko, A. (1966), *J. Amer. Ceram. Soc.*, **49**, 562.
Paul, A. (1975), *J. Mater. Sci.*, **10**, 422.
Isard, J. O. (1959), *J. Soc. Glass Technol.*, **33**, 113.
Kühl, C., Rudow, H. and Weyl, W. A. (1938), *Sprechsaal*, **71**, 118.
Tress, H. J. (1960), *Phys. Chem. Glasses*, **1**, 196.
Paul, A. and Douglas, R. W. (1966), *Phys. Chem. Glasses*, **7**, 1.
Lenhart, A. and Schaeffer, H. A. (1987), in *Effects of Modes of Formation on the Structure of Glass-2nd Int. Conf.*, Trans. Tech., p. 335.
Manring, W. H. and Hopkins, R. W. (1958), *Glass Ind.*, **39**, 139.
Simpson, W. and Myers, D. D. (1978), *Glass Technol.* **19**, 82.

CHAPTER 8

Acid–Base Concepts in Glass

8.1 INTRODUCTION

The concept of acids and bases has undergone numerous revisions and is still the subject of considerable controversy. Of the several definitions of acids and bases available, each is designed to be satisfactory for a group of allied substances only. Arrhenius [1] defined an acid as a compound that dissociates in solution to give hydrogen ions, and a base as a compound yielding hydroxyl ions in solution. The process of neutralization is regarded simply as the combination of H^+ and OH^- to form water. Arrhenius' acids–base theory relied exclusively on the phenomena observed in aqueous solution and is of limited applicability in non-aqueous systems.

The Brönsted and Lowry [2] theory offered a broader definition of acids and bases to determine acid–base functions in a protonic solvent. According to this theory an acid is defined as a species which has a tendency to yield a proton, and a base as a substance which can accept a proton. This definition can be represented by the equation:

$$A^{n+} \rightleftharpoons B^{(n-1)+} + H^+$$

which in fact is a hypothetical scheme for defining an acid and its conjugate base, and does not in any way represent the actual reaction which can occur to yield a free proton. The reaction can take place only when the acid is brought in contact with some base which has a higher proton affinity than the conjugate base $B^{(n-1)+}$. Thus, the actual acid–base reaction can be represented by the equation:

$$\text{Acid}(1) + \text{Base}(2) \rightleftharpoons \text{Acid}(2) + \text{Base}(1).$$

The equilibrium constant of the above reaction is determined by the relative affinities of the bases for the proton. The essential nature of the neutralization, according to the proton transfer concept, must be viewed in terms of a complete proton transfer reaction. The strengths of acids and bases in the Brönsted and Lowry sense depend on their environments, and thus a substance which functions as an acid in one solvent does not necessarily behave in the same manner in other solvents. For example, urea is a weak base in water, a strong base in acetic acid, and an acid in ammonia.

There are many substances which are capable of acting either as an acid or as a base in terms of proton donating or accepting capacities, and these are called amphioprotic. Water acts as a proton donor (acid) towards ammonia but a eptor (base) towards acetic acid.

$$H_2O + NH_3 \rightleftharpoons NH_4^+ + OH^-$$
$$CH_3COOH + H_2O \rightleftharpoons H_3O^+ + CH_3COO^-$$
$$A_1 + B_2 \rightleftharpoons A_2 + B_1$$

ɔnsted–Lowry picture still does not represent the full generality of phenomena. Familiar acid–base properties have been observed in ontaining no proton at all. The proton exchange theory does take into ɪe experimental fact that there are many substances besides the ion which exhibit typical basic properties, but fails to recognize ntary data with regard to acids. For obvious reasons the proton :id–base theory offers practically no application in glass.

l–base theory advanced by Lewis [3] considers acid–base functions and ɔcesses independently of the solvent. According to Lewis an acid is any ɪt is capable of accepting a pair of electrons to form a covalent bond by th the electron donor, and a base is any species that can donate a pair of to form a covalent bond with the acid. Neutralization is then the of a coordinate covalent bond between the acid and base. Formation of ɪte covalent bond is regarded as a first step, even though ionization may ɪtly occur. A typical example of a Lewis acid is boron trifluoride:

$$\begin{array}{c} F \diagdown \\ B\!-\!F \\ F \diagup \end{array} + \begin{array}{c} H \\ :N\!-\!H \\ H \end{array} \rightleftharpoons \begin{array}{c} F \diagdown \\ F\!-\!B \\ F \diagup \end{array} \leftarrow \begin{array}{c} H \\ N\!-\!H \\ H \end{array}$$

ion between boron trifluoride and ammonia takes place independently vent even in the gaseous phase. In this case ionization does not occur ralization; however, in other cases formation of a covalent bond may be or accompanied by ionization.

nphoteric substances are those which can both accept and donate pairs in the formation of a coordinate covalent bond. The close ɪip between acid–base function and oxidation–reduction reactions is arent, in that both oxidizing agents and acids tend to accept electrons. accepts electron pairs held by a base and forms coordinate covalent ɪ sharing, while the oxidizing agent keeps the electrons to itself. The ndependent Lewis theory, although it can overcome many difficulties in chemistry, cannot be suitably applied to a glass melt.

ɔase relationships in non-aqueous solvents have also been formulated in the solvent system concept postulated by Jander [4]. The solvent system envisages autoionization of 'water-like' solvents into two ions. 'Water-ɪents are those solvents having strong solvent power, a low intrinsic vity and high conductivity of solutions arising from the ability to bring ectrolytic dissociation of dissolved substances in it. According to the ɪystem theory, an acid is defined as a species capable of increasing the ɔoncentration of the cation characteristic of the pure solvent, whereas a reases the concentration of the corresponding anion formed via the ization' reaction. The greater the concentration of the negative ion due to the addition of the solute, the greater is the basicity of the solution. Thus:

$$\text{Acid} \qquad \text{Base}$$

$$H_2O + H_2O \rightleftharpoons H_3O^+ + OH^-$$

$$CH_3COOH + CH_3COOH \rightleftharpoons CH_3COOH_2^+ + CH_3COO^-$$

$$AsCl_3 + AsCl_3 \rightleftharpoons AsCl_2^+ + AsCl_4^-$$

Thus each solvent is the parent system of a series of acids and bases. Based on the specific conductivity of the pure solvent, the degree of autoionization varies over a wide range in different liquids, as the autoionization constants below will show:

$$AsCl_3 \quad 10^{-7}$$
$$H_2O \quad 10^{-14} \text{ at } 25°C$$
$$NH_3 \quad 10^{-33} \text{ at } 50°C$$

8.2 ACID–BASE RELATIONSHIPS IN GLASSES

Though extensive investigations have been made by numerous authors on acid–base phenomena in aqueous and non-aqueous chemistry, very little has been done to determine acid–base relationships in glass. Initially, it was believed that acidic oxides are glass formers, and basic oxides merely modify the properties of glass. Weyl in his monograph *Coloured Glasses* considers acidic glass-forming cations as those having high ionic potential, i.e. small ionic radii and higher charges, while basic glass constituents are considered to be cations possessing smaller ionic potential. Weyl concludes that a sharp distinction between acids and bases in glass is not possible, in spite of the fact that the terms acidic and basic glasses are widely used. According to him the quesion, 'how acidic is a glass?' cannot be answered simply.

Lüx and Rogler [5] studied acid–base relationships in mixed alkali–borate glasses. According to these authors, substitution of B_2O_3 by the molar equivalent of Li_2O in a sodium–borate glass makes the melt more acidic. This leads to the conclusion that in certain composition ranges Li_2O behaves as a stronger acid than even B_2O_3, while in certain composition ranges, like binary lithium–silicate, this obviously acts as a base. However, the method used by these authors to determine the acid–base character of glass was rather crude and it is doubtful if the neutral point observed by them using Cr^{3+}–Cr^{6+} as a colour indicator actually indicated any constant end point of the acid–base neutralization in the melt.

Paul and Douglas [6] have defined acids and bases in borate and silicate glasses in a manner analogous to Jander's *solvent system concept*, in a non-protonic solvent. The autoionization in silica and boric oxide were assumed to be:

$$2\,SiO_2 \rightleftharpoons SiO_{3/2}^+ + SiO_{5/2}^- \qquad (8.1)$$

$$B_2O_3 \rightleftharpoons BO^+ + BO_2^- \qquad (8.2)$$

The conventional way of representing the reaction between silica and alkali oxide is:

$$2\,Si(O_{1/2})_4 + R_2O \rightleftharpoons 2\,Si(O_{1/2})_3\,O^-\,R^+ \qquad (8.3)$$

represents an oxygen shared between two silicons.
g to the autoionization hypothesis, using convenient symbols,
.1) can be written as:

$$2\,Si(O_{1/2})_4 \rightleftharpoons Si(O_{1/2})_3{}^+ + Si(O_{1/2})_3\,O^- \tag{8.4}$$

this in terms of *bridging* and *non-bridging* oxygens

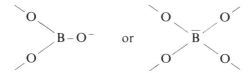

equation (8.4) suggests that basicity of the melt increases with the
f concentration of the $Si(O_{12})_3O^-$ group. The addition of alkali oxide
ncreases this negative ion characteristic of the autoionization product,
ncreasing the basicity of the glass. The complementary reaction of
(8.3) may be written as:

$$Si(O_{1/2})_3\,O^-R^+ \rightleftharpoons Si(O_{1/2})_3\,O^- + R^+ \tag{8.5}$$

: same molar concentration of alkali oxides, the alkali silicate group
nizes to a greater degree will produce a more basic melt. As the bond
between alkali and oxygen decreases with increase in the size of the alkali
extent of ionization of the alkali silicate group increases with the increase
radii of the alkali ion. Thus the basicity of a binary alkali–silicate, with
ar concentrations of alkali oxides, will increase in the order:

$$Li_2O < Na_2O < K_2O < Rb_2O < Cs_2O.$$

ture, however, is not so simple in borate melts because $BO_2{}^-$ as shown in
n (8.2) may represent a three-coordinated boron with one non-bridging
or a four-coordinated boron tetrahedron:

$$\begin{array}{c} O \\ \diagdown \\ O \diagup \end{array} B-O^- \qquad or \qquad \begin{array}{cc} O \diagdown & O \diagup \\ & B \\ O \diagup & O \diagdown \end{array}$$

g symbolically

$$2\,B(O_{1/2})_3 = B(O_{1/2})_2{}^+ + B(O_{1/2})_2\,O^-$$
$$= B(O_{1/2})_2{}^+ + B(O_{1/2})_4{}^- \tag{8.6}$$

clear magnetic resonance studies by Bray and O'Keefe[7] in alkali–borate
s have shown beyond doubt that when alkali oxide is added to boric oxide,
rst reaction product is tetrahedral boron. On further addition of alkali oxide,
nelt is eventually found to contain three-coordinated boron with a non-
ging oxygen. The sequence of reaction can thus be represented schematically

$$\underset{2\,B(O_{1/2})_3 \rightarrow 2\,B(O_{1/2})_4{}^- \rightarrow 2\,B(O_{1/2})_2\,O^-}{\overline{\text{Increasing alkali oxide} \longrightarrow}}$$

The reaction sequence may therefore be regarded as analogous to acid–base titration where acidic $B(O_{1/2})_3$ triangles are stepwise neutralized by the alkali oxide, the intermediate product of neutralization being $B(O_{1/2})_4{}^-$ tetrahedral which may be termed as mildly basic. Further addition of alkali oxide produces more basic $B(O_{1/2})_2\,O^-$ groups; higher basicity may be assigned to this group over the tetrahedral group, owing to the fact that excess electrons available with the former group have been utilized in the formation of sp^3 hybrid orbitals thereby reducing the electron-donating capacity of the individual oxygen ligands, whereas, the latter being an sp^2 hybrid, the electron-donating capacity of the oxygen ligands in this group is more than that of the tetrahedral group.

8.3   OXYGEN ION ACTIVITY

It would appear from this discussion that there is a close relationship between acidity–basicity and the concept of non-bridging oxygens, and oxygen ion activity. Franz and Scholze [8], and Douglas *et al.* [9] defined the basicity of glass in terms of oxygen ion activity – the greater the oxygen ion activity, the greater the basicity of the glass. From his investigation of solubility of water in glasses, Scholze suggests that one can draw a pO scale in glass analogous to a pH scale in aqueous solution. However, none of the entities such as *oxygen ion activity, non-bridging oxygen* or *non-bridging oxygen ion activity* are directly measurable and these have so far been regarded as convenient ways of discussing availability of electrons or active oxygens.

The activity of the oxygen ion in a glass enters into the equations representing the potential taken up by a reversible oxygen electrode immersed in the molten glass, the equilibrium between a gas such as H_2O, SO_2, or CO_2 and the melt at high temperature, the coordination equilibrium of transition metal ions in glass, and the redox equilibrium in glass. Molar refractivity of oxygen was also thought by some workers to represent oxygen ion activity in crystals and glasses.

8.3.1   Molar refractivity of oxygen

Fajans *et al.* used the Lorentz–Lorenz molar refraction of ions, R as a measure of the 'looseness' of the electrons in a system.

$$R = \frac{n^2 - 1}{n^2 + 2} \cdot \frac{M}{d} = \frac{n^2 - 1}{n^2 + 2} \cdot v$$

where n = refractive index (n_D), M = Molecular weight, taken from the formula, d = density, and v = molar volume.

The refractive index of a glass, and its density, can be measured easily. The refractivity can then be calculated. A comparison of the ionic refractivities of the O^{2-} ions in different binary and ternary silicates (Table 4.3), reveals that the basicity, as measured by the refractivity of the O^{2-} ions, increases with decreasing field strength of the cation or when a mixture of basic oxides is replaced by a single basic oxide.

.f. measurement

ion activity of two melts can, in principle, be compared by measuring
ice of the electrochemical potential of oxygen in the melts. This
be the only direct method by which oxygen ion activities can be
independently from interference from other reactions. In order to
ieasurement, it is necessary to devise a reversible oxygen electrode, and
a bridge between the glasses being compared which has negligible
otential.

workers have attempted to carry out such measurements. The work of
<o and Rochow [10] would appear to be most useful in the present
though the general finding of all experimenters who have made this
:asurement is that the oxygen ion activity increases with the basicity of

:nko and Rochow used two identical porous porcelain cups which were
e sintering temperature (1200°C). The insides of the cups were lined
ium foil, and they were placed in a flat platinum dish. The dish, and one
s were then filled with finely powdered standard glass, while the other
the glass containing the dissolved oxide under investigation. This
ent permitted the diffusion of the melts through the walls of the
hus establishing electrolytic contact with the glass in the platinum dish
:d as an electrolytic bridge. These authors used two platinum electrodes:
:rsed in pure fused $PbO.SiO_2$, the other immersed in a fused $PbO.SiO_2$
different oxides had been added. Neglecting the junction potentials and
ng the oxygen ion activity of the standard glass $PbO.SiO_2$ as the
state', the oxygen ion activity of the test melt is given by the Nernst

$$\text{e.m.f.} = \frac{RT}{nF} \ln O^{2-} \text{ (melt).} \qquad (8.7)$$

nko and Rochow found that the e.m.f. of the cell varied with the nature
itity of the alkali oxide. A typical result at 900°C is shown in Figure 8.1
ows that e.m.f. or the oxygen ion activity is increased either by increasing
i content in the melt or by replacing lithia by soda or soda by potash, i.e.
reasing basicity of the melt.

Gaseous solubility

er vapour solubility

iction between water vapour and oxygen ions in a glass at high
.tures may be represented as:

$$H_2O \text{ (gas)} + O^{2-} \text{ (glass)} \rightleftharpoons 2OH^- \text{ (glass)}$$

sence of the OH^- group in glass from dissolved water, either, as $\equiv Si-OH$
$-OH----O-Si \equiv$ has been established by infrared spectroscopy, and in
n its presence has been deduced from the dependence of the solubility on
are root of the partial pressure of water vapour with which the glass had

251

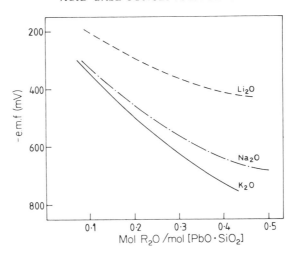

Fig. 8.1 Electromotive force as a function of alkali oxide concentration in PbO–SiO$_2$ melt at 900°C.

been brought to equilibrium. Franz and Scholze [8] wrote:

$$K_1 = \frac{[\text{OH}^-]^2}{[f_{\text{H}_2\text{O}}]\text{O}^{2-}} \qquad (8.8)$$

where $f\text{H}_2\text{O}$ is the fugacity of water vapour. From their experimental data it was known that:

$$[\text{OH}^-] = L \cdot \sqrt{p\text{H}_2\text{O}}$$

or

$$[\text{OH}^-]^2 = L^2 \cdot [p\text{H}_2\text{O}]$$

When $p\text{H}_2\text{O}$ is unity, L becomes $x\text{H}_2\text{O}$, the mole fraction of water dissolved. Again, $f\text{H}_2\text{O} = K_2 \cdot \text{H}_2\text{O}$.

Putting these values of $f\text{H}_2\text{O}$ and $[\text{OH}^-]^2$ in equation (8.8)

$$K_1 = \frac{L^2 \cdot p\text{H}_2\text{O}}{K_2 \cdot p\text{H}_2\text{O} \cdot [\text{O}^{2-}]}$$

Franz and Scholze assumed that K_1 was independent of composition and wrot

$$\text{O}^{2-} = K_3 \cdot x^2\text{H}_2\text{O},$$

where

$$K_3 = \frac{1}{K_1} \cdot \frac{1}{K_2} \qquad (8$$

K_3 in equation (8.9) cannot be calculated, but can be eliminated by comparing oxygen ion activity in two different melts. For convenience, the oxygen activity in pure silica was taken as the 'standard state', and the activities in bin

252

ate melts were calculated using the following equation:

$$\log [O^{2-}] \text{ (glass)} = 2 \log \frac{x H_2O \text{ (glass)}}{x H_2O \text{ (silica)}}$$

$_2O$ is the amount of water vapour dissolved per gram of glass. ygen ion activities so deduced in binary alkali silicate glasses at 1700° C n in Figure 8.2. These activities also increase with the basicity of the melt.

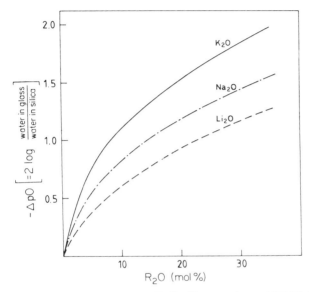

. 8.2 Oxygen ion activity in binary alkali–silicate melts at 1700°C from water solubility measurement.

Carbon dioxide solubility

solubility of CO_2 in a melt may be represented as:

$$CO_2 \text{ (gas)} + O^{2-} \text{ (glass)} \rightleftharpoons CO_3^{2-} \text{ (glass)}$$

$$K = \frac{[CO_3^{2-}]}{[CO_2][O^{2-}]} \tag{8.10}$$

rce [11] measured the CO_2 solubility in some binary alkali–borate and li–silicate melts at different temperatures. He assumed that for low con-trations of CO_3^{2-} ions the activity of CO_3^{2-} could be replaced by the centrations and that $f CO_2 = pCO_2$ which was equal to unity (i.e. one osphere in his experiments).

$$\text{Thus } K = \frac{\% CO_3^{2-} \text{ (glass)}}{[O^{2-}]}$$

$$O^{2-} = \bar{K} . \% CO_3^{2-}$$

253

$\bar{K}$ cannot be calculated, but Pearce assumed that it was independent of composition and thus could be eliminated by comparing the oxygen ion activity in two oxide melts. Pearce took the oxygen ion activity in pure Na_2O as the 'standard state'.

Thus

$$\log O^{2-} \text{ in glass} = \log \frac{\% CO_3^{2-} \text{ in glass}}{\% CO_3^{2-} \text{ in } Na_2O}$$

The oxygen ion activity was calculated in binary alkali–borate and alkali–silicate melts in the temperature range: $900°–1200°C$. Some of his results are shown in Figure 8.3. Again, these activities increase with the basicity of the melt.

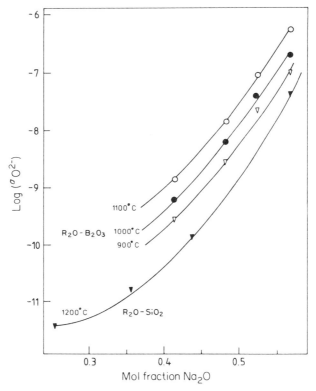

Fig. 8.3 Oxygen ion activity in sodium–silicate and sodium–borate melts from carbon dioxide solubility measurement.

It is difficult to compare Pearce's results with those of Franz and Scholze because the temperature and composition ranges do not overlap much; it does appear, however, that there would be nearly a factor of two difference in the relative values of the oxygen ion activities of a melt, when compared by the two methods.

lity of sulphur dioxide

ssolves in glass either as sulphide or as sulphate, depending upon the
ssure of oxygen in the furnace atmosphere. As shown by Fincham and
n [12], the reactions involved during sulphur dissolution may be
d as:

$$\tfrac{1}{2}S_2\,(gas) + O^{2-}\,(melt) \overset{K_{12}}{\rightleftharpoons} \tfrac{1}{2}O_2 + S^{2-}\,(melt) \tag{8.12}$$

$$[O^{2-}] = \frac{[O_2]^{1/2} \cdot [S^{2-}]}{K_{12} \cdot [S_2]^{1/2}}$$

$$\tfrac{1}{2}S_2\,(gas) + \tfrac{3}{2}O_2 + O^{2-}\,(melt)$$

$$\overset{K_{13}}{\rightleftharpoons} SO_4{}^{2-}\,(melt) \tag{8.13}$$

$$O^{2-} = \frac{SO_4{}^{2-}}{S_2{}^{1/2} \cdot O_2{}^{3/2} \cdot K_{13}}$$

be mentioned that Fincham and Richardson did not attempt to
the oxygen ion activity by assuming that K_{12} or K_{13} was constant in
hardson [13] clearly recognized that the reaction constant K varies with
ion, as did Toop and Samis [14].

he above equations, the solubility of sulphur in a glass will increase at
ant pressure of oxygen and sulphur as the oxygen ion activity in the melt
. Experimentally it has been found by Fincham and Richardson [12],
an and Pearce [15] and Zaman [16] that the equilibrium solubility of
ncreases with the basicity of the glass.

ll the experiments on the solubility of gases in melts could be interpreted
ing an increase of oxygen ion activity with increasing basicity of the melt.

RANSITION METAL IONS AS ACID–BASE INDICATORS IN GLASS

dance with the autoionization concept, it is evident that the basicity of
r silicate glass is dependent on the 'concentration', or more precisely on
vities', of $B(O_{1/2})_2\,O^-$ or $Si(O_{1/2})_3\,O^-$ groups formed in the melt as a
 acid–base reaction. The possibility of using transition metal ions as
rs to detect the acid–base relationship in borate and silicate glasses thus
apparent, owing to the fact that the coordination symmetry of these ions
 with the basicity of the melt and a sharp change at the neutralization
uld be expected. Moreover, these ions, owing to their incompletely filled
ell, give coloured paramagnetic ions; thus the type and nature of the
es formed can be studied satisfactorily by taking spectroscopic and
c measurements. The author used chromium (VI), vanadium (V),
n (VI), cobalt (II) and nickel (II) indicators to estimate the basicity of the
lkali–borate glasses, in which a change of cobalt or nickel coordination
ctahedral to tetrahedral symmetry, the conversion of chromium and

vanadium complexes from 'borochromate' or 'borovanadate' to chromate or vanadate respectively and the change of acidic uranyl group to some basic modification with increasing basicity were demonstrated by optical absorption measurements. The results were interpreted in terms of autoionization concept of acid–base and some equations to show the possible reactions taking place in the melt with the addition of alkali.

8.4.1 Chromium (VI) as an acid–base indicator in glass

In aqueous solution, several oxy-anions of chromium (VI) can exist, depending upon the pH. The equilibria amongst the most commonly encountered chromium (VI) complex ions may be summarized in order of decreasing pH as:

$$CrO_4^{2-} + H^+ \rightleftharpoons HCrO_4^- \tag{8.14}$$

$$Cr_2O_7^{2-} + H_2O \rightleftharpoons 2\,HCrO_4^- \tag{8.15}$$

$$HCrO_4^- + H^+ \rightleftharpoons H_2CrO_4 \tag{8.16}$$

Above a pH of 8, only CrO_4^{2-} is present, whilst in the pH range 2–6 equilibrium equation (8.15) predominates. In more acid solutions (below $pH = 1$) the presence of other species such as H_2CrO_4 and $HCr_2O_7^-$ has been suggested.

When an alkali oxide is added to a boric oxide melt, the following equilibria may exit amongst the various boron–oxygen groupings in order of increasing basicity:

$$2\,[B(O_{1/2})_3] + R_2O \rightleftharpoons 2\,[B(O_{1/2})_4]^- + 2R^+ \tag{8.17}$$

$$2\,[B(O_{1/2})_3] + X[B(O_{1/2})_4]^- + XR^+ + R_2O$$
$$\rightleftharpoons (2+X)\,[B(O_{1/2})_2O]^- + (2+X)R^+ \tag{8.18}$$

$$2\,[B(O_{1/2})_4]^- + R_2O \rightleftharpoons 2\,[B(O_{1/2})\,(O)_2]^{2-}$$
$$+\,2R^+ \qquad\qquad +\,4R^+ \tag{8.19}$$

$$[B(O_{1/2})_4]^- + R_2O \rightleftharpoons [B(O)_3]^{4-} + 3R^+$$
$$R^+ \tag{8.20}$$

Equations (8.17–20) are written in terms of one boron atom, and no attempt has been made to indicate mechanisms or to take account of the more complex groupings found, for example, by Krogh-Moe [17]. They are suggested by the results of nuclear magnetic resonance experiments; by increasing the alkali oxide concentration in binary borate melt the concentration of the $[B(O_{1/2})_4]^-$ group is found to decrease after reaching a maximum and different types of borate groups with non-bridging oxygens could be identified.

Figure 8.4 shows the absorption spectra of chromium (VI) in a series of binary sodium borate glasses. The spectra of glasses containing 33.3 and 27.6 mol % Na_2O (curves 2 and 3) resemble closely the absorption of chromate solution (curve 1), whereas for glass containing 20 mol % Na_2O (curve 4) the absorption resembles closely those of $[HCrO_4]^-$ (curve 5). The absorption curves for other glasses also follow the same trend. Parallel experiments made with alkaline-earth–borate glasses gave similar results.

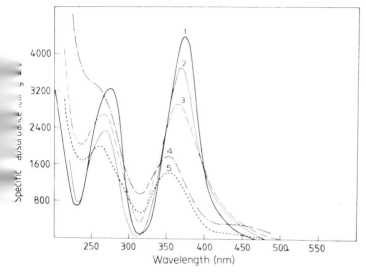

.4 Ultraviolet absorption of chromium (VI) in aqueous solution and in sodium–borate glasses.
(1) alkaline chromate solution (pH 10)
(2) Na_2O, $2 B_2O_3$ glass
(3) 27 6 Na_2O, 72 4 B_2O_3 glass
(4) Na_2O, $4 B_2O_3$ glass
(5) acidic chromate solution (pH 4)

ulating that only chromate and borochromate groups exist in these the following type of reaction for alkali chromate may be suggested:

$$CrO_4^{2-} + 2 B(O_{1/2})_3 \rightleftharpoons [CrO_3 - O - B(O_{1/2})_2]^- + B(O_{1/2})_2O^- \quad (8.21)$$
$$\text{Borochromate}$$

romate groups in high boric oxide glasses react with $B(O_{1/2})_3$ triangles g borochromate complex groups. The $B(O_{1/2})_2O^-$ groups with non-ng oxygen can only be maintained in high alkali glasses; they react with the ₂)₃ triangles, producing $B(O_{1/2})_4^-$ tetrahedra. Further addition of R_2O ter a critical concentration of alkali oxide has been reached, produce a non-ng oxygen ion, equation (8.18), on a $B(O_{1/2})_3$ group and this in turn will rt a borochromate group to a chromate group by causing the reaction of ion (8.21) to go to the left. The order in which reactions (8.17–21) will occur, he critical concentrations of R_2O at which (8.21) will begin to move to the epend upon the free energy change of the individual reactions, and the ty coefficients of the species involved in these reactions. ure 8.5 contains plots of extinction ratio, R against the concentration of oxide in different $R_2O–B_2O_3$ glasses. The absorbance ratio was calculated the ratio of the absorbance at the maximum around 360 nm to that around

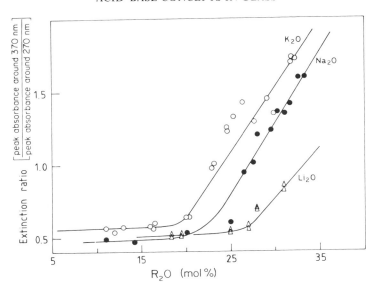

Fig. 8.5 Chromium (VI) absorbance ratio in different alkali–borate glasses.

260 nm for the same piece of glass. The critical concentration of the basic oxides was determined from the point where there is a sudden increase in the slope of the curve shown in Figure 8.5. With a constant molar concentration of basic oxides, the relative basicities of the melts increase in the order:

$$CaO < SrO < Li_2O < BaO < Na_2O < K_2O$$

The critical compositions so determined range from 35 per cent CaO to 19 per cent K_2O. A plot of critical composition against $Z/(r_0 + r_1)^2$, where Z is the valency, r_0 is the radius of the oxygen ion and r_1 is the radius of the alkali- or alkaline-earth ion, produces a straight line. As usual, the alkalis and alkaline-earths fall on two separate curves. There is no 'a priori' reason as to why these plots are straight lines, but the order in which the ions place themselves would be expected, from the fact that the basicity of these glasses would be determined by an ionization equilibrium equivalent to equation (8.6).

The critical compositions of 19, 22 and 26 mol % for K, Na and Li, at which the concentrations of $B(O_{1/2})_2O^-$ groups have become sufficient to cause the sudden increase in CrO_4^{2-} groups, apparently differ from the current view of the occurrence of such groups in borate glasses. However, as the specific absorption coefficients of chromate and borochromate groups are very high, and at the same time quite different, the chromium (VI) absorption spectrum is very sensitive to the changes in the concentration of non-bridging oxygens in the melt, and the concentrations which cause the change to CrO_4^{2-} may be much smaller than can be detected by nuclear magnetic resonance or infrared studies.

nadium (V) as an acid–base indicator in glass

(V) complexes have been extensively studied in basic and acidic olutions. In basic solutions mononuclear vanadate ions, VO_4^{3-}, are d these aggregate into binuclear and trinuclear species as the basicity of m is reduced:

$$[VO_4]^{3-} + H^+ \rightleftharpoons [VO_3(OH)]^{2-} \qquad\qquad K = 10^{12.6}$$
$$)_3(OH)]^{2-} + H^+ \rightleftharpoons [V_2O_6(OH)]^{3-} + H_2O \qquad\qquad K = 10^{10.6}$$
$$)_3(OH)]^{2-} + H^+ \rightleftharpoons [VO_2(OH)_2]^- \qquad\qquad K = 10^{7.7}$$
$$)_3(OH)]^{2-} + 3H^+ \rightleftharpoons [V_2O_9]^{3-} + 3H_2O \qquad\qquad K = 10^{30.7}$$

ical absorption curves of vanadium (V) in K_2O–B_2O_3 glasses are shown e 8.6. These curves resemble very closely the absorption curves of n (V) in aqueous buffer solutions also shown in Figure 8.6. In aqueous , the first step towards neutralization of the $[VO_4]^{3-}$ complex is:

$$[VO_4]^{3-} + H_2O \rightleftharpoons [VO_3(OH)]^{2-} + OH^- \qquad\qquad (8.22)$$

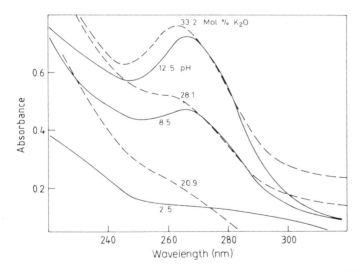

g. 8.6 Ultraviolet absorption of vanadium (V) in aqueous solution and in some potassium–borate glasses.

n the change in ultraviolet absorption spectra this change seems to take place n pH7 to 8. At pH < 7, complicated equilibria occur and no simple rpretation of the ultraviolet spectrum is possible. The trend of the change of adium (V) absorption in alkali–borate glasses resembles very closely the sformation represented in equation (8.22). Since the non-bridging oxygen ion lass has a similar function to that of the OH^- ion in aqueous solutions, the

259

equilibria involved in alkali–borate melts may be tentatively represented as follows:

$$[VO_4]^{3-} + 2B(O_{1/2})_3 = [VO_3 - O - B(O_{1/2})_2]^{2-} + B(O_{1/2})_2O^- \qquad (8.23)$$
$$\text{borovanadate}$$

The 'borovanadate' group is converted to the metavanadate group when the activity (basicity) of non-bridging oxygen is sufficient to shift the reaction (8.23) towards the left.

Since the concentration of vanadium (V) in these glasses was not estimated, the extinction coefficient of V^{5+} for the various glasses could not be compared. But the variation of log absorbance with frequency is independent of the concentration of vanadium (V) and thickness of the specimen (for these are constant for a single piece of glass), and thus a satisfactory comparison can be made by calculating the area under the 'hump' in such a plot. This area from 225 nm ($44\,440\ cm^{-1}$) to 295 nm ($33\,900\ cm^{-1}$) for different glasses is plotted against mol % R_2O in Figure 8.7, and that versus pH is also shown in this figure. The critical concentrations of the alkali oxides were determined from the point where there was a sudden increase in the slope of the curves in Figure 8.7, and were found to be 24, 26 and 30 mol % for K_2O, Na_2O and Li_2O respectively. For aqueous buffer solutions, the analogous change is observed at pH $\sim$ 7.

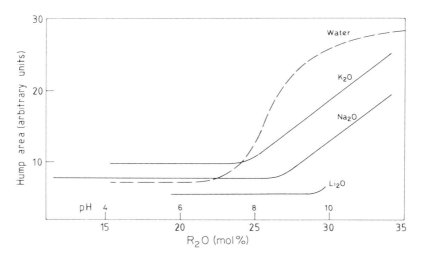

Fig. 8.7 Change of hump area with alkali oxide content of alkali–borate glasses and pH of aqueous solution.

8.4.3 Uranium (VI) as an acid–base indicator in glass

According to Weyl, uranium (VI) can be present in glass either as a uranyl group $[UO_2]^{2+}$ or as a uranate group $[UO_4]^{2-}$ depending upon the basicity of

260

Weyl suggests that basic glasses would favour the formation of a
:oup while a uranyl group will be favoured by an acidic glass. In
ate glasses containing uranium, the change of colour from pale
.o deeper yellow, and the decrease in fluorescence with increased
has been attributed to the shift of the uranyl–uranate equilibria
he formation of a uranate group.
►tical absorption spectra, with increasing pH, of uranium in uranyl
aqueous solution, are shown in Figure 8.8. Uranyl nitrate hexahydrate
was used and the pH was increased by gradual addition of Na_2CO_3.
itrate in aqueous solution gives an acidic reaction, and the solution
H more than 4.0 could not be prepared owing to precipitation. The
)n of uranyl ions to uranate could not, therefore, be demonstrated in
solution; neither was it possible to obtain the characteristic absorption
f the uranate ion in aqueous solution, owing to insolubility of uranate in

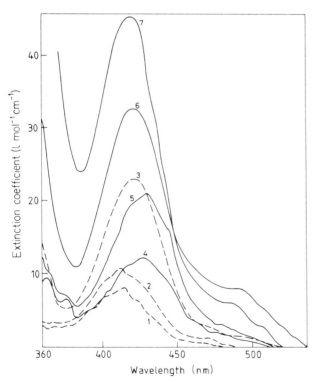

3.8 Optical absorption of uranium(VI) in aqueous solutions and in
sodium–borate glasses.

(1) aqueous solution pH 2.85   (5) 3 Na_2O, 17 B_2O_3 glass
(2) aqueous solution pH 3.32   (6) Na_2O, 3 B_2O_3 glass
(3) aqueous solution pH 3.92   (7) Na_2O, 2 B_2O_3 glass.
(4) Na_2O, 9 B_2O_3 glass

The characteristic feature of uranyl compounds is that uranium has two closely held oxygen neighbours; other ligands are situated at a greater distance, being held together by secondary uranium–ligand bonds. The values of U–O bond lengths in uranyl ion, $[UO_2]^{2+}$ are 1.67 Å in synthetic uranyl carbonate and 1.71 Å in sodium uranyl acetate. In a plane normal to the axis of sodium uranyl acetate, six oxygens of acetate groups are linked by secondary U–O bonds at greater distances.

The preference for two oxygen atoms by uranium in forming a colinear O–U–O is also exhibited by certain 'uranates'. These have crystal structures related to those of the uranyl compounds, and cannot really be regarded as true uranates, as uranates have discrete $[UO_4]^{2-}$ ions with uranium in tetrahedral configuration, analogous to the chromate ion $[CrO_4]^{2-}$.

The reactions of uranyl nitrate in aqueous solution may be represented as:

$$UO_2(NO_3)_2 + H_2O \rightleftharpoons UO_2(NO_3)OH + HNO_3.$$

$$UO_2(NO_3)OH + H_2O \rightleftharpoons UO_2(OH)_2 + HNO_3.$$

The absorption spectra of uranium (VI) in borate glasses are very similar to those of uranyl spectra in aqueous solution as shown in Figure 8.8. The absorption curves of low alkali containing glasses (10–15 mol $\%$ Na_2O), like aqueous uranyl spectra at lower pH (2.85–3.32), show indications of fine structure; the fine structures in both cases disappear with increase in alkali content. The peak height of the broad absorption band at 540 nm in glass increases with increased alkali content, as does the peak at 415 nm in aqueous uranyl solution with increased pH.

From an examination of the absorption spectra in solution it appears quite unlikely that uranium (VI) can occur as the uranate group $[UO_4]^{2-}$ in basic glasses. It appears that in glass, uranium (VI) is present essentially as a uranyl group for all alkali contents; for high alkali containing glasses the uranyl group in a uranyl complex is probably present as $[UO_2(O)_2]^{2-}$ with considerable increase in structural asymmetry which is responsible for the increase in the intensity of absorption.

The possible reactions of uranium in the borate melts may be represented as:

$$O-B(O_{1/2})_2$$
$$|$$
$$[UO_2(O)_2]^{2-} + 2B(O_{1/2})_3 \rightleftharpoons UO_2(O) + B(O_{1/2})_2O^-$$

basic form

$$\begin{bmatrix} O-B(O_{1/2})_2 \\ | \\ UO_2(O) \end{bmatrix}^- + 2B(O_{1/2})_3 \rightleftharpoons \begin{matrix} O-B(O_{1/2})_2 \\ | \\ UO_2 \\ | \\ O-B(O_{1/2})_2 \end{matrix} + [B(O_{1/2})_2O]^-$$

boro-uranyl complex

formation of non-bridging oxygen in the melt, the reaction would
כ the left. In low alkali borate glasses the $[B(O_{1/2})_2O]^-$ group would
with $B(O_{1/2})_3$ to produce tetrahedral boron. Thus in high alkali
g borate glasses the basic form of uranyl complex may exist, while in the
כ-borate glasses the existence of a boro–uranyl complex is a possibility.

obalt (II) as an acid–base indicator in glass

I) may be present in glass in octahedral as well as in tetrahedral
y, the ratio being dependent upon the basicity of the glass. The optical
כn characteristics of octahedral and tetrahedral cobalt (II) are distinctly
. As octahedral cobalt (II) reacts with basic glass to form tetrahedral
i), an acid–base titration similar to that of chromium (VI) may be studied
–borate glasses using cobalt (II) as the indicator.
typical absorption curves of cobalt (II) in sodium–borate glasses are
ר Figure 8.9. These curves resemble very closely the absorption curves of
I) in aqueous thiocyanate solutions also shown in Figure 8.9.
ıary alkali–borate glasses, cobalt (II) in octahedral symmetry absorbs
throughout the visible region, having absorption maxima at approxi-
500 and 550 nm. Cobalt (II) in tetrahedral symmetry has absorption
around 540, 600 and 635 nm. The overall absorption in both the cases is
ln aqueous solution the extinction coefficient of cobalt (II) in octahedral
ry at the most intense absorption maxima is about $4 \, mol^{-1} \, cm^{-1}$ (Figure
ve 1), while for cobalt (II) in tetrahedral symmetry it is about 500 and so the
intensity of absorption gives some guide to the coordination.
odium borate glasses when Na_2O is 13 mol %, the molar extinction
ent is only 12, but in the glass containing 33 mol % Na_2O it rises to 124. In
um–borate glasses, the maximum extinction becomes 140 when K_2O is
כl %, but for a glass containing 15 mol % K_2O it is only 15. If it is assumed
e same orders of magnitude of maximum extinction as those in aqueous
ın are retained in the glasses, then the greater conversion of octahedral
(II) to tetrahedral cobalt (II) in potassium glass is reflected in the greater
se in maximum extinction. In Figure 8.10 the specific absorbance area in the
400–700 nm, in arbitrary units, has been plotted against mol % R_2O.
ges in slope occur around 18, 19 and 22 mol % for K_2O, Na_2O and Li_2O
tively. However, the change is much less pronounced with lithium than
odium or potassium.
absorption spectra of cobalt (II) in all these alkali borate glasses and
us thiocyanate solutions were analysed into the minimum number of
onent gaussian bands. A typical analysis is shown in Figure 8.11. It is clear,
this analysis, that the absorption band centred at about $16\,000 \, cm^{-1}$ which
e to tetrahedral cobalt (II) $^4A_2 \rightarrow {}^4T_1$ (P) transition is absent in low alkali
es but appears as soon as the critical alkali oxide concentration is reached in
lass.
gure 8.12 show the absorption of cobalt (II) in a series of 20 mol % Na_2O,
ol % SiO_2 glasses where different amounts of SiO_2 have been replaced by
valent amounts of Al_2O_3. On resolution of the absorption spectra it became
that even in glasses containing 20 mol % Al_2O_3 a part of cobalt (II) is still in

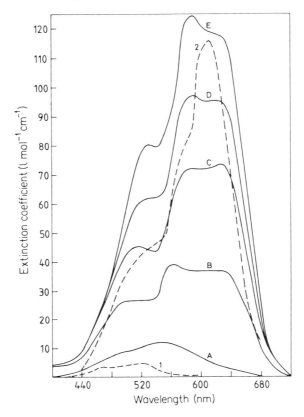

Fig. 8.9 Optical absorption of cobalt(II) in aqueous solutions and in some sodium–borate glasses.

(1) water
(2) 30 wt. % KCNS in water
(A) 13 Na_2O, 87 B_2O_3 glass
(B) 22.5 Na_2O, 77.5 B_2O_3 glass
(C) 26.5 Na_2O, 73.5 B_2O_3 glass
(D) 30.2 Na_2O, 69.8 B_2O_3 glass
(E) 33 Na_2O, 67 B_2O_3 glass.

tetrahedral configuration. This is due to the fact that in a glass melt both cobalt octahedra and aluminium octahedra compete for non-bridging oxygen; and although the decrease in free energy is more in the case of tetrahedral aluminium formation, in the presence of cobalt(II) this reaction is by no means complete. Thus, it is expected that with larger concentrations of Al_2O_3 in the melt, ultimately all the cobalt(II) will be converted to the octahedral symmetry. Unfortunately, due to very high melting temperature, glasses containing more than 20 mol% Al_2O_3 could not be prepared.

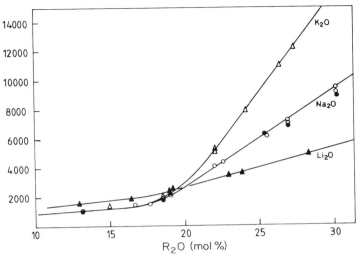

.10 Variation of the absorption intensity of cobalt(II) with alkali oxide content in binary alkali–borate glasses.

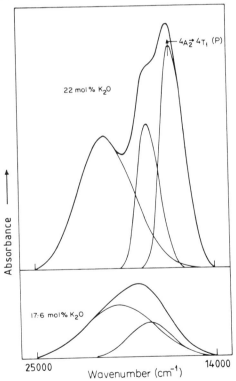

Fig. 8.11 Resolution of cobalt(II) absorption bands in two potassium–borate glasses.

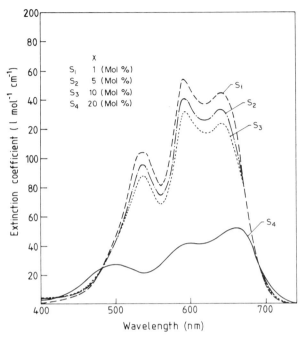

Fig. 8.12 Optical absorption of cobalt (II) in $20 \, Na_2O$, $x \, Al_2O_3$, $(80-x) \, SiO_2$ glasses.

8.4.5 Nickel (II) as an acid–base indicator in glass

Nickel (II) changes its coordination from six to four with increasing basicity of the glass. The absorption spectra of nickel (II) in different potassium–borate glasses are shown in Figure 8.13. The spectra of nickel (II) in low-alkali–borate glasses (curves 1 and 2 in Figure 8.13) are similar to that of $[Ni(H_2O)_6]^{2+}$ and may be interpreted in terms of ligand field theory in an analogous manner. When K_2O concentration is gradually increased beyond 19 mol%, newer bands characteristic of nickel (II) in square planar and tetrahedral symmetries appear, indicating a shift of the nickel (II) equilibria in the direction:

$$Octahedral \rightarrow Square \ planar \rightarrow Tetrahedral$$
$$\xrightarrow{\hspace{4cm}}$$
$$Increasing \ basicity$$

The composition of the 410 nm band is complicated. On close inspection even th spectra of low-alkali glasses, for example curves 2 and 3 in Figure 8.13, reveal shoulder on the long-wavelength side which may be represented as a separate s of bands buried in the more intense neighbouring band. It has been suggeste recently that the components of the 410 nm band arise from different transitio and cannot be assigned to a single energy level with spin-orbit splitting.

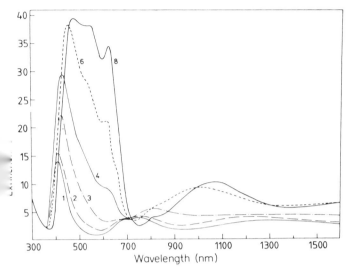

.13 Optical absorption of nickel (II) in binary potassium–borate glasses.

Curve	Mol % K_2O	Curve	Mol % K_2O
1	16.2	4	25.3
2	19.6	6	26.8
3	22.2	8	29.8

O is increased in boric oxide glasses, a shoulder develops on the long-
th side of the $^3A_2(F) \rightarrow {}^3T_1(P)$ transition (see Chapter 9) which is caused
velopment of the 515 nm peak. Square planar nickel (II) complexes give
similar peak and the extent of broadening of the 410 nm peak thus may be
ed as an indication of the degree to which tetragonal nickel (II) is present
glasses. The other remarkable difference caused by this tetragonal state is
nsification of the 410 nm band.

the R_2O concentration is further increased, an extra band at 615 nm is
veloped, and the appearance of the overall spectrum becomes quite
t from that of the low-alkali glasses. The newly developed 615 nm band in
ich glasses may be assigned to the $^3T_1(F) \rightarrow {}^3T_1(P)$ transition of nickel (II)
rahedral field.

e band positions of tetrahedral, square planar, and octahedral nickel (II)
considerably in solution, as well as in glass, the only spectroscopic way to
e them (at least semi-quantitatively) is to measure the total absorption
der all the important peaks between 350 and 1300 nm. It is clear from
8.14 that when the alkali contents reach about 19 mol % there is a change
rdination which, because of the increase in intensity, is from hexa-
nation to tetra-coordination. A further change occurs in glasses very rich in
which, from the development of the absorption at about 650 nm (shown in
8.13), can be identified as a change from square planar to tetrahedral. Thus,

267

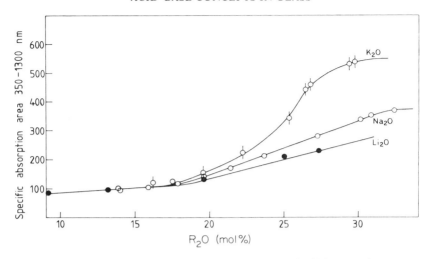

Fig. 8.14 Intensity of nickel(II) absorption in binary alkali–borate glasses as a function of the alkali oxide content.

two equilibria exist:

(less than 19 mol % R_2O)
Nickel (II) octahedral → Nickel (II) square planar

(Alkali-rich soda and potash glasses)
Nickel (II) square planar → Nickel (II) tetrahedral

The results of this investigation indicate that the formation of tetra-coordinated nickel (II) begins at about 19 ± 1 mol % R_2O in alkali–borate glasses. Beyond this alkali concentration, however, the concentration of tetra-coordinated nickel (II) increases very rapidly in potassium glasses, moderately in sodium glasses and only very slowly in lithium glasses. It appears that the tetra-coordination is initially square planar, but, certainly in potassium glasses, it appears to change to tetrahedral when the alkali content increases above 30 mol %. These changes are clearly associated with the basicity of the glasses, and may be discussed as in the case of cobalt (II).

8.5 OXIDATION–REDUCTION EQUILIBRIUM IN GLASS

The relation between basicity and oxidation–reduction equilibrium in glass will now be discussed. When a redox oxide is introduced into a glass melt, it distributes into different states of oxidation depending upon the time and temperature of melting, the glass composition, the furnace atmosphere, and the batch composition. In a given set of conditions, after sufficient time of melting, a melt comes to equilibrium with the partial pressure of oxygen in the ambient atmosphere and the relative concentrations of the different oxidation states reach equilibrium values (Chapter 7).

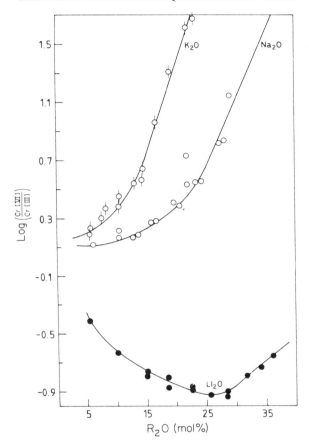

g. 8.15 Chromium (III)–chromium (VI) equilibrium in alkali–borate melts at 1000°C.

he redox reaction for chromium (III)–chromium (VI) in a glass melt may be ten as:

$$Cr^{3+} + \tfrac{3}{4}O_2 \rightleftharpoons Cr^{6+} + \tfrac{3}{2}O^{2-} \tag{8.26}$$

ne experimental results in silicate and borate glasses are shown in Figures 7.5 8.15. Equation (8.26) appears to indicate the wrong direction for the omium (III)–chromium (VI) equilibrium found experimentally. From the vious acid–base discussion, it can be recalled that chromium (VI) is present in ss as chromate and acid–chromate groups and not as free Cr^{6+} ion ionically nded with oxide ligands of glass, Thus the equations relating reactions between components of the system and chromium oxide may be written as:

$$\tfrac{1}{2}Cr_2O_3 + \tfrac{3}{4}O_2 \rightleftharpoons CrO_3 \tag{8.27}$$

$$2\,Si(O_{1/2})_4 + R_2O \rightleftharpoons 2[Si(O_{12})_3\,O]^- + 2R^+ \tag{8.28}$$

269

$$2[Si(O_{1/2})_3O]^- + CrO_3 \rightleftharpoons [CrO_4]^{2-} + 2Si(O_{1/2})_4$$

$$\tfrac{1}{2}Cr_2O_3 + \tfrac{3}{4}O_2 + R_2O \rightleftharpoons [CrO_4]^{2-} + 2R^+ \tag{8.29}$$

The more basic the alkali oxide and the greater its concentration, the more reaction (8.28) will move to the right and more $[CrO_4]^{2-}$ will be formed. Thus the tendency of the ratio [chromium (VI)/chromium (III)] to increase with alkali content and in the order: Li < Na < K, is predicted assuming that the basicity of the alkali oxide increases with the atomic weight.

In alkali–borate glasses, experimental results are different from those of alkali–silicate glasses. By plotting log [chromium (VI)/chromium (III)] against mol % R_2O, a distinct curvature is obtained; the point of inflection around 17, 22 and 26 mol% for K_2O, Na_2O and Li_2O respectively corresponds approximately to the chromate-borochromate transitions in these glasses.

8.6 FILLED SHELL IONS WITH ns^2 CONFIGURATION AS ACID–BASE INDICATOR IN GLASS

Thallium (I), lead (II), and bismuth (III) ions all have $(Xe)6s^2$ electronic configuration. In an alkali–halide matrix these ions produce three well known absorption bands A, B and C (after Seitz [20]) which involve $(s)^2 \rightarrow (s)(p)$ electronic transitions. The highest-energy C band is attributed to the allowed transition: $^1S_0 \rightarrow {}^1P_1$, the B band to the $^1S_0 \rightarrow {}^3P_2$ transition, and the A band which occurs at the lowest energy to the $^1S_0 \rightarrow {}^3P_1$ transition. The position of the A band in any medium depends on the donor capacity or Lewis basicity of the solvent. This can be explained as follows: in simple terms, the decrease in energy of the s → p transition can be thought of as a consequence of the covalency effect in which the electrons received from the ligands are accommodated in σ and π bonding molecular orbitals of the system. Some of this electron density is located between the inner electron core of the metal ion and its 6s orbital. This is shown schematically for lead (II) ion in Figure 8.16. One of the factors affecting the energy of the s → p transition is the force of attraction that the 6s electron experiences from the nucleus. Much of the positive pull of the nucleus is screened

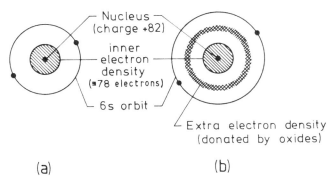

(a)   (b)

Fig. 8.16 Schematic diagram of (a) free Pb^{2+} ion, and (b) Pb^{2+} ion after receiving negative charge from neighbouring oxide ions (after Duffy and Ingram).

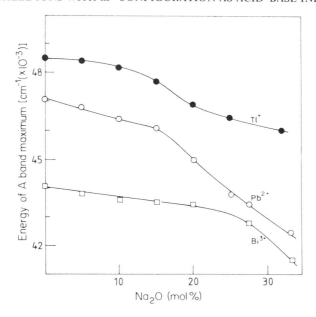

8.17 Energy of A band maximum for different ns^2 ions in binary sodium–borate glasses.

inner electron core, but the electron density donated by the ligands serves ease this screening further and this in turn allows the 6s electron to escape easily to the 6p level (even though the 6p level is also screened). Electron ion by the ligands to the probe ion therefore brings about a reduction in the energy difference, and hence a reduction in energy in the ultraviolet ption band, compared with the free Pb^{2+} ion. Figure 8.17 shows the on of A band due to thallium (I), lead (II), and bismuth (III) in different)–B_2O_3 glasses. As usual, the basicity seems to increase with soda content of lass, and inflections are obtained around 15–25 mol% Na_2O. The energy e A band maximum has been used to calculate 'optical basicity' values of rent glasses by some workers [21].

owever, the positions of the A band of these ions are known to change even e they are in the same halide system, but have different structures. For nple, in the chloride system the A band due to lead (II) occurs around 00 cm^{-1} in NaCl, whereas the transition energy reduces to about 35 000 cm^{-1} n the coordination number increases from 6 to 8 in CsCl. When the rdination number does not change, the nature of the cation seems to have no ificant effect on the transition energy (Table 8.1). The site symmetry and rdination number of these probe ions in glass are not clearly known. From alt (II) and nickel (II) indicators we have already seen that the coordination nber changes with glass composition. Thus it is not obvious from the results of ure 8.17 whether the changes in $^1s_0 \rightarrow {}^3p_1$ transition energy are entirely due to

271

Table 8.1

The maxima of the $^1s_0 \rightarrow {}^3p_1$ and $^1s_0 \rightarrow {}^1p_1$ absorption bands in some lead-doped halides

	Coordination number	Transition energy (cm^{-1})	
		$^1s_0 \rightarrow {}^3p_1$	$^1s_0 \rightarrow {}^1p_1$
PbCl$_2$	9	37 080	52 160
NaCl:Pb	6	37 080	52 160
KCl:Pb	6	37 000	51 340
RbCl:Pb	6	37 330	51 350
CsCl:Pb	8	35 370	47 270
CdCl$_2$:Pb	6	37 490	46 460
PbBr$_2$	9	30 160	44 010
NaBr:Pb	6	33 170	
KBr:Pb	6	33 580	45 640
RbBr:Pb	6	33 740	44 830
CsBr:Pb	8	32 190	44 010
CdBr$_2$	6	31 460	38 310

From: De Gruijtez, (1973), *J. Solid State Chem.* **6**, 151.

changes of covalency between the indicator ion and oxygen ligands of glass or due to changes of coordination number of the indicator ion with respect to oxygen.

8.7 VANADYL ION AS AN ACID–BASE INDICATOR IN GLASS

Vanadium (IV) usually occurs as VO^{2+} group in common oxide glasses. The bonding scheme of the VO^{2+} group according to Ballhausen and Grey [22] is as follows: A very strong σ-bond is formed between the $(2p_z + 2s)$ hybrid of the oxygen and the $(3d_z^2 + 4s)$ hybrid of the vanadium ion. Furthermore, since the $2p_x$ and $2p_y$ orbitals of the oxygen make a strong π-bond with the $3d_{xz}, 3d_{yz}$ orbitals on the metal ion, vanadium (IV) form the very stable VO^{2+} complex. The $(3d_z^2 - 4s)$ hybrid, together with the orbitals $3d_{x^2-y^2}$ and $4p_x$, $4p_y$ and $4p_z$ are then just capable of five σ-bonds directed in a tetragonal pyramid with the vanadium ion located at its base. Thus vanadium (IV) usually occurs in a tetragonally distorted octahedral unit.

In non-donor solvents like acidic low-alkali borate glasses, vanadium (IV) occurs predominantly as VO^{2+} group. As the electron donor capacity of the solvent increases, or in other words as the glass become more basic, the covalency of the in-plane σ-bonds increases and simultaneously the degree of π-bonding with the vanadyl oxygen decreases. Thus vanadium (IV) assumes more regular octahedral symmetry with increasing basicity of the glass.

Vanadium (IV) at low concentrations in glass produces strong well resolved electron spin resonance (e.s.r.) spectra, a typical example being shown in Figure 8.18. This spectrum can be satisfactorily described by means of an axial spin Hamiltonian of the form:

$$\mathcal{H} = g_{\parallel}\beta S_z H_z + g_{\perp}\beta (S_x H_x + S_y H_y) + A_{\parallel}S_z I_z + A_{\perp}(S_x I_x + S_y I_y)$$

272

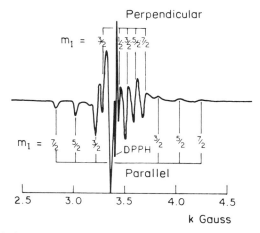

Figure 8.18 Typical e.s.r. spectrum of vanadium (IV) in a sodium–germanate glass.

β is the Bohr magneton, $g_\parallel$ and $g_\perp$ are the parallel and perpendicular principal components of the g-tensor; $A_\parallel$ and $A_\perp$ are the parallel and perpendicular components of the hyperfine-coupling tensor; H_x, H_y and H_z are the components of the magnetic field, and S_x, S_y, S_z and I_x, I_y, I_z are the components of the spin operators of the electron and nucleus respectively.

In a glass the resonance condition arising from the above Hamiltonian is to be summed over all values of the angle, θ, which defines the orientation of the magnetic field with respect to the symmetry axis of the axial ion site. The absorption curve in this case is composed of a set of envelopes, one for each $2I + 1$ value of the nuclear magnetic quantum number, m, whose extreme points occur at field strengths given by:

$$H_\parallel = \frac{2H_0}{g_\parallel} - \frac{A_\parallel}{g_\parallel \beta} \cdot m \qquad \text{for} \quad \theta = 0$$

$$H_\perp = \frac{2H_0}{g_\perp} - \frac{A_\perp}{g_\perp \beta} \cdot m \qquad \text{for} \quad \theta = \pi/2$$

The extreme points which are relatively pure, can readily be identified on the experimental curve as shown in Figure 8.18. The linear plots of $H_\parallel$ and $H_\perp$ versus the nuclear quantum number, m can be used to estimate the values of $g_\parallel$, $g_\perp$, $A_\parallel$ and $A_\perp$ a glass.

The optical transition energies ($b_2 \to e_\pi^*$ and $b_2 \to b_1^*$) of vanadium (IV) are related with the g factors by the following relations:

$$g_\perp = g_e \left[1 - \frac{\lambda \gamma^2}{E(b_2 \to e_\pi)} \right]$$

$$g_\parallel = g_e \left[1 - \frac{4\lambda \alpha^2}{E(b_2 \to b_1^*)} \right]$$

273

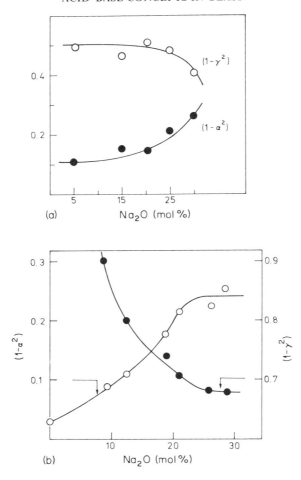

Fig. 8.19 (a) Covalency change of vanadium (IV) in sodium–borate glasses.
(b) Covalency change of vanadium (IV) in sodium–germanate glasses.

where λ is the spin-orbit coupling constant ($165\,\mathrm{cm}^{-1}$) and the expressions $(1 - \alpha^2)$ and $(1 - \gamma^2)$ are degrees of covalency; the covalency increases as $(1 - \alpha^2)$ and $(1 - \gamma^2)$ increase, the former gives an indication of the influence of the σ-bonding between the vanadium ion and the equilateral ligands, while the latter indicates the influence of the π-bondings with the vanadyl oxygen.

The parameters $(1 - \alpha^2)$ and $(1 - \gamma^2)$ for a series of Na_2O–B_2O_3 and Na_2O–GeO_2 glasses are shown in Figure 8.19. As expected, $(1 - \alpha^2)$ increases and $(1 - \gamma^2)$ decreases with increasing soda content (basicity), and points of inflection are obtained in both plots around 15 to 20 mol% Na_2O, which is consistent with other measured properties of these glass systems.

REFERENCES

REFERENCES

nius, S. (1887), *Z. Physik. Chem.*, **1**, 631.

ted, J. (1923), *Rec. Trav. Chim.*, **42**, 718.

y, T. (1923), *Chem. Ind.*, (London), **42**, 43.

, G. N. (1923), *Valence and Structure of atoms and molecules*, ACS monograph, ical Catalog Co., New York.

r, G. (1938), *Naturwissen Schaften*, **26**, 779.

H. and Rogler, E. (1942), *Z. Anorgan. Allegem. Chem.*, **250**, 159.

A. and Douglas, R. W. (1967), *Phys. Chem. Glasses*, **8**, 151.

P. J. and O'Keefe, J. G. (1963), *Phys. Chem. Glasses*, **4**, 37.

, V. H. and Scholze, H. (1963), *Glastech. Ber.*, **36**, 347.

las, R. W., Nath, P. and Paul, A. (1965), *Phys. Chem. Glasses*, **6**, 216.

chenko, R. and Rochow, E. G. (1954), *J. Amer. Chem. Soc.*, **76**, 3291.

e, M. L. (1964; 1965), *Jr. Amer. Ceram. Soc.* **47**, 342; **48**, 175.

am, C. J. B. and Richardson, F. D. (1954), *Proc. Roy. Soc.*, **A223**, 40.

rdson, F. D. (1955), *The Vitreous State*, Glass Delegacy of the University of ield, **22**, 82.

, G. E. and Samis, C. S. (1962), *Trans. AIME.*, **224**, 878.

dogan, E. J. and Pearce, M. L. (1963), *Trans. AIME.*, **227**, 940.

an, M. S. (1968), Ph.D. Thesis, University of Sheffield.

gh-Moe, J. (1960), *Phys. Chem. Glasses*, **1**, 26.

l, W. A. (1951), *Coloured Glasses*, Society of Glass Technology.

, A., Unpublished work.

, F. (1938), *J. Chem. Phys.*, **6**, 150.

ensen, C. K. (1962), *Absorption Spectra and Chemical Bonding in Complexes*, amon Press, New York, N. Y., p. 138.

fy, J. A. and Tngram, M. D. (1971), *J. Amer. Chem. Soc.*, **93**, 6448.

hausen, C. J. and Gray, H. B. (1962), *Inorg. Chem.*, **1**, 111.

uki, H. and Akagi, S. (1972), *Phys. Chem. Glasses*, **13**, 15.

l, A. and Assabghy, F. (1975), *J. Mater. Sci.*, **10**, 613.

275

CHAPTER 9

Coloured Glasses

Glass is usually coloured by dissolving transition-metal ions in it. In this chapter we shall discuss the elementary principles of *ligand field* and *molecular orbital* theory with which the absorption bands produced by different transition-metal ions can be interpreted.

9.1 ATOMIC STRUCTURE AND THE PERIODIC CLASSIFICATION OF TRANSITION METALS

In a modified Bohr atom, the electrons are arranged in shells ($n = 1, 2$, etc.) within which there are a number of sub-shells (s, p, d, f) comprising respectively 1, 3, 5 and 7 orbitals, characterized by the same energy. Each orbital can accommodate two electrons which according to *Pauli's exclusion principle* should be of opposite spin.

The first eighteen electrons fill successively the 1s, 2s, 2p, 3s and 3p levels. In potassium the 19th electron goes not to the 3d, but into the 4s sub-shell and the 20th electron in Ca also goes to the 4s sub-shell. After this the electrons continue to fill up the incomplete 3d sub-shell. In the elements from Sc to Ni the 3d sub-shell is only partly filled and associated with it are some characteristic properties of these metals such as variable valency and the formation of coloured paramagnetic ions. Copper may be included in the first series of transition elements because in the divalent state it has an incomplete 3d shell and the characteristic properties of a transition metal. In the next set of transition elements, Y to Pd, the filling of the 4d sub-shell takes place. The filling of the 4f level does not begin, however, until Ce is reached, where the 4th shell expands from 18 to 32 electrons in the elements from Ce to Lu inclusive. Since the two outer sub-shells are similarly constituted in all of these fourteen elements (the rare-earths) their chemical properties are very similar. It is likely that the filling of the 5f sub-shell in the elements from Th or Pa onwards leads to a second series of chemically similar elements analogous to the rare-earths. Thus there are four series of transition elements of general electronic structure as follows:

Sc to Ni	1st series	$1s^2\ 2s^2\ 2p^6\ 3s^2\ 3p^6\ 3d^n\ 4s^2$
Y to Pd	2nd series	$1s^2\ 2s^2\ 2p^6\ 3s^2\ 3p^6\ 3d^{10}\ 4s^2\ 4p^6\ 4d^n\ 5s^{1-2}$
Ce to Lu	3rd series	(1s to 4d – complete) $4f^n\ 5s^2\ 5p^6$
Pa to——	4th series	(1s to 4f – complete) $5s^2\ 5p^6\ 5d^{10}\ 5f^n\ 6s^2$

THEORIES OF CHEMICAL BONDING IN TRANSITION METAL COMPLEXES

...ing of coordination chemistry is usually dated from the discovery of ...mines by Tassaert (see Graddon [1]) in 1798. Tassaert observed that ...al solutions of cobalt chloride deposited the orange compound ...H_3 on standing overnight, and recognized in this a new type of ...ubstance, formed by the combination of two already fully-saturated ...s, but possessing properties quite different from either. In 1890 Werner ...lished the first of his papers on complex compounds in which he ...the concept of *Primary* and *Secondary* valencies. According to Werner ...ry valencies are those involved in satisfying the chemical equivalence of ...and the secondary valencies are those by which the coordinated ...are attached. From experimental evidence with different luteo-cobaltic ...precipitated with $AgNO_3$ solutions it is apparent that one of the ...s between primary and secondary valencies is that the primary valency ...nization of the bound atoms whereas the secondary valency does not. ...sence of six groups attached by secondary valencies to the metal atom in ...x compound raised a similar stereochemical problem to that posed to ...hemists by the benzene molecule. However, it has been confirmed by ...s X-ray analysis results, and from the number of possible isomers of ...compounds, that these six groups are arranged at the apices of an ...on.

...are three major approaches to the study of bonding in transitional metal ...es:

...igand field theory, which is a modified form of crystal field theory in ...hich allowance is made for orbital overlap.
...alency bond theory.
...Iolecular orbital theory.

...lency bond theory has not yet been widely used in interpreting absorption ...of coloured glasses, only ligand field and molecular orbital theories will be ...ed here.

Crystal field and ligand field theory

...p and d orbitals of a free ion are shown in Figure 9.1. In the crystal field ...the salt of a transition-metal ion is considered as an aggregate of ions ...r dipolar molecules which interact with each other electrostatically but do ...change electrons.
...us consider a metal ion M^{m+} having one 3d electron, lying at the centre of ...tahedral set of anions x^-. In the free ion, this d-electron would have had ...probability of being in any one of the five degenerate 'd' orbitals, but in the ...ic field created by the ligands the five d orbitals split into two groups, as ...n in Figure 9.2(a).
...the ligands present a negative charge (ligands are always negative ions or ...ted dipoles) towards the central ion, it follows that an electron will tend to ...I the $d_{x^2-y^2}$ and d_{z^2} orbitals the lobes of which point towards the ligand, and

277

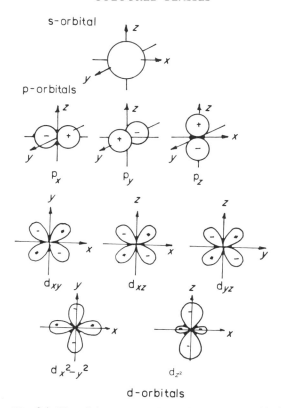

Fig. 9.1 Pictorial representation of s, p and d orbitals.

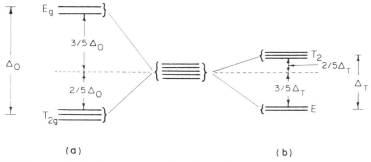

Fig. 9.2 Energy level diagrams showing the splitting of a set of d orbitals by (a) octahedral field, and (b) tetrahedral field.

occupy preferentially the d_{xy}, d_{xz} and d_{yz} orbitals (see Figure 9.3a). The effect of an octahedral field, then, is to remove the degeneracy of the five 'd' orbitals, so that these split into lower groups of three, known as T_{2g} or γ_5 orbitals, and an upper group of two known as E_g or γ_3 orbitals, (Figure 9.2a).

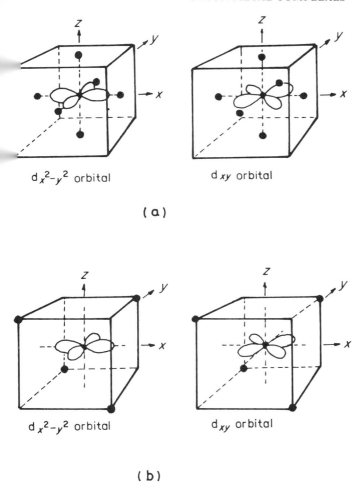

Fig. 9.3 (a) The $d_{x^2-y^2}$ and d_{xy} orbitals in an octahedral field.
(b) The $d_{x^2-y^2}$ and d_{xy} orbitals in a tetrahedral field.

n the transition-metal ion is in tetrahedral symmetry as shown in Figure
e situation is reversed. The lobes of $d_{x^2-y^2}$ or d_{z^2} orbital now lie in the
on between the ligands, whilst the lobes of the d_{zy} orbital, though not
g directly towards the ligands, lie closer to them. Thus the T_{2g} orbitals are
ilized with respect to E_g orbitals (Figure 9.2b).

y often the octahedral arrangements of the ligands around the transition-
ion are distorted – rather than perfect. The pattern of splitting of the d
ls in tetragonally distorted octahedral complexes and in planar complexes
e described as follows. If, from an octahedral complex Mx_6, the two *trans*
ls along the z-axis are slowly withdrawn, new energy differences among the

d orbitals arise. First of all, the degeneracy of the E_g orbital is lifted, the d_{z^2} orbital becoming more stable than the $d_{x^2-y^2}$ orbital. At the same time, the three-fold degeneracy of the T_{2g} orbitals is also lifted, the d_{yz} and d_{xz} orbitals remain equivalent to one another, but they become more stable than the d_{xy} orbital. For the opposite type of tetragonal distortion, that is when the two *trans* ligands lie closer to the metal ion than do the other four, the relative energies of the split components will be inverted: In the limiting case, when the *trans* ligands are completely removed, a square planar, four coordinated complex is obtained.

(a)   High spin and low spin complexes of the first transition series
Let us consider a hypothetical molecule having two orbitals separated by an energy difference ΔE and that two electrons are to occupy these orbitals. If one electron is placed in each orbital, their spins will be parallel, and the energy is given by $2E^\circ + \Delta E$, where E° is the inherent energy of an electron. If both the electrons are placed in the lower orbital, their spins will have to be coupled to satisfy the exclusion principle, and the total energy will be $2E^\circ + P$, where P stands for the energy required to cause pairing of two electrons in the same orbital. Thus, whether this system will have high spin distribution depends on whether ΔE is greater or less than P. Applying this sort of analysis to complexes of octahedral and tetrahedral symmetry, results as listed in Table 9.1 can be obtained.

Table 9.1
Ligand field stabilizations of high-spin and low-spin transition metal ions in octahedral and tetrahedral symmetries

	Octahedral		Tetrahedral	
Ions	*High-spin*	*Low-spin*	*High-spin*	*Low-spin*
d^3 Cr^{3+}, V^{2+}			$T_{2g}^1 E_g^2$ $-4/5\Delta_t$	E_g^3 $-9/5\Delta_t + P$
d^4 Mn^{3+}, Cr^{2+}	$T_{2g}^3 E_g^1$ $-3/5\Delta_o$	$T_{2g}^4 E_g^0$ $-8/5\Delta_o + P$	$T_{2g}^2 E_g^2$ $-2/5\Delta_t$	$T_{2g}^0 E_g^4$ $-12/5\Delta_t + 2P$
d^5 Mn^{2+}, Fe^{3+}	$T_{2g}^3 E_g^2$ 0	$T_{2g}^5 E_g^0$ $-2\Delta_o + 2P$	$T_{2g}^3 E_g^2$ 0	$T_{2g}^1 E_g^4$ $-2\Delta_t + 2P$
d^6 Fe^{2+}, Co^{3+}	$T_{2g}^4 E_g^2$ $-2/5\Delta_o + P$	$T_{2g}^6 E_g^0$ $-12/5\Delta_o + 3P$	$T_{2g}^3 E_g^3$ $-3/5\Delta_t + P$	$T_{2g}^2 E_g^4$ $-8/5\Delta_t + 2P$
d^7 Co^{2+}	$T_{2g}^5 E_g^2$ $-4/5\Delta_o + 2P$	$T_{2g}^6 E_g^1$ $-9/5\Delta_o + 3P$		

Ions having d^1, d^2, d^8, d^9 and d^{10} have only one possibility for each regarding the occupancy of the orbitals by the electrons.
$-\Delta_t = 4/9\Delta_o$ where Δ_o is the energy of separation in octahedral field, and Δ_t is the separation in tetrahedral field.

ng energy, P, is composed of two terms. One is the inherent repulsion
ch must be overcome when forcing two electrons to occupy the same
e might expect this to be fairly constant from one element to another
ably independent of other factors. The larger, more diffuse d orbitals
vier transition metals (5d) might more readily accommodate two
larges than the smaller 3d orbitals, but otherwise little variation is
he second factor of importance is the loss of exchange energy, E_{ex} (the
und's rule) which occurs as electrons with parallel spins are forced to
arallel spins. The exchange energy is proportional to the number of
ctrons of the same spin that can be arranged from n parallel electrons:

$$E_{ex} = \frac{n(n-1)}{2} K$$

st loss of exchange energy occurs when the d^5 configuration is forced to
e the apparent stability of the half-filled d sub-shell.
ctron–electron interactions are commonly expressed in terms of the
ameters and may be obtained from the spectra of the free, gaseous ions.
ical values of P are listed in Table 9.2.

Table 9.2
Pairing energies for some 3d metal ions (cm^{-1})

Ion	P_{coul}	P_{ex}	P
Cr^{2+}	5 950	14 475	20 425
Mn^{3+}	7 350	17 865	25 215
Cr^+	5 625	12 062	17 687
Mn^{2+}	7 610	16 215	23 825
Fe^{3+}	10 050	19 825	29 875
Mn^+	6 145	8 418	14 563
Fe^{2+}	7 460	11 690	19 150
Co^{3+}	9 450	14 175	23 625
Fe^+	7 350	10 330	17 680
Co^{2+}	8 400	12 400	20 800

ear from Table 9.1 that, depending upon the relative values of ΔE and P,
edral symmetry d^4, d^5, d^6, d^7 may change from high spin to low spin
es and in the case of tetrahedral symmetry this inter-conversion is
in d^3, d^4, d^5 and d^6 ions.
when the strictly octahedral environment does not permit the existence of
in state, as in the d^8 case of Ni^{2+}, distortion of the octahedron will cause
splitting of the degenerate orbitals, which may become great enough to
ne pairing energies and cause electron pairing.

states derived from electronic configurations
ctron in an atom has a set of quantum numbers, n, l, m_l, m_s; it is the last
hich are of concern here. Just as l is used, as $\sqrt{[l(l+1)]}$, to indicate the

orbital angular momentum of a single electron, there is a quantum number, L, such that $\sqrt{[L(L+1)]}$ gives the total orbital angular momentum of the atom. The symbol M_L is used to represnt a component of L in a reference direction and is analogous to m_l for a single electron. Similarly, we use a quantum number, S, to represent the total electron spin angular momentum, given by $\sqrt{[S(S+1)]}$, in analogy to the quantum number, s, for a single electron. There is the difference here that s is limited to the value $1/2$, whereas S may take any integral or half-integral value beginning with 0. Components of S in a reference direction are designated by M_s, analogous to m_s.

Symbols for the states of atoms are analogous to the symbols for the orbitals of single electrons. Thus the capital letters S, P, D, F, G, H, . . . are used to designate states with $L = 0, 1, 2, 3, 4, 5, ...$ The complete symbol for a state also indicates the total spin, but not directly in terms of the value of S. Rather, the number of different M_S values, which is called the spin multiplicity, is used. Thus, for a state with $S = 1$, the spin multiplicity is 3 because there are three M_s values, $1, 0, -1$. In general the multiplicity is equal to $2S + 1$, and is indicated as a left superscript to the symbol for L. The following examples should make the usage clear.

For $M_L = 4$, $S = \frac{1}{2}$, the symbol is 2_G

For $M_L = 2$, $S = \frac{3}{2}$, the symbol is 4_D

For $M_L = O$, $S = 1$, the symbol is 3_S

In speaking or writing of states with spin multiplicities of $1, 2, 3, 4, 5, 6 . . .$ we call them respectively, singlets, doublets, triplets, quartets, quintets, sextets, . . . Thus, the three states shown above would be called doublet G, quartet D and triplet S, respectively.

As in the case of a single electron, we may sometimes be interested in the total angular momentum, that is the vector sum of L and S. For the entire atom this is designated J. When required, J values are appended to the symbol as right subscripts. For example, a 4_D state may have any of the following J values, the appropriate symbols being indicated:

L	M_S	J	Symbol
2	$\frac{3}{2}$	$\frac{7}{2}$	$4_{D_{7/2}}$
2	$\frac{1}{2}$	$\frac{5}{2}$	$2_{D_{5/2}}$
2	$-\frac{1}{2}$	$\frac{3}{2}$	$2_{D_{3/2}}$
2	$-\frac{3}{2}$	$\frac{1}{2}$	$4_{D_{1/2}}$

In order to determine what states may actually occur for a given atom or ion, we begin with the following definitions:

$$M_L = m_l^{(1)} + m_l^{(2)} + m_l^{(3)} + \ldots + m_l^{(n)}$$

$$M_S = m_s^{(1)} + m_s^{(2)} + m_s^{(3)} + \ldots + m_s^{(n)}$$

in which $m_l^{(i)}$ and $m_s^{(i)}$ stand for the m_l and m_s values of the ith electron in an at having a total of n electrons.

In general it is not necessary to pay specific attention to all the electrons in atom when calculating M_L and M_S since those groups of electrons wh fill any one set of orbitals (s, p, d, etc.) collectively contribute zero to M_S. For instance, a complete set of p electrons includes two with with $m_l = 1$ and two with $m_1 = -1$, the sum, $0 + 0 + 1 + 1 - 1 - 1$, At the same time, half of the electrons have $m_s = +1/2$ and the other $-1/2$, making M_S equal to zero. This is obviously a generalization for any therefore we need only concern ourselves with partly filled

rtly filled shell, there is always more than one way of assigning m_l and to the various electrons. All ways must be considered except those which prohibited by the exclusion principle or are physically redundant, as ained presently. For convenience we shall use symbols in which + and ripts represent $m_s = +1/2$ and $m_s = -1/2$ respectively. Thus, when ectron has $m_l = 1$, $m_s = +1/2$, the second electron has $m_l = 2$, $m_s = $ e third electron has $m_l = 0$, $m_s = +1/2$, etc., we shall write $^+$, ...). Such a specification of m_l and m_s values of all electrons will be icrostate.

ow consider the two configurations 2p3p and $2p^2$. In the first case, our o assign quantum numbers m_l and m_s to the two electrons is unrestricted clusion principle, since the electrons already differ in their principal numbers. Thus, microstates such as $(1^+, 1^+)$ and $(0^-, 0^-)$ are permitted. not permitted for the $2p^2$ configuration, however. Secondly, since the ons of the 2p3p configuration can be distinguished by their n quantum two microstates such as $(1^+, 0^-)$ and $(0^-, 1^+)$ are physically different. , for the $2p^2$ configuration, such a pair are actually identical since there is al distinction between *the first electron* and *the second electron*. For the figuration there are thus $6 \times 6 = 36$ different microstates, while for onfiguration six of these are nullified by the exclusion principle and the g 30 consist of pairs which are physically redundant. Hence, there are but states for the $2p^2$ configuration.

9.3(a) lists the microstates for the $2p^2$ configuration, in which they are according to their M_L and M_S values. It is now our problem to deduce s array the possible values for L and S. We first note that the maximum imum values of M_L are 2 and -2, each of which is associated with M_S ese must be the two extreme M_L values derived from a state with $L = 2$ 0, namely a ^1D state. Also belonging to this ^1D state must be microstates $= O$ and $M_L = 1, 0$ and -1. If we now delete a set of five microstates iate to the ^1D state, we are left with those shown in Table 9.3(b). Note that important which of the two or three microstates we have removed box which originally contained several, since the microstates occupying e box actually mix among themselves to give new ones. However, the of microstates per box is fixed, whatever their exact description may be. g now at Table 9.3(b) we see that there are microstates with $M_L = 1, 0, -1$ h of the M_S values, 1, 0, -1. Nine such microstates constitute the nents of a 3p state. When they are removed, there remains only a single ate with $M_L = 0$ and $M_S = 0$. This must be associated with a ^1S state of the ration. Thus, the permitted states of the $2p^2$ configuration – or any np^2 ration – are ^1D, ^3P and ^1S. It is to be noted that the sum of the racies of these states must be equal to the number of microstates. The ^1S

Table 9.3 Tabulation of microstates for $2P^2$ configuration

(a)

M_L	M_S 1	0	-1
2		$(1^+,1^-)$	
1	$(1^+,0^+)$	$(1^+,0^-)\,(1^-,0^+)$	$(1^-,0^-)$
0	$(1^+,-1^+)$	$(1^+,-1^-)\,(0^+,0^-)\,(1^-,-1^+)$	$(1^-,-1^-)$
-1	$(-1^+,0^+)$	$(-1^+,0^-)\,(-1^-,0^+)$	$(-1^-,0^-)$
-2		$(-1^+,-1^-)$	

(b)

M_L	M_S 1	0	-1
2			
1	$(1^+,0^+)$	$(1^-,0^+)$	$(1^-,0^-)$
0	$(1^+,-1^+)$	$(1^-,-1^+)\,(0^+,0^-)$	$(1^-,-1^-)$
-1	$(-1^+,0^+)$	$(-1^-,0^+)$	$(-1^-,0^-)$
-2			

state has neither spin nor orbital degeneracy; its degeneracy number is therefore 1. The ^1D state has no spin degeneracy but is orbitally $2L+1$–5-fold degenerate. The ^3P state has 3-fold spin degeneracy and 3-fold orbital degeneracy giving it a total degeneracy number of $3 \times 3 = 9$. The sum of these degeneracy numbers is indeed 15.

For the 2p3p configuration the allowed states are again of the types of S, P and D, but now there is a singlet and a triplet of each kind. This can be demonstrated by making a table of the microstates and proceeding as before. It can be seen perhaps more easily by noting that for every combination of $m_l^{(1)}$ and $m_l^{(2)}$ there are four microstates, with spin assignments $++$, $+-$, $-+$ and $--$. One of these, either $+-$ or $-+$ can be taken as belonging to a singlet state and the other three then belong to a triplet state. It will be noted that the sum of the degeneracy numbers for the six states ^3D, ^1D, ^3P, ^1P, ^3S, ^1S is 36, the number of microstates.

While the method shown for determining the states of an electron configuration will obviously become very cumbersome as the number of electrons increases beyond 5, there is, fortunately, a relationship which makes tractable many of the problems with still larger configurations. This relationship is called the *hole formalism*, and with it a partially filled shell of n electrons can be treated either as n electrons or as $(N-n)$ positrons, where N represents the total capacity of the shell. As far as electrostatic interactions among themselves are concerned, it makes no difference whether they are all positively charged or all negatively charged since the energies of interaction are all proportional to the product of two charges. It is actually rather easy to see that the hole formalism must be true, for, whenever we select a microstate for n electrons in a shell for capacity N, there remains a set of m_l and m_s values which could be used by $(N-n)$ electrons.

eral states derived from a particular configuration have different
However, purely theoretical evaluation of these energy differences is
y nor accurate, since they are expressed as certain integrals represent-
on–electron repulsions, which cannot be precisely evaluated by
on. However, when there are many terms arising from a configuration,
y possible to express all of the energy differences in terms of only a few
Thus, when the energies of just a few of the states have been measured,
may be estimated with fair accuracy, though not exactly, because the
scheme itself is only an approximation.

ntioned before, each state of type $2S + 1L$ actually consists of a group
tes with different values of the quantum number J. The energy
es between these substates are generally an order of magnitude less than
y differences between the various states themselves, and usually they can
ed. However, in certain cases, e.g. in understanding the magnetic
es of the lanthanides and in nearly all problems with the very heavy
(those of the 3rd transition series and the actinides), these energy
ces are of great significance and cannot be ignored. Indeed, for the very
ements they become comparable in magnitude to the energy differences
the $2S + 1L$ states. When this happens, the LS coupling method
s inherently unreliable.

ow energy states that can arise in atoms or ions having incomplete d
(d^1 to d^9) by Russell–Saunders coupling is given in Table 9.4. Racah [5]
culated the energy of different states by a group theoretical method and
ed them in terms of three parameters A, B and C. In practice only the latter
rameters are needed to describe the relative separation of levels, as the
eter A merely shifts all levels of an ion equally. The third column of
9.4 gives the lower free ion term spacings in terms of B and C, with the
d state energy taken as zero. The parameter B is usually sufficient to
te the difference in energy between states of the same spin multiplicity,
both B and C parameters are necessary for terms of different spin
plicity.

ore proceeding further with the investigation of d^n states, let us first look at
teraction of an octahedral field with s^1, p^1, d^1 and f^1 configuration.

s orbital is completely symmetrical and hence is unaffected by all fields. The
itals are not split by an octahedral field since all interact equally, but fields of
r symmetry can cause a splitting of the p orbitals. In octahedral fields the d
als, as discussed before, split into $^2T_{2g}$ state (representing the $T_{2g}^1E_g^0$
guration) and ^2E_g state (representing the $T_{2g}^0E_g^1$ configuration). The set of 7f
als is split by an octahedral field into three levels: a triply degenerate level
below, a triply degenerate level 2Dq above, and a single level 12Dq above
oresplit average energy (barycentre) as shown in Figures 9.4 and 9.5(b). We
thus say that these orbitals and their states transform in an octahedral field
ollows:

e	*Transforms in an octahedral field as*	*States*
	-->	^2S (A_{1g})
	-->	^2P ($^2T_{1g}$)
	-->	$^2T_{2g}$, ^2E_g
	-->	$^2T_{1g}$, $^2T_{2g}$, $^2A_{2g}$

Table 9.4
Lower free ion term spacings

Configuration	State	Energy
$d^1 \equiv d^9$	^2D	0
	^3F	0
	^1D	$5B + 2C$
$d^2 \equiv d^8$	^3P	$15B$
	^1G	$12B + 2C$
	^1S	$22B + 7C$
	^4F	0
	^4P	$15B$
$d^3 \equiv d^7$	^2G	$4B + 3C$
	^2H	$9B + 3C$
	^2P	$9B + 3C$
	^5D	0
	^3H	$4B + 4C$
	^3_aP	$16B + 5\frac{1}{2}C - 1/2(912B^2 - 24BC + 9C^2)^{1/2}$
$d^4 \equiv d^6$	^3_aF	$16B + 5\frac{1}{2}C - 3/2(68B^2 + 4BC + C^2)^{1/2}$
	^3G	$9B + 4C$
	^1I	$6B + 6C$
	^6S	0
	^4G	$10B + 5C$
	^4P	$7B + 5C$
d^5	^4D	$17B + 5C$
	^2I	$11B + 8C$
	^4F	$22B + 7C$

A complete set of transformations for S, P, D, F, G, H and I states is given in Table 9.5.

Table 9.5
Splitting of d^n terms in an octahedral field

Terms	Components in an octahedral field
S	A_{1g}
P	T_{1g}
D	$E_g + T_{2g}$
F	$A_{2g} + T_{1g} + T_{2g}$
G	$A_{1g} + E_g + T_{1g} + T_{2g}$
H	$E_g + T_{1g} + T_{1g} + T_{2g}$
I	$A_{1g} + A_{2g} + E_g + T_{1g} + T_{2g} + T_{2g}$

Now, let us turn to the problem of d^n configurations, namely the d^2. In the

286

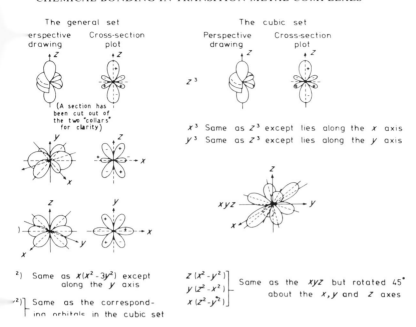

The general set		The cubic set	
erspective drawing	Cross-section plot	Perspective drawing	Cross-section plot

z^3

(A section has been cut out of the two "collars" for clarity)

x^3 Same as z^3 except lies along the x axis
y^3 Same as z^3 except lies along the y axis

xyz

$^2)$ Same as $x(x^2-3y^2)$ except along the y axis

$^2)$ Same as the corresponding orbitals in the cubic set

$\left. \begin{array}{l} z(x^2-y^2) \\ y(z^2-x^2) \\ x(z^2-y^2) \end{array} \right]$ Same as the xyz but rotated $45°$ about the x, y and z axes

9.4 The f orbitals: polar plots of the angular part of the wave functions of the f orbitals.

ce of an external field this produces two states: a low energy ^3F and a higher ^3P. The ground state, in this case, consists of two electrons in different ls. If an octahedral field is imposed the two orbitals must be two of the T_{2g} ils. Now if we excite one of these electrons to the higher energy E_g orbitals, d that there are two possibilities. If the electron being promoted comes from : or yz orbitals it will cost less energy to promote it to the d_{z^2} orbital than to x^2-y^2 orbital. The reason for this is that the electron-electron repulsion s the two resulting configurations $(xy)^1 (x^2-y^2)^1$ and $(xy)^1 (z^2)^1$ to be at rent energies. The source of this energy difference is simply that the $(xy)^1$ $-y^2)^1$ configuration has both electrons confined to the vicinity of the xy e; that is neither orbital has a z component. The alternative configuration vs the electrons to separate somewhat and spread along all three coordinate . Finally, if two electrons are promoted (configuration E_g^2) we obtain a th, high-energy state.

o return to the subject of the transformation of orbitals in an octahedral field, recall that p orbitals remain unchanged but that f orbitals are split into three ls: T_{1g}, T_{2g} and A_{2g}. Now recall that the two triplet states arising from the d^2 figuration are the ^3F and ^3P. These states behave in an octahedral field exactly the F and P states arising from the f^1 and p^1 configurations discussed ive. That is, the ^3F state is split into $^3T_{1g}$ (F), $^3T_{2g}$, and $^3A_{2g}$ states and the is unsplit and becomes the $^3T_{1g}$ (P) state. The splitting of d^1 (^2D) and

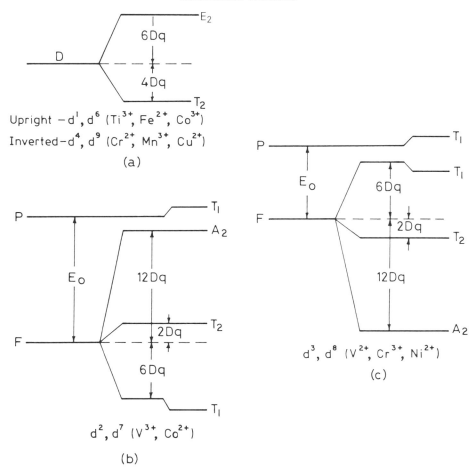

Fig. 9.5   The splitting of d orbitals (a) for D-state, (b), (c) for P and F states. The diagrams are appropriate for octahedral fields and when inverted apply to tetrahedral or cubic coordination.

$d^2(^3F, {}^3P)$ configurations is shown in Figure 9.5. The low-energy states for all d^n configurations are S, P, D, or F states: d^1 and d^9, ^2D; d^2 and d^8, ^3F and ^3P; d^3 and d^7, ^4F and ^4P, d^4 and d^6, ^5D, and d^5, ^6S. The S and P states are not split in an octahedral field and the D and F states split as shown in Figures 9.5(a)–9.5(c).

By means of the hole formalism all of the splittings can be interpreted in terms of Figure 9.5. The d^9 configuration of Cu^{2+} could be treated as an inverted d^1 configuration; that is; the single 'hole' which tends to 'float' to the top in a $t^6_{2g}e^3_g$ configuration is the equivalent of an electron in the inverted $e^1_g t^0_{2g}$ configuration. So the ^2D term of Cu^{2+} may be treated by Figure 9.5 inverted; that is the ground state in an octahedral field is ^2E_g with an energy lying 10Dq below the first excited

288

The d^8 configuration of Ni^{2+} (^3F and ^3P) can be treated in a similar
an inverted d^2. Consideration of the splitting of F and P states in
provides us with the ^3F split into $^3A_{1g}$ lying 10 Dq below the $^3T_{2g}$
3 Dq below the $^3T_{1g}$. Note that the inversion applies only to the F states
ie F state is always (whether from d^2 or d^8) lower in energy than the 3 P.
milar manner, by considering a d^3 configuration to a spherically
:al d^5 with two holes, and the d^4 to be a d^5 with one hole, these
ions can also be determined.
now examine the absorption spectra of some Cr^{3+} salts and compare
h those expected from theory. The spectra of several complexes of
re shown in Figure 9.6. Each spectrum shows at least two well-defined
n bands in the visible region. A third band is sometimes discernible,
it is often obscured by the very intense *charge transfer band* at a higher

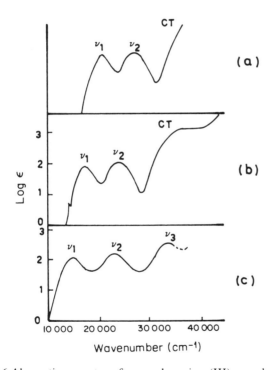

Fig. 9.6 Absorption spectra of some chromium (III) complexes.
(a) $[Cr(en)_3]^{3+}$ (b) $[Cr(ox)_3]^{3-}$ (c) $[CrF_6]^{3-}$

se two visible bands can be assigned on the basis of Figure 9.5. We note that
$^+$ is a d^3 ion with a ^4F ground state. Under the influence of an octahedral field
latter will split into a $^4A_{2g}$ ground state and two excited states, $^4T_{2g}$ and $^4T_{1g}$.
: transitions are thus $^4A_{2g} \rightarrow {}^4T_{2g}(\nu_1)$ and $^4A_{2g} \rightarrow {}^4T_{1g}(\nu_2)$. The first transition

is seen to be 10 Dq from Figure 9.5. The second transition would be 18 Dq were it not for the interaction of the $^4T_{1g}$ (F) and $^4T_{1g}$ (P) terms. Since the wave functions for these two states have identical symmetry they will 'mix'; the amount of mixing being inversely proportional to the difference in energy between the ^4F and ^4P levels. This mixing is analogous to molecular orbital formation in which two orbitals of appropriate symmetry mix (form linear combinations) to yield two new orbitals, one at a lower energy and one at a higher energy than the contributing orbitals. As a result of the mixing, the $^4T_{1g}$ (F) will lie somewhat lower in energy and the $^4T_{1g}$ (P) will lie somewhat higher in energy than they would if mixing had not occurred.

We can compare the experimental results from spectra with those expected from theory. In general, the observed transitions will be $^4A_{2g} \rightarrow ^4T_{2g}$, $^4A_{2g} \rightarrow ^4T_{1g}$ (F), and $^4A_{2g} \rightarrow ^4T_{1g}$ (P). The transitions arising from the ^2G state are either not observed or very weak, since they are spin forbidden. When we compare the experimental results for $[CrF_6]^{3-}$, $[Cr(Ox)_3]^{3-}$ and $[Cr(en)_3]^{3+}$ with the theoretical expectations, we obtain some interesting results. The $^4A_{2g} \rightarrow ^4T_{2g}$ transition (v_1 equal to 10 Dq) increases in progressing from L = F$^-$ to L = en, as expected on the basis of the spectrochemical series. From Figure 9.5, the $^4A_{2g} \rightarrow ^4T_{1g}$ (F) transition is expected to be 18 Dq or 80 per cent greater than the $^4A_{2g} \rightarrow ^4T_{2g}$. The transition $^4A_{2g} \rightarrow ^4T_{1g}$ (P) is expected, on the basis of the free ion (Table 9.6) to be $15B + 12$ Dq or 15×918 cm^{-1} + 12 Dq (for free Cr^{3+} ion $B = 918$ cm^{-1}). It can be seen that these expectations are not realized very closely by the experimental results. Two corrections must be made in order to improve the interpretation and correlation of the spectra. First, the extent of the mixing of the F and P terms was not included. This will be discussed presently, but we must first note that even if this is known exactly, the experimental results cannot be duplicated using the free-ion value of B, even if it, too, is known exactly. The apparent value of B in complexes is always smaller than that of the free ion. This phenomenon is known as the nephelauxetic effect and is attributed to delocalization of the metal electrons over molecular orbitals that encompass not only the metal but the ligands as well. As a result of this delocalization or 'cloud expanding' the average inter-electronic repulsion is

Table 9.6
Spectral correlations for octahedral Chromium (III) complexes

Energy level diagram	Predicted energy (Figure 9.9)	Experimentally observed energy (cm^{-1})		
		$(CrF_6)^{3-}$	$[Cr(Ox)_3]^{3-}$	$[Cr(en)_3]^{3+}$
$^4T_{1g}$(P)	12 Dq + 15 B + X	34 400	—	—
$^4T_{1g}$(F)	18 Dq − X	22 700	23 900	28 500
$^4T_{2g}$	10 Dq	14 900	17 500	21 850
$^4A_{2g}$	0	0	0	0
		Dq = 1 490	Dq = 1 750	Dq = 2 185
		B′ = 827	B′ = 641	B′ = 641

ıd B' (representing B in the free ion) is smaller. The nephelauxetic ratio, by:

$$\beta = \frac{B'}{B}$$

s less than one and decreases with increasing delocalization. Estimates ıe obtained from the nephelauxetic parameters h_x for the ligand and k_M ɔtal:

$$(1 - \beta) = h_x \cdot k_M$$

ɔ transitions are observed, it is a simple matter to assign a value to B' following equation must hold:

$$15B' = \nu_3 + \nu_2 - 3\nu_1$$

is the absorption occuring at the lowest frequency. For example, the ℶ' in the fluoro complex is:

$$B' = \frac{34\,400 + 22\,700 - 3(14\,900)}{15} = 827\,\text{cm}^{-1}$$

q can be measured directly as in the case of Cr^{3+} spectra $(10\,Dq = \nu_1)$ ıluated by means of the above equations, it is quite simple to estimate all ınsition energies. For high-spin octahedral d^n species the appropriate ː to calculate energy differences are given in Table 9.7.

Table 9.7
For octahedral symmetry

Configuration	Transition	Energy
d^1, d^6	$T_2 \to E$	$10\,Dq$
d^4, d^9	$E \to T_2$	$10\,Dq$
d^2, d^7	$T_1(F) \to T_1(P)$	$(225\,B^2 + 180\,B \cdot Dq + 100\,Dq^2)^{1/2}$
	$\to A_2(F)$	$-\dfrac{15}{2}B + 15\,Dq + \tfrac{1}{2}(x)^{1/2}$
	$\to T_2(F)$	$-\dfrac{15}{2}B + 5\,Dq + 1/2(x)^{1/2}$
d^3, d^8	$A_2(F) \to T_1(F)$	$\dfrac{15}{2}B + 15\,Dq - 1/2(x)^{1/2}$
	$\to T_1(P)$	$\dfrac{15}{2}B + 15\,Dq + 1/2(x)^{1/2}$
	$\to T_2(F)$	$10\,Dq$

where $x = (225\,B^2 + 180\,B \cdot Dq + 100\,Dq^2)$.

uld be noted that the same equations can be used for tetrahedral y also if only it is remembered that splitting scheme of d^n (octahedral) is ıt to that of $d^{(10-n)}$ (tetrahedral). Using these equations, more accurate ʒ can be made of the transitions and the spectrum fitted quite suitably.

Before proceeding further, let us make a few generalizations regarding the energy level equations of Table 9.7.

(1) States with identical designations never cross.
(2) The ligand field states have the same spin multiplicity as the free ion states from which they originate.
(3) Non-repeating states have their energies linearly dependent on ligand field strength, whereas when two or more states occur with identical designation they show non-linear dependence of energy on field strength. This is because such states interact with each other and also with the ligand field.
(4) Transitions are spin-forbidden between states of different spin multiplicity.

The use of adjusted B and C parameters is the basis of the *ligand field theory* in which B and C are treated as empirical, adjustable parameters which are fitted to experimentally obtained spectral data. To explain this point, let us interpret the electronic spectra of a typical transition metal ion, say V^{3+} in aqueous solution. V^{3+} corresponds to a d^2 configuration. From Table 9.7 the ground state is $^3T_{1g}$ (F) and the possible spin-allowed transitions that may be expected are

$$^3T_{1g}\,(F) \rightarrow\, ^3T_{2g}\,(F),\, ^3T_{1g}\,(F) \rightarrow\, ^3T_{1g}\,(P),\ \text{and}\ ^3T_{1g}\,(F) \rightarrow\, ^3A_{2g}\,(F).$$

Two absorption bands have been experimentally observed, and these are assigned to $^3T_{1g}$ (F) $\rightarrow$ $^3T_{2g}$ (F), and $^3T_{1g}$ (F) $\rightarrow$ $^3T_{1g}$ (P) transitions. From Table 9.7 the energies associated with these transitions are:

$$^3T_{1g}(F) \rightarrow\, ^3T_{2g}(F) = (v_1) = 1/2[10Dq - 15B + (225B^2 + 180\,BDq + 100\,Dq^2)^{1/2}]$$

$$^3T_{1g}(F) \rightarrow\, ^3T_{1g}(P) = (v_2) = [225B^2 + 180\,B.Dq + 100\,Dq^2]^{1/2}$$

In order to take into consideration the parameterization of B, the above terms may be expressed as:

$$\frac{E[^3T_{1g}(F) \rightarrow\, ^3T_{2g}(F)]}{B} = \frac{1}{2}\left[\frac{10Dq}{B} - 15 + \left(225 + \frac{180\,Dq}{B} + \frac{100\,Dq^2}{B^2}\right)^{1/2}\right] \quad (9.1)$$

$$\frac{E[^3T_{1g}(F) \rightarrow\, ^3T_{1g}(P)]}{B} = \left[225 + \frac{180\,Dq}{B} + \frac{100\,Dq^2}{B^2}\right]^{1/2} \quad (9.2)$$

The spectrum of $[V(H_2O)_6]^{3+}$ indicate absorption maxima at $17\,200$ cm^{-1} and $25\,600$ cm^{-1} corresponding to first and second transitions respectively (Figure 9.13). Thus,

$$\frac{17\,200}{25\,600} = 0.672 = \frac{1/2[10\,Dq/B - 15 + (225 + 180\,Dq/B + 100\,Dq^2/B^2)^{1/2}]}{[225 + 180\,Dq/B + 100\,Dq^2/B^2]^{1/2}}$$

Now, either of two different methods may be adopted to determine the value of $10\,Dq$ and B. In the first case, by a trial and error method, the appropriate value of $10\,Dq/B$ is obtained which corresponds to the above mentioned quotient of 0.672. This gives a $10\,Dq/B$ value of 28.

tions (9.1) and (9.2), E/B have been expressed as functions of $10\,Dq/B$, ore

$$\frac{E[^3T_{1g}(F) \to {}^3T_{2g}(F)]}{B} = 25.9$$

$$\frac{E[^3T_{1g}(F) \to {}^3T_{1g}(P)]}{B} = 38.6.$$

the experimental transitions, the value of B comes out to be $665\,cm^{-1}$, ce $10\,Dq = 18\,600\,cm^{-1}$. The approach of the second method is exactly except that $10\,Dq/B$ is obtained from a master plot of

$$\frac{E[^3T_{1g}(F) \to {}^3T_{2g}(F)]}{E[^3T_{1g}(F) \to {}^3T_{1g}(P)]} \text{ vs. } \frac{10\,Dq}{B}.$$

ore, this procedure is applicable for the analysis of a large number of data.

The molecular orbital theory

asic assumption in molecular orbital theory (MO) is that if two nuclei are oned at an equilibrium distance and electrons are added they will go into ular orbitals which are in many ways analogous to the atomic orbitals. In om there are s, p, d, f . . . . orbitals determined by various sets of quantum ers. The Pauli exclusion principle and Hund's principle of maximum plicity are obeyed in these molecular orbitals as well as in atomic orbitals. hen we attempt to solve the Schrödinger equation to obtain the various cular orbitals, we run into the same problem as for atoms heavier than ogen; we cannot solve the Schrödinger equation exactly and therefore must e some approximations concerning the form of the wave function for the ecular orbitals.

f the various methods of approximating the correct molecular orbitals, we ll discuss here only one, the linear combination of atomic orbitals (LCAO) thod. In this approach we assume that we can approximate the correct lecular orbitals by combining the atomic orbitals of the atoms to form the lecule. The rationale is that most of the time the electrons will be nearer, and ace 'controlled' by, one or the other of the two nuclei; when this is so, the lecular orbital should be very nearly the same as the atomic orbital for that om. We therefore combine the atomic orbitals ψ_A and ψ_B, to obtain two olecular orbitals:

$$\psi_b = \psi_A + \psi_B \tag{9.3}$$

$$\psi_a = \psi_A - \psi_B \tag{9.4}$$

he one-electron molecular orbitals thus formed consist of a bonding molecular rbital (ψ_b) and an *antibonding* molecular orbital (ψ_a). If we allow a single lectron to occupy the bonding molecular orbital (as in H_2^+, for example), the pproximate wave function for the molecule is

$$\psi = \psi_{b(1)} = \psi_{A(1)} + \psi_{B(1)} \tag{9.5}$$

For a two-electron system such as H_2, the total wave function is the product of the wave functions for each electron:

$$\psi = \psi_{b(1)} \cdot \psi_{b(2)} = [\psi_{A(1)} + \psi_{B(1)}][\psi_{A(2)} + \psi_{B(2)}] \tag{9.6}$$

$$= \psi_{A(1)} \cdot \psi_{A(2)} + \psi_{B(1)} \cdot \psi_{B(2)} + \psi_{A(1)} \psi_{B(2)} + \psi_{A(2)} \cdot \psi_{B(1)} \tag{9.7}$$

In equation (9.7) the ionic terms $\psi_{A(1)} \cdot \psi_{A(2)}$ and $\psi_{B(1)} \cdot \psi_{B(2)}$ are weighted as heavily as the covalent ones. This is not surprising since we did not take into account the repulsion of electrons in obtaining equation (9.6). This is a general result; simple molecular orbitals obtained in this way from LCAO tend to exaggerate the ionicity of molecules and the chief problem in adjusting this simple method to make the results more realistic is the matter of taking into account *electron correlation*. However, it is possible to optimize the wave function by the addition of correcting terms. Some typical results for the hydrogen molecule are shown in Table 9.8.

Table 9.8
Energies and equilibrium distances for molecular orbital wave functions of hydrogen molecule

Types of wave function	Energy (cal)	Distance (A)
Uncorrected, $\psi = \psi_A + \psi_B$	62 265	0.85
Addition of shielding	80 483	0.73
Addition of electron–electron repulsions	94 781	0.71
Observed values	109 471	0.741

The two orbitals, ψ_b and ψ_a differ from each other as follows. In the bonding molecular orbital, the wave functions for the component atoms reinforce each other in the region between the nuclei but in the antibonding molecular orbital they cancel, forming a node between the nuclei. To find the electron distribution in a hydrogen molecule we have to square the wave functions:

$$\psi_b^2 = \psi_A^2 + 2\psi_A\psi_B + \psi_B^2 \tag{9.8}$$

$$\psi_a^2 = \psi_A^2 - 2\psi_A\psi_B + \psi_B^2 \tag{9.9}$$

The difference between the two probability functions lies in the cross term $2\psi_A\psi_B$. The integral $\int \psi_A\psi_B d\tau$ is known as the *overlap integral*, S, and is very important in bonding theory. In the bonding orbital, the overlap is positive and the electron density between the nuclei is increased, whereas in the antibonding orbital the electron density between the nuclei is decreased. In the former case, the nuclei are shielded from each other and the attraction of both nuclei for the electrons is enhanced. This results in a lowering of the energy of the molecule and is therefore a bonding situation. In the second case, the nuclei are partially exposed to each other and the electrons tend to be in those regions of space in which mutual attraction by both nuclei is severely reduced. This is a repulsive or antibonding situation.

now normalize the molecular orbitals. Since $\psi^2 d\tau = 1$ for the probability of finding an electron somewhere in space, equation (9.8) becomes:

$$N_b^2 \psi_b^2 d\tau = N_b^2 (\int \psi_A^2 d\tau + \int \psi_B^2 d\tau + 2\int \psi_A \psi_B d\tau) = 1,$$

is the normalizing constant. If we let S be the overlap integral, $\int \psi_A \psi_B d\tau$,

$$\int \psi_b^2 = \left(\psi_A^2 d\tau + \psi_B^2 d\tau + 2S \right)$$

e atomic wave functions ψ_A and ψ_B were previously normalized, $\psi_A^2 d\tau$ and ich equal one. Hence:

$$N_b^2 = 1/(2 + 2S) \text{ or } N_b = \sqrt{[1/(2 + 2S)]},$$

nilarly,

$$N_a = \sqrt{[1/(2 - 2S)]}.$$

ur molecular wave functions become:

$$\psi_b = \frac{1}{\sqrt{(2 + 2S)}} (\psi_A + \psi_B)$$

$$\psi_a = \frac{1}{\sqrt{(2 - 2S)}} (\psi_A - \psi_B)$$

e relative energies (neglecting overlap, for S is numerically rather small) of wo molecular orbitals are shown in Figure 9.7. The bonding orbital is lized relative to the energy of the isolated atoms by the quantity ΔE. The onding orbital is destabilized by almost an equivalent amount (strictly, tly more for $(2 + 2S) > (2 - 2S)$). The quantity ΔE is termed the *exchange* gy.

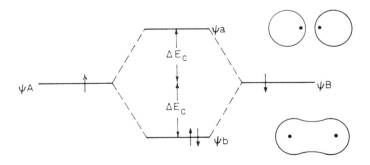

Fig. 9.7 Energy levels for the H_2 molecule with neglect of overlap. The quantity ΔE_c represents the difference in energy between the energy levels of the separated atoms and the bonding molecular orbital.

295

As may be seen from equations (9.8) and (9.9) the only difference between the electron distribution in the bonding and antibonding molecular orbitals and the atomic orbitals is in those regions of space for which both ψ_A and ψ_B have appreciable values, so that their product ($S = \psi_A \psi_B d\tau$) has an appreciable non-zero value. Furthermore, for bonding $S > 0$, and for antibonding, $S < 0$. The condition $S = 0$ is termed *nonbonding* and corresponds to no interaction between the orbitals. Now we may make the generalization that the strength of a bond will be roughly proportional to the extent of the overlap of the atomic orbitals. This is known as the *overlap criterion of bond strength* and indicates that bonds will form in such a way as to maximize overlap.

In s orbitals the sign of the wave function is the same everywhere and so there is no problem with matching the sign of the wave functions to achieve positive overlap. With p and d orbitals, however, there are several possible ways of arranging the orbitals, some resulting in positive overlap, some in negative overlap, and some in which the overlap is exactly zero. Bonding can take place only when the overlap is positive.

Some of the possible combinations of atomic orbitals are shown in Figure 9.8. Those orbitals which are cylindrically symmetrical about the internuclear axis are

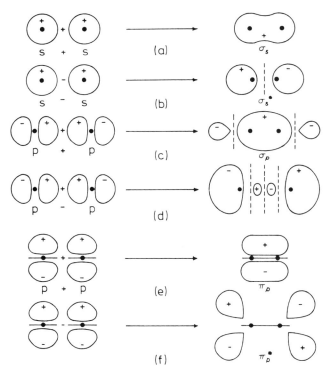

Fig. 9.8 Symmetry of molecular orbitals formed from atomic orbitals illustrating sigma (a)–(d) and pi (e), (f) orbitals, and bonding (a), (c), (e) and antibonding (b), (d), (f) orbitals.

orbitals, analogous to an s orbital, the atomic orbital of highest
. If the internuclear axis lies in a nodal plane, a π-bond results. In delta
the internuclear axis lies in two mutually perpendicular planes. All
ng orbitals possess an additional nodal plane perpendicular to the
ear axis and lying between the nuclei.

are two criteria which must be satisfied for molecular orbitals to be more
n the contributing atomic orbitals. One is that the overlap between the
rbitals must be positive. Furthermore, in order for there to be effective
on between orbitals on different atoms, the energies of the two atomic
must be approximately the same.

proceeding with transition metal complexes involving d orbitals, let us
the simple BeH_2 molecule as a complex of Be^{2+} and $2H^-$. There are two
available on the beryllium for bonding: the 2s (with A_{1g} symmetry) and
(with A_{2u} symmetry). Since the resulting molecular orbitals will be linear
ations of the atomic orbitals of metal and ligand with the same symmetry,
ropriate to construct linear combinations of the ligand orbitals to match
metry of the metal orbitals. An A_{1g} ligand group orbital (LGO) can be
cted by adding the 1 s wave functions of the hydrogen atoms:

$$\psi_{a1g} = \psi_H + \psi_{H'}$$

GO has the required symmetry since, like the beryllium 2s orbital, the wave
n is positive everywhere. The second LGO can be constructed with the
de symmetry, A_{2u}:

$$\psi_{a2u} = \psi_H - \psi_{H'}$$

ppropriate molecular orbitals may now be written:

$$\sigma_g = \psi_{2s} + \psi_{a1g}$$
$$\sigma_u = \psi_{2pz} + \psi_{a2u}$$

e case of first row transition metals, out of the five d orbitals, the d_z^2 and
$_{y^2}$ are directed towards the ligands providing positive overlap. The three
ining, the d_{xy}, d_{xz}, and d_{yz}, are directed between the ligands and the net
lap is zero. Note that these two groups of d orbitals are the same as we have
in ligand field theory, namely, T_{2g} and E_g.
n LGO can now be constructed as follows:

$$\psi_{LGO}, x^2 - y^2 = 1/2\,(\psi_{\sigma x} + \psi_{\sigma - x} - \psi_{\sigma y} - \psi_{\sigma - y})$$

expressed in somewhat simpler symbolism:

$$\Sigma_{x^2 - y^2} = 1/2(\sigma_x + \sigma_{-x} - \sigma_y - \sigma_{-y})$$

ere Σ and σ represent the wave functions for the ligand group orbital and the
ntributing atomic orbitals, respectively. The second E_g orbital of the metal is the
orbital and the appropriate Σ_{z^2} LGO can be written:

$$\Sigma_{z^2} = \frac{1}{2\sqrt{3}}(2\sigma_z + 2\sigma_{-z} - \sigma_x - \sigma_{-x} - \sigma_y - \sigma_{-y})$$

he pictorial representation of the E_g orbitals is given in Figure 9.9.

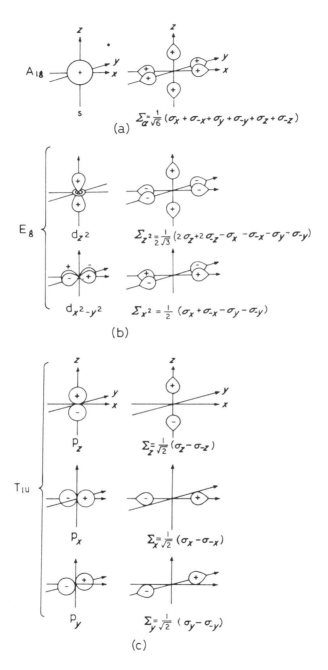

Fig. 9.9 Ligand group orbitals (LGO's) and matching atomic orbitals (AO's) of the same symmetry.

tal T_{2g} orbitals cannot form sigma bonds. The spherical 4s orbital has
netry and a ligand group orbital of A_{1g} symmetry can readily be
ed (Figure 9.9a). In a similar manner, LGOs can be written for the 4p
Γ_{1u} symmetry, Figure 9.9c).
is possible to set up an energy diagram for an octahedral complex such
$[H_3)_6]^{3+}$. The overlap of the 4s and 4p orbitals with the ligands is
bly better than that of the 3d orbitals. (In general, d orbitals tend to be
diffuse and, as a result, overlap of d orbitals is quantitatively poor, even
alitatively favourable). As a result, the A_{1g} and T_{1u} molecular orbitals
west in energy and the corresponding antibonding orbitals, A_{1g}^* and
e highest in energy. The E_g and E_g^* orbitals arising from the 3d orbitals
aced less from their barycentre because of poorer overlap. The T_{2g}
re nonbonding (in sigma-only system) and not displaced. The resulting
vel diagram is shown in Figure 9.10.

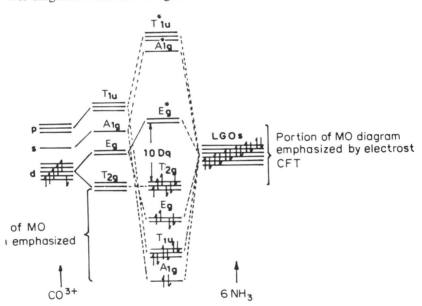

9.10 MO energy level diagram of $[Co(NH_3)_6]^{3+}$ illustrating electron distribution and relation between MOT (Molecular Orbital Theory), CFT
(Crystal Field Theory), and VBT (Valence Bond Theory).

trons may now be added to the molecular orbitals of the complex in order
easing energy. In a complex such as $[Co(NH_3)_6]^{3+}$, there will be a total of
n electrons, twelve from the nitrogen lone pairs and six from the $3d^6$
iration of the Co^{3+} ion. The electron configuration will then be:
$_uE_g^4T_{2g}^6$ (Figure 9.10). Note that the complex is diamagnetic because the
ns pair in the T_{2g} level rather than entering the higher energy E_g^* level. If
ergy difference is small, as in $[CoF_6]^{3-}$ the electrons will be distributed
^{2}. Thus, both molecular orbital theory and ligand field theory account for

nd spectral properties of octahedral complex ions by supposing the
o sets of orbitals separated by an energy gap, 10Dq. If this energy is
e pairing energy, low-spin complexes will be formed. Likewise, the
of complexes are attributed to electronic transitions such as T_{2g}
...- qualitative results of ligand field theory and molecular orbital
theory are quite similar although the fundamental assumptions, that is, purely
electrostatic perturbations versus orbital mixing, seem considerably different.

Molecular orbital theory to non-octahedral complexes
Although MO treatment can be applied to planar, tetrahedral, linear, etc.
symmetries as well, the simplicity of the energy level diagram is rapidly lost. Even
for tetrahedral complexes the analysis is complicated, due to the lack of a centre of
symmetry to keep various kinds of orbitals symmetrically distinct from one
another. For instance, the three p orbitals and the three d orbitals (d_{xy}, d_{xz}, d_{yz})
are in the same symmetry class in a tetrahedral complex and then they mix with
one another. This makes it harder to produce simple prediction.

The bonding capabilities of various metal orbitals in several types of complex
are shown in Table 9.9.

Table 9.9
Bonding orbitals for some highly symmetrical molecules

Stereo chemistry	σ-orbitals	π-orbitals
Linear L–M–L	(s, p_z)	(p_x, p_y) (d_{xz}, d_{yz})
Planar equilateral triangle	(s, p_x, p_y) (d_{z^2}, p_x, p_y)	(p_z, d_{xz}, d_{yz})
Tetrahedral ML$_4$	(s, p_x, p_y, p_z) $(s, d_{xy}, d_{yz} \cdot d_{xz})$	(p_x, p_y, p_z) (p_z, d_{xz}, d_{yz}) $(d_{yz}, d_{xz}, d_{x^2-y^2}, d_{z^2}, d_{xy})$
Square planar ML$_4$	$(d_{x^2-y^2}, s, p_x, p_y)$ $(d_{x^2-y^2}, d_{z^2}, p_x, p_y)$	$(d_{z^2}, d_{x^2-y^2})$ (p_z, x_z, y_z)
Trigonal bipyramid ML$_6$	$(s, p_x, p_y, p_z, d_{z^2})$	(p_x, p_y, p_z) $(d_{x^2-y^2}, d_{xy})$ (d_{xz}, d_{yz})

9.3 APPLICATION OF BONDING THEORIES IN INTERPRETING d–d ABSORPTION SPECTRUM

So far we have discussed the energy levels that arise when an atom or ion having a
finite number of electrons is placed in a chemical environment producing an
electrostatic field (CFT or LFT) or forming bonding, non-bonding, and antibonding orbitals at different energies (MOT). In the case of most of the transition
metal ions the separation between the ground state and the nearest excited state is
much larger than kT. Thus, at ordinary temperatures it is only the ground state
that is populated. If electromagnetic radiation is now allowed to fall on this
material, a photon will be absorbed when the photon energy ($h v$) is equivalent to
the separation energy:

$$E(\text{excited}) - E(\text{ground}) = \Delta E = h\nu,$$

and thus an absorption band will arise in the spectrum.

ualitative predictions of ligand field theory

elds three important symmetries are found, octahedral, tetrahedral and
ıbic. To a first approximation, and provided the ligand–metal ion
are the same in all three symmetries, one can express the ligand field
ɔarameters in terms of the octahedral ligand field splitting parameter,

$$10 \, Dq_{Td} = -4/9 \; 10 \, Dq_{Oh}$$

$$10 \, Dq_{cubic} = -8/9 \; 10 \, Dq_{Oh}$$

therefore obtain an approximate value of $10 \, Dq_{Td}$ if that of $10 \, Dq_{Oh}$ is
n this way the spectrum of a tetrahedral complex can be predicted from
ı octahedral complex of the same ion. The relationship may also be used
ξuish whether two different coordinations exist.

ntensities of electronic spectra

ısities of absorption bands due to transition metal ions in glass vary over
ders of magnitude. A low intensity usually indicates that some formally
ın transition is occurring and that several mechanisms may contribute to
idden transition. The variation of intensity with wavelength is usually
d by the extinction coefficient, ε, defined by the relation, $D = \varepsilon C_1$, where
optical density per centimetre thickness of the sample, and C_i is the
ration in mols per litre of the absorbing species i. D is the experimentally
:d quantity and the absorption spectrophotometers can be adjusted to
: this directly. D is defined by the relationship, $D = \log_{10} I_0/I$, where I_0 is
nsity of incident light and I is the intensity of transmitted light. The
ɔn coefficient is governed largely by the probability of the electronic
ɔn and the polarity of the excited state. In order that interaction may take
photon must strike a chromophore approximately within the space of the
lar dimensions, and the transition probability, g, will be the proportion of
ξet hits which lead to absorption. Thus:

$$\frac{-\partial I}{I} = \frac{1}{3} gCNA \left(\frac{\partial t}{1000} \right)$$

$\mathcal{C}$ is the molar concentration, N is Avogadro's number, t is the thickness in
ıs the cross-sectional target area (obtainable from X-ray diffraction data)
} is a statistical factor to allow for random orientation. On integration and
ıg numerical constants,

$$\log \frac{I_0}{I} / Ct = 0.87 \times 10^{20}. \, gA$$

ıing the average cross-sectional area to be $10 \, Å$ (a value which has been
for many simple molecules) for a transition of unit probability

$$\varepsilon \simeq 10^5$$

ighest extinction coefficients observed in glass are of this order. The
ity of a given absorption band may be measured by the maximum

301

extinction coefficient and the broadness by the halfwidth (that is the width of the band at half maximum intensity). Often the assumption is made that absorption bands have a Gaussian shape, that is they may be described by the relation,

$$\varepsilon = \varepsilon_0 \exp - \alpha (v - v_0)^2,$$

where ε is the extinction coefficient at any particular wave-number v, ε_0 is the maximum extinction coefficient at the wavenumber v_0 and α is a constant which is related to the halfwidth Γ by the relation,

$$\Gamma = 1/2 \sqrt{\left(\frac{\log_e 2}{\alpha} \right)}$$

The oscillator strength of a band, f, may be defined by,

$$f = (4.32 \times 10^{-9}) \int_{v_1}^{v_2} \varepsilon \, dv$$

and is directly proportional to the area under an absorption band. For a Gaussian band the integral is approximately equal to $\varepsilon \cdot \Gamma$, and this provides a convenient method for the estimation of the oscillator strength.

9.3.3 Selection rules

The electric dipole mechanism is the only one of importance for the absorption of light by complex ions. The transition moment Q is defined by,

$$Q = \langle \psi_1 | \mathbf{r} | \psi_2 \rangle$$

where ψ_1 and ψ_2 are the wave functions of the two energy states between which electrons are excited when absorbing radiative energy whose frequency corresponds to the energy difference between the two states, while $\mathbf{r}$ is the radius vector which has the symmetry of an electric dipole. For polarized light the absorption may be anisotropic, the absorption in the z direction for example is a function of Q where $Q = \langle \psi_1 | \mathbf{z} | \psi_2 \rangle$. For an absorption to occur A must be non-zero. By the use of group theory one can readily define two selection rules:

(1) *Spin selection rules*: Transitions for which $\Delta S \neq 0$ are said to be spin-forbidden, where ΔS is the change in multiplicity between the two energy states, and

(2) *Laporte selection rule*: As discussed before the energy levels or states are usually denoted by symmetry symbols. This is because the wave functions for those energy levels transform according to the irreducible representations of the group to which they belong. The Laporte rule states that transitions are allowed only between states with opposite parity, that is between states labelled g and u, but not between states which are both u or both g. The wave functions for the energy levels in ligand field theory are all constructed from metal ion d orbitals which are all g in character, and therefore transitions between these states, in principle, are Laporte forbidden.

..laxation of selection rules

...ove selection rules were rigorously obeyed, transition metal ion would show no absorption spectra. Several mechanisms operate which ...selection rules to be relaxed and spectra to be observed.

..selection rule

...vave function of an atomic complex involves orbital, spin, vibrational, and transitional wave functions. In treating these complexes the ..ation is made that the wave functions may be solved separately without ..action between them. Coupling between spin and orbital angular ..m does occur, however, which makes it impossible to factorize the wave ..ccurately into spin and orbital wave function products. This means that ..election rule is not completely valid, and spin-forbidden bands do occur, ..weakly. The intensity of these bands increases as spin-orbit coupling ..increase, particularly in moving from left to right in the transition metal ..d on going from the first to third series in the periodic table.

..orte selection rule

...ence of molecular vibrations in a complex allows coupling to occur ..electronic and vibrational energy levels in both the ground and excited ..ch coupled energy levels are denoted vibronic energy levels, and for them ..st certain that a g electronic level is combined with a u vibrational level. ..ting vibronic level will be u, and it is almost certain that both the ground ..ed states have g and u energy levels.

..nsity stealing

...idden level lies very near in energy to a fully allowed transition which ..nds to a very intense band, there is in general a vibrational level which ..s with both the forbidden and allowed electronic levels. In this way the ..of the forbidden level is increased further than it would have been if the ..level were not present. Such allowed levels are normally charge transfer

..xing d–p wave functions in non-centrosymmetric complexes

...entrosymmetric complex such as tetrahedral, when treated by molecular ..theory, has energy levels whose wave functions are constructed not ..from d orbitals. Excited states exist whose wave functions are com- ..ns of d and p wave functions. Such wave functions are u, and therefore ..ons are possible between the g ground state composed of d and p wave ..ns.

..stortion from octahedral symmetry

...plex may exist whose symmetry is not octahedral – an elongation or ..ssion along the z axis, for example, gives a symmetry of D_{4h}. Although the ..e rule still holds, investigation of the symmetry properties indicates that ..n in the xy plane of the complex is preferred to that in the xz or yz planes. ..ne absorption properties are anisotropic, and the resulting absorption is ..ed.

The above mechanisms for relaxation of the selection rules results in a wide variation in observed intensities of absorption bands in glass, which are given in Table 9.10.

Table 9.10
Representative values for the intensities of bands of various types in transition metal complexes.

Type of transition	Approximate f	Approximate ε
Spin-forbidden, Laporte-forbidden	10^{-7}	0.1
Spin-allowed, Laporte-forbidden	10^{-5}	10
Spin-allowed, d–p mixing	10^{-3}	100
Spin-allowed, intensity stealing	10^{-2}	1 000
Spin-allowed, Laporte-allowed (charge transfer)	10^{-1}	10 000

9.3.5 Band widths

If an absorption corresponds to an electronic transition between two single energy levels, then the observed bands should be extremely sharp. In practice, however, band widths in glass are often of the order of $1000 \, \text{cm}^{-1}$ and rarely less than $100 \, \text{cm}^{-1}$. This implies that the energy levels vary over a range of energies, and mechanisms for which this may occur are discussed below.

(a) *Vibrations*
The ligand field splitting parameters depend upon the ligand–metal ion separation in a critical manner, and molecular vibrations therefore modulate the ligand field. Those energy levels in the Orgel diagrams which vary with 10 Dq are those most broadened in this way, while those energy levels which run parallel to the abscissa are least affected by vibrations and are therefore sharp. Widths of up to $1000 \, \text{cm}^{-1}$ may be accounted for in this way.

(b) *Spin-orbit coupling*
Levels with the symmetry label T are split by spin-orbit coupling, which may vary from 100 to $1000 \, \text{cm}^{-1}$ for ions in the first transition series. Therefore transitions between levels, such as the T type, are split in this way. Ideally this should be observed as a fine structure, but it is rarely observed at room temperature.

9.3.6 Band shape

The mechanisms for band broadening give some qualitative information as to the shape of the absorption bands.

(a) *Vibrational interaction*
Vibrational interaction is expected to lead to an asymmetrically shaped band. For bands in which vibronic coupling is the origin of most of the intensity, symmetry is expected only at very low temperatures. At higher temperatures, a tail towards

ncy (the infrared) is expected, and the position of the band maximum
:hange.

orbit coupling
does not split a term symmetrically and asymmetric bands are

symmetry complexes
:lds in low symmetry complexes lift the degeneracies of E and T type
which the transition moments of each component may differ. If the
; small, overlap of the partially separated bands may lead to a resultant
symmetric shape.
in the spectra of transition metal ions are often quite symmetrical in
d can often be fitted to a Gaussian relation. It is possible that all three
;cussed above occur simultaneously to give an average band which is
:. Structure and asymmetry are more obvious in low energy bands
perturbing mechanisms represent a larger fraction of the energy of the
is.

ihn–Teller effect

ı important mechanism by which absorption bands are broadened and
rom symmetry. The Jahn–Teller theorem states that if, as first
ed, a molecular complex is seen to give rise to an orbitally degenerate
itate, it will be found to have distorted itself so as to remove the
cy.
ergy by which the ground state is split may be of the order of a few times
ng Boltzmann's constant), which is approximately $250 \, \text{cm}^{-1}$ at room
ure. Splitting of higher energy states usually occurs simultaneously. A
which is assumed to be octahedral is considered to be distorted either by
:ning or shortening of the bond lengths along the z axis.
ihn–Teller effect has been demonstrated satisfactorily in glass only for
though the asymmetry of the Ti^{3+} absorption spectrum is also attri-
this splitting. In Cu^{2+} it is of the order of $2000 \, \text{cm}^{-1}$. No idea as to the
f the distortion can be predicted, apart from the requirement that a centre
ietry be retained.
ic Jahn–Teller effect occurs when the ground state is degenerate. In those
ere the excited states are degenerate the theorem applies as well, although
:ases the effect is a complicated dynamical one because the short lifetime
ectronically excited state does not permit the attainment of a stable
ium configuration.

ABSORPTION SPECTRA OF TRANSITION METAL IONS

iscussing all these elementary principles of ligand field and molecular
:heories let us now examine the absorption spectra of some representative
ɔn-metal ions in glasses.

(a) $3d^1$ *system. Absorption spectra of Ti^{3+} in glass*

The energy level diagram for a d^1 system is shown in Figure 9.11(a) and the absorption spectrum of Ti^{3+} in a series of borate and phosphate glasses is shown in Figure 9.12. The main absorption band (around $18\,000\,cm^{-1}$ in phosphate glasses and $21\,000\,cm^{-1}$ in borate glasses) is due to $^2T_2 \rightarrow {}^2E$ transition. Due to Jahn–Teller effect, however, the octahedron of ligands is subjected to small tetragonal distortion giving rise to the shoulder around $14\,000\,cm^{-1}$. In borate glasses also this shoulder can be resolved as a component Gaussian band, as shown dotted in Figure 9.12 for the Na_2O, $9B_2O_3$ glass.

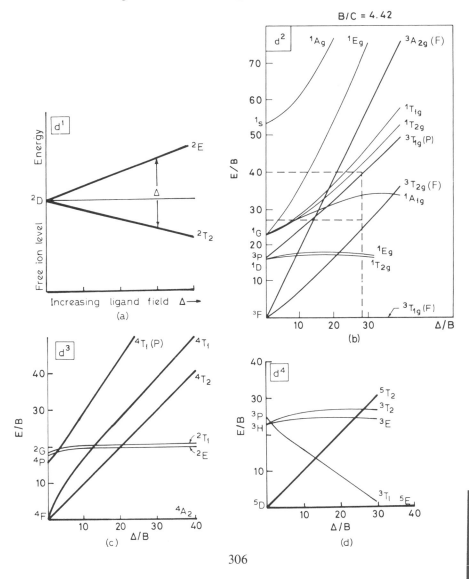

306

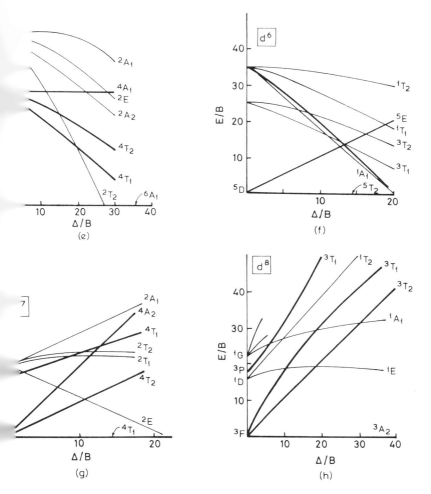

9.11 Semiquantitative energy level diagrams for d^n configurations in octahedral field.

d^2 system. Absorption spectra of V^{3+} in glass

energy level diagram for a d^2 system in octahedral symmetry is shown in 9.11(b) (for tetrahedral configuration consult the octahedral diagram of em Figure 9.11 h). From this Figure, we see the three bands corresponding allowed transitions:

$$^3T_1\,(F) \longrightarrow {}^3T_2$$
$$\longrightarrow {}^3A_2$$
$$\longrightarrow {}^3T_1\,(P),$$

307

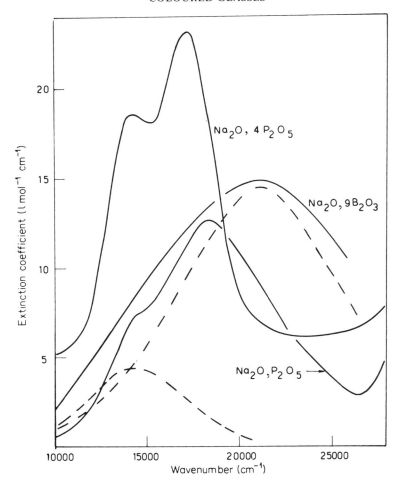

Fig. 9.12 Optical absorption of titanium (III) in some glasses.
The dotted lines are the resolved gaussian bands.

together with a few spin-forbidden transitions. The third band, due to its forbidden two-electron jump, is expected to be of low intensity. The absorption spectrum of V^{3+} in aqueous solution and in some glasses is shown in Figure 9.13. Aqueous solution of V^{3+} show two wide, fairly weak bands at $17\,200\,cm^{-1}$ and $25\,600\,cm^{-1}$. There are also very weak bands between $20\,000$ and $30\,000\,cm^{-1}$ (not shown in the figure); these weak bands are from spin-forbidden transitions to the excited singlet terms. The two relatively stronger bands are due to spin-allowed transitions. If the band at $17\,200\,cm^{-1}$ is assigned to $^3T_1(F) \rightarrow {}^3T_2(F)$ transition, then it may be seen from Figure 9.11 (b) that the transition $^3T_1(F) \rightarrow {}^3A_2(F)$ should lie at about $38\,000\,cm^{-1}$, and evidently cannot be at $25\,600\,cm^{-1}$. $^3T_1(P)$ is the only other term which is spin allowed but involves a

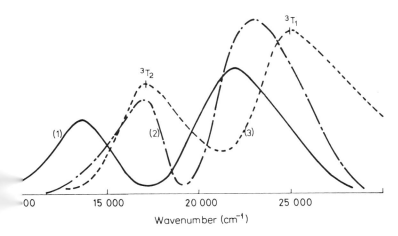

Fig. 9.13 Optical absorption of vanadium (III) in:
(1) $2 Na_2O, 5 B_2O_3, 3 P_2O_5$ glass
(2) $2 Na_2O, 7 B_2O_3, P_2O_5$ glass
(3) 1 molar perchloric acid.

n two-electron jump, and may lie below the $^3A_2(F)$ term. Figgis [6] has
the $^3T_1(F) \rightarrow {}^3T_1(P)$ transition to the band at $25\,600\,cm^{-1}$. On the
these two transitions the band corresponding to $^3T_1(F) \rightarrow {}^3A_2(F)$
n should be at $36\,000\,cm^{-1}$. Although no band has been found for V^{3+}
us solution in this region, V^{3+} in Al_2O_3 has been found to show a weak
ound $36\,000\,cm^{-1}$.
ss V^{3+} shows two bands around $15\,000\,cm^{-1}$ and $23\,000\,cm^{-1}$ and in
to aqueous solutions may be assigned to $^3T_1(F) \rightarrow {}^3T_2(F)$ and $^3T_1(F)$
P) transitions respectively.

^{3} *system. Absorption of Cr^{3+} in glass*
ergy level diagram for a d^3 ion in octahedral symmetry is shown in
9.11(c). Due to very large ligand field stabilization energy Cr^{3+} exists only
hedral complex in conventional glasses. From the energy level diagram the
ng transitions are expected:

Spin allowed	*Spin forbidden*
$^4A_2 \rightarrow {}^4T_2$	$^4A_2 \rightarrow {}^2E$
$\rightarrow {}^4T_1(F)$	$\rightarrow {}^2T_1$
$\rightarrow {}^4T_1(P)^*$	$\rightarrow {}^2T_2$
	$\rightarrow {}^2A_1$

cates a two-electron jump

bsorption spectra of Cr^{3+} in a $Na_2O, 2P_2O_5$ and in a $Na_2O, 2B_2O_3$
are shown in Figure 9.14. Soda-phosphate glass is chosen, for in this glass
is the only stable oxidation state, and glasses can easily be made without any

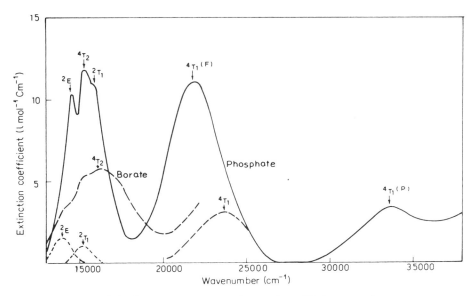

Fig. 9.14 Optical absorption of chromium (III) in some glasses.

trace of Cr^{6+} in them. In Figure 9.14 the assigned transitions are labelled with the bands. The excited states $^4T_2$, $^4T_1(F)$ and $^4T_1(P)$ slope sharply upward from the ground state and thus all the main absorption bands are broad. The spin forbidden states ^2E and $^2T_1$ are almost parallel to the ground state and thus the corresponding bands can be seen to be quite narrow in the Figure 7.14. For further details on chromium (III) absorption in solution and in glass, see the excellent paper by Bates and Douglas [7].

(*d*) *3d⁴ system. Absorption of Mn³⁺ in glass*

(d) 3d⁴ system. Absorption of Mn³⁺ in glass

The energy level diagram for a d^4 ion in octahedral ligand field is shown in Figure 9.11(d). Figure 9.15 contains the absorption spectra of Mn^{3+} in two $Rb_2O-B_2O_3$ glasses. The main absorption band around $20\,000\,cm^{-1}$ originates from $^5E \rightarrow {}^5T_2$ transition, the other bands hidden under this main absorption band are probably due to the splitting of the ground state due to Jahn–Teller distortion [8].

(*e*) *3d⁵ system. Absorption of Mn²⁺ in glass*

The energy level diagram for a d^5 ion in either octahedral or tetrahedral ligand field is shown in Figure 9.11(e) and the absorption spectrum of Mn^{2+} in a phosphate and in a silicate glass is shown in Figure 9.16. The absorption spectrum of Mn^{2+} provide a good illustration of the usefulness of ligand field theory to interpret the absorption spectra of transition-metal ions. In Figure 9.11(e) the first two excited states $^4T_1(G)$ and $^4T_2(G)$ sharply change their energy with ligand field, and this corresponds to the two broad bands around $17\,000$ and $22\,000\,cm^{-1}$.

310

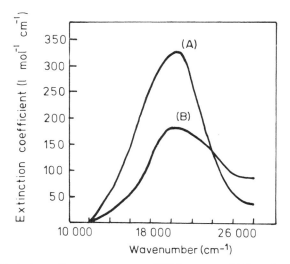

15 Optical absorption of manganese(III) in some rubidium–borate glasses.
(A) 30.3 mol % Rb_2O
(B) 20.0 mol % Rb_2O

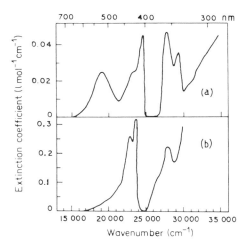

.16 (a) Optical absorption spectrum of manganese(II) in a $3K_2O$, $7P_2O_5$ glass.
(b) Optical absorption spectrum of manganese(II) in a $3K_2O$, $7SiO_2$ glass.

xcited states ^4E and $^4A_1$ are degenerate and parallel to the ground state
·espond to a sharp band at $25\,000$ cm^{-1}. The sharp bands around 27 800
400 cm^{-1} correspond to the excited levels $^4T_2(D)$ and $^4E(D)$. The
ion spectrum of the tetrahedral Mn^{2+} in a potassium silicate glass shows

311

six absorption bands in two groups of three. Just as there are in octahedral symmetry, but they are here much closer together. This is to be expected since the 10 Dq value for the tetrahedral complex is only 4/9 of the octahedral. The most intense and sharp band at lower energy is due to the field independent ^4E and $^4A_1$ levels [27].

(f) 3d^6 system. Absorption of Fe^{2+} in glass
The energy level diagram for a d^6 system in octahedral ligand field is shown in Figure 9.11(f), and the absorption spectrum of Fe^{2+} in aqueous solution and in different phosphate glasses is shown in Figure 9.17. The main broad absorption band around 9000 cm^{-1} is due to the $^5T_2 \rightarrow {}^5E$ transition; the shoulder around 5000 cm^{-1} is due to distortion splitting [28].

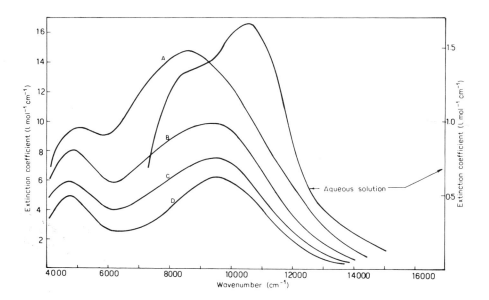

Fig. 9.17 Optical absorption spectra of iron (II) in aqueous solution and in some binary alkaline–earth–phosphate glasses.
(A) MgO–P$_2$O$_5$
(B) CaO–P$_2$O$_5$
(C) SrO–P$_2$O$_5$
(D) BaO–P$_2$O$_5$.

(g) 3d^7 system. Absorption of Co^{2+} in glass
The energy level diagram for a d^7 ion in octahedral symmetry is shown in Figure 9.11(g) (for tetrahedral energy level diagram consult Figure 9.11(c). Absorption spectra of Co^{2+} in two different Na$_2$O–B$_2$O$_3$ glasses is shown in

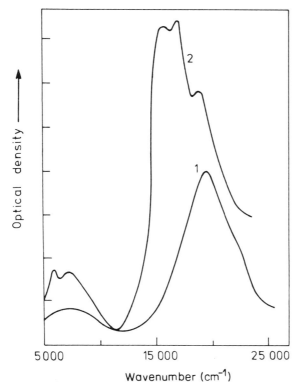

9.18 Optical absorption spectra of cobalt (II) in binary sodium–borate glasses.
(1) 10 mol % Na_2O (2) 30 mol % Na_2O.

: 9.18. From the energy level diagram, octahedral Co^{2+} is expected to have
spin-allowed transitions:

$$^4T_1(F) \underline{\qquad} ^4T_2$$
$$\underline{\qquad} ^4T_1(P)$$
$$\underline{\qquad} ^4A_2$$

.rve 1 of Figure 9.18 where Co^{2+} is in octahedral symmetry, the main band is
.o the $^4T_1(F) \rightarrow ^4T_1(P)$ transition, the high energy shoulder is a consequence
▪in-orbit coupling in the $^4T_1(P)$ state. The $^4T_1(F) \rightarrow ^4T_2$ transition occurs
nd $8000\,cm^{-1}$. Using these two assignments the $^4T_1(F) \rightarrow ^4A_2$ transition
calculated by Cotton and Wilkinson [9] to be around $18\,000\,cm^{-1}$. However,
transition is expected to be weaker because it involves the forbidden two-
▪tron jump. This weakness combined with the closeness of the $^4T_1(F)$
$T_1(P)$ band results in the $^4T_1(F) \rightarrow ^4A_2$ transition being unresolved.
he main absorption band in tetrahedral Co^{2+} (curve 2 of Figure 9.18) is due
the $^4A_2 \rightarrow ^4T_1(P)$ transition, although two other absorption bands cor-
▪ponding to $^4A_2 \rightarrow ^4T_2$ and $^4A_2 \rightarrow ^4T_1(F)$ occur in the infrared. The splitting

313

of the $^4A_2 \rightarrow {}^4T_1$ (P) band is caused by spin-orbit coupling which both splits the $^4T_1$ (P) state and allows the transitions to the neighbouring doublet states to gain in intensity.

(h) $3d^8$ system. Absorption of Ni^{2+} in glass

The energy level diagram for a d^8 ion in octahedral symmetry is shown in Figure 9.11 (h). This diagram predicts three spin-allowed transitions, and these were observed in water and ammonia solutions as shown in Figure 9.19. Spin-orbit coupling, which mixes the $^3T_1$ (F) and ^1E states, splits the middle band in the $[Ni(H_2O)_6]^{2+}$ complex, but in a stronger ligand field with ammonia these levels are too far away to produce any significant mixing.

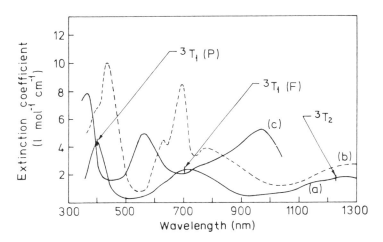

Fig. 9.19 Optical absorption of nickel(II)
(a) in water
(b) in concentrated HCl
(c) in 80% ammonia.

The spectra of low-alkali–borate glasses containing nickel (Figure 9.20, Curve 1) is similar to that of $[Ni(H_2O)_6]^{2+}$ and may be interpreted in terms of ligand field theory in an analogous manner. When the R_2O concentration is gradually increased beyond 20 mol %, newer bands characteristic of Ni^{2+} in square planar and tetrahedral symmetries appear [29].

The absorption band at about $15\,400\,cm^{-1}$ (650 nm) of Ni^{2+} in tetrahedral complex (curve 3 of Figure 9.20) is assigned to the $^3T_1$ (F) $\rightarrow {}^3T_1$ (P) transition, and the one at about $7700\,cm^{-1}$ to the $^3T_1$ (F) $\rightarrow {}^3A_2$ transition.

Another band corresponding to $^3T_1$ (F) $\rightarrow {}^3T_2$ is expected in the infrared region, and indeed, a band at about $5000\,cm^{-1}$ has been observed in potash-rich borate glasses [10]. Spin-orbit coupling which removes the degeneracy of the $^3T_1$ (P) state splits the visible band.

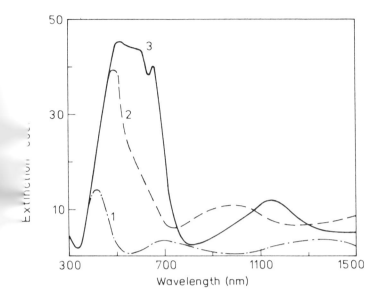

.20 Optical absorption of nickel (II) in some binary potassium–borate glasses.
(1) 16.2 mol % K_2O
(2) 25.3 mol % K_2O
(3) 29.8 mol % K_2O

[9] *system. Absorption of Cu^{2+} in glass*
 produces blue–green colour in glass. The absorption spectrum of Cu^{2+} in
 alkali borate glasses is shown in Figure 9.21. The absorption spectrum
 s of a broad band around 12 500 cm^{-1} and is attributed to the transition
 $^2T_2$. The band is asymmetric on account of splitting by a low symmetry
 field component.

9.5   CHARGE-TRANSFER BANDS

oportion to the vast amount of experimental and theoretical work devoted
easuring and analysing Laporte-forbidden d–d transitions of transition
 ions in glass over the last 30 years, the Laporte-allowed charge-transfer
 s have been rather neglected. At the moment there is no commonly accepted
retical scheme for analysing such spectra and extracting theoretical para-
ers from them in the way which has become so common in ligand-field
troscopy; there is not a single charge-transfer spectrum of any metal complex
se detailed assignment, band by band, is universally agreed upon. For this
e are a number of reasons, both technical and theoretical: the bands are nearly
ays intense, and quite frequently rather broad; and they sometimes occur far
ugh into the ultraviolet to become entangled with the fundamental absorp-
n edge of the glass. Theoretical analyses encounter the difficulty that one can

315

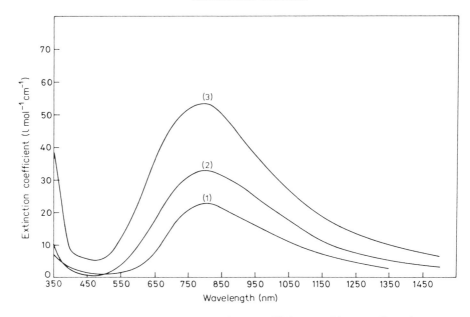

Fig. 9.21 Optical absorption spectra of copper (II) in some binary sodium–borate glasses.
(1) 5.6 mol % Na_2O
(2) 20.2 mol % Na_2O
(3) 34.9 mol % Na_2O

no longer use a perturbation approach based on the assumption that both ground and excited orbitals are largely localized on the same atomic centre. However, if correctly analysed, these spectra contain a wealth of information which in some ways is more relevant to the structure of glass than that furnished by intrasubshell transitions because it concerns orbitals of both metal and ligand character, some of which are deeply engaged in molecular bonding.

In the molecular orbital sense, electron transitions between electronic levels of different atoms may always take place, although sometimes very high energy is required. For simplicity and relevance to glass technology, we shall restrict ourselves to charge-transfer transitions occurring with energies comparable to those of the visible and near-ultraviolet region of the electromagnetic spectrum. For a transition to give rise to visible colour there must exist an energy separation between the relevant states of some $14\,000$–$30\,000$ cm^{-1}. The relative energies of the highest field and lowest empty orbitals of an atom may be qualitatively discussed in terms of its ionization potential and electron affinity, respectively. Thus, atoms or ions which have low ionization potentials, i.e. are readily oxidizable, have filled orbitals of relatively high energy. Conversely, atoms or ions which have high electron affinities, i.e. are readily reduced, have relatively low-lying empty orbitals. If we form a 'complex', for example, between a metal which is readily oxidizable, and a ligand which is readily reducible, we can expect the

316

)aration between the highest filled level on the metal and the lowest
:l on the ligand to be relatively small. If the separation between these
o small, say less than 5000–$10\,000\,cm^{-1}$, then commonly there will be a
ron transfer between the species, resulting in oxidation of the metal and
of the ligand. Provided the energy separation is great enough for the
be stable, we shall see a charge-transfer absorption at relatively low
corresponding to a transition, which may be schematically written as

$$M^{n+} - L \rightarrow M^{(n+1)+} - L^{-}$$

nsitions are called 'metal to ligand' charge-transfer. Conversely, if we
a readily reducible metal with a readily oxidizable ligand, we can expect
gand to metal' charge transfer, which may be written as

$$M^{n+} - L^{-} \rightarrow M^{(n-1)+} - L$$

at these transitions do not involve the complete transfer of an electron
e atom to another; rather, in a molecular orbital sense, they represent the
on of an electron from a molecular orbital primarily located on one atom
lecular orbital primarily located on another atom.

Ligand to metal charge-transfer

a expect to see such transitions in or close to the visible region if we bind
ible ligands to reducible metals i.e. metals in the higher oxidation states.
ore readily oxidizable ligands are those with low ionization potentials, such
ide, I^{-}, and sulphide, S^{2-} in particular. In a periodic group such as, for
le, the halides we can expect that with a given metal, the ligand to metal
e-transfer band will decrease in energy in the order $F^{-} > Cl^{-} > Br^{-} > I^{-}$,
ding to the ease of oxidation of these ions. Electron-transfer spectra of some
)-tetrahalides are shown in Figure 9.22.

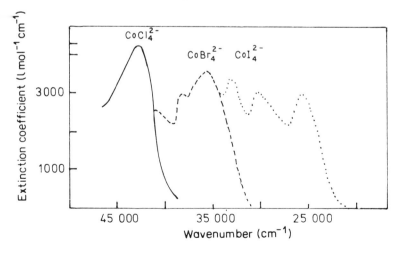

Fig. 9.22 The charge transfer spectra of the anions $CoCl_4^{2-}$, $CoBr_4^{2-}$ and CoI_4^{2-}

In a series of species such as ML_2, ML_4, ML_6, etc., increasing the number of ligands donating charge to the central metal atom is expected to destabilize the metal orbitals, by virtue of a spherical repulsion term between the metal and ligand electron density. One would then expect that the ligand to metal charge-transfer band would decrease in energy as the coordination number is decreased. In general this has been found to be true in simple coordination compounds of some transition-metals of the first series [30].

There is not much information concerning the variation of charge-transfer energies with stereochemistry, maintaining the coordination number constant. The square and tetrahedral $[CuCl_4]^{2-}$ complexes have essentially the same charge-transfer spectra. However, this may be viewed as a coincidence since the acceptor-orbital energies on the copper atom in the two cases are somewhat different.

In discussing charge-transfer transitions, we have to ask ourselves how close do the oxidizing and reducing species have to approach each other? For an electron transition to occur, there must be some non-zero overlap between the ground and the excited wave functions. If the species are too far apart, the transition probability (intensity) approaches zero. The charge-transfer complexes in glasses reported so far are all inner-sphere, in that a direct bond (molecular orbital) between the metal and the ligand of interest is formed. However, it is possible to see transitions in outer-sphere complexes, such as ion pairs, where there is no direct bond between the metal and ligand.

Species of the type $[ML_6]^{n+}$ nX^- have been studied and $X \to M$ charge-transfer bands have been reported both in crystalline materials and in solution. These have energies which may be higher or lower than the corresponding inner coordination charge-transfer band, but are invariably weaker [11]. Such transitions, however, have never been reported in glass.

9.5.2 Metal to ligand charge transfer

In general, the principles outlined above apply in reverse to metal to ligand charge transfer. In this case, for an homologous series of complexes, the charge-transfer band will:

(a) move to lower energy as the oxidation state of the metal decreases and as the ligand becomes more electronegative.
(b) move to higher energy as the coordination number decreases.

A relationship between the energy of such a metal to ligand transition and the affinity of the ligand to accept an electron (as indicated, for example, by its electrode-potential for reduction) would be anticipated, and has been reported in solution [12]. Ligands with empty orbitals of appropriate energy and symmetry to give rise to such charge-transfers close to the visible region, are much less common than are those which give rise to ligand to metal charge-transfer. In consequence metal to ligand charge transfer is less well defined, particularly in glass.

As stated earlier, electronically allowed charge-transfer bands are generally very intense, with molar extinction coefficients ranging from $10^3-10^5 \, l \, mol^{-1} \, cm^{-1}$. Since the metal d orbital which 'accepts' the electron is even in character, it follows

onor orbital primarily on the ligand must have odd parity. In an
complex such as ML_6, if L is a species such as ammonia with only one
lectron pair used for σ-bonding, then there will be a set of σ-bonding
symmetry A_{1g}, E_g and T_{1u}. Only the last of these can be the source of a
.nsfer band. If the ligand has several available pairs, such as halides or
then in addition to this σ-bonding set, there will also be a π-bonding
ymmetry T_{1g}, T_{2g}, T_{1u} and T_{2u}. The last two levels are sources of
.nsfer transitions but usually lie very close together. Separate trans-
.m these two levels can only sometimes be identified.
:eptor orbitals on the metal are the T_{2g} and E_g orbitals. Thus in general,
:s of ligand to metal charge-transfer transitions are possible. Qualit-
. order of ascending energy, they are:

(a) $\pi \rightarrow T_{2g}$
(b) $\pi \rightarrow E_g$ }
(c) $\sigma \rightarrow T_{2g}$ } close
(d) $\sigma \rightarrow E_g$

.rst of these transitions, (a), involves charge-transfer from one essentially
.nding molecular orbital to another essentially non-bonding molecular
Under such circumstances, the equilibrium internuclear distances in the
state are effectively the same as in the ground state, such conditions giving
.arrow and, in this case, relatively weak absorption bands. The remaining
.ransitions all involve a bonding or anti-bonding orbital as the donor or
)r. Bands (b) and (d) are generally broad and strong, while band (c), for
:s of poor overlap, is weak and rarely observed.
.oretically, with a metal ion in which there are fewer than six electrons in the
bitals, all four transitions can be observed. However, if the T_{2g} set is filled
.onfiguration), transitions (a) and (c) will not occur. transition (d) generally
; at very high energies. Similar comments may be made about complexes
other symmetries. Charge-transfer bands often exhibit considerable
.ure; this is particularly true for the heavier ligands and the heavier metals
: spin-orbit coupling (which can lift the degeneracy of the otherwise
.ally degenerate states) is important (see Figure 9.22). It also should be noted
.. transition such as (a), which corresponds to $T_{1u} \rightarrow T_{2g}$, is really a group of
.itions (since $T_{2g} \times T_{1u} = A_{2u} + E_u + T_{1u} + T_{2u}$) which usually lie close together.
)wever, the relative simplicity of the spectra of complexes involving only the
.er atoms suggests that spin-orbit coupling rather than inter-electronic
.lsion is primarily responsible for band structure.
.o calculate the energy of charge-transfer transition, it is necessary to make
.ections for any changes in inter-electronic repulsion which may arise as a
.lt of 'transferring' the electron from the ligand to the central ion. This may be
.e by considering the difference in the spin pairing energy of the d level before
. after the addition of one electron, that is, for the configuration l^q and l^{q+1}.
: spin pairing energy, E, is given by:

$$E = \left[\frac{3}{4} q \left(1 - \frac{q+1}{4l+1} \right) - S(s+1) \right] D$$

.ere S is the total spin and D is the 'spin pairing parameter' which for d electrons

319

equals $\dfrac{7}{6}\left[\dfrac{5B}{2}+C\right] = 7B$ approximately (B and C are Racah parameters). For the d electrons, $l = 2$, and depending upon the configuration S can take the value from 0 to 5/2.

If one records the energy of the onset of charge transfer in a series of complexes MF_6^{n-}, MCl_6^{n-}, and MI_6^{n-} the difference in these energies for any pair of halogens, with constant n, has been found to be essentially independent of the metal, M. This led Jorgensen [13] to postulate the concept of 'optical electronegativity'. A donor or acceptor orbital on a metal or ligand was assigned an optical electronegativity which was, incidentally, connected by an arbitrary constant to the Pauling electronegativity scale. Using the difference in the optical electronegativities of the donor and acceptor orbitals and a correction, where necessary, for changes in the mean inter-electronic repulsion energy, it is possible to reproduce the observed charge-transfer energies with reasonable accuracy. On the other hand, by using observed charge-transfer energies it is possible to map optical electronegativity values for glasses of different compositions.

In oxide glasses some charge transfer bands occur in the near-ultraviolet region. Due to their very high intensity and broad nature, charge transfer bands are very important in making ultraviolet absorbing or ultraviolet transmitting glasses.

9.5.3 Charge-transfer bands due to iron in glass

Iron usually occurs in both the ferrous and ferric states in glass. Both Fe^{3+} and Fe^{2+} have been reported to have strong charge-transfer bands in the near-ultraviolet region [14]. However, the nature of the ferrous charge-transfer band in the near-ultraviolet has not yet been carefully examined. Ultraviolet absorption of Fe^{3+} in some binary alkali–silicate, alkaline-earth–phosphate glasses, and in aqueous solution are shown in Figure 9.23. In silicate glasses Fe^{3+} is in four-fold coordination, whereas in borate and phosphate glasses Fe^{3+} may occur in four-fold as well as in six-fold coordination, depending upon the composition of the glass. Correspondingly the charge-transfer band of Fe^{3+} in phosphate, and particularly in borate glasses changes markedly with the glass composition.

The ultraviolet absorption of Fe^{3+} in glass is characterized by a steep absorption edge which in some cases, such as in low-alkali–borates, extends into the visible region. This edge has been shown to be part of the strong charge-transfer band, and is found to obey Urbach's Rule, indicating that this low-energy absorption may be due to some photon–photon interaction.

The extinction coefficient of Fe^{3+} in glass varies considerably from system to system. In alkali–borates, on changing from 10 to 30 mol % Na_2O, the intensity of the charge-transfer band increases by a factor of almost two and the width decreases correspondingly.

In borate and silicate glasses, Fe^{2+} contributes to the absorption in the ultraviolet, but with a much lower intensity than that of Fe^{3+} in this region. Probably the peak of the Fe^{2+} charge-transfer band lies further down the ultraviolet region (higher energy), as would be expected from the lower oxidation state of Fe^{2+} relative to Fe^{3+}.

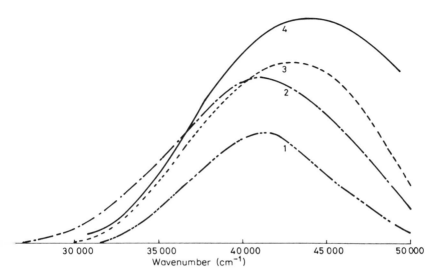

30 000   35 000   40 000   45 000   50 000
Wavenumber (cm^{-1})

Fig. 9.23 Ultraviolet absorption of iron (III) in:
(1) 4.6 molar perchloric acid
(2) $MgO-P_2O_5$ glass
(3) $CaO-P_2O_5$ glass
(4) Na_2O-SiO_2 glass

Change-transfer bands of Cr^{6+} in glasses

nium usually occurs as Cr^{3+} and Cr^{6+} in oxide glasses, and of these, Cr^{3+}
ot produce any near-ultraviolet charge-transfer band. Ultraviolet absorp-
f Cr^{6+} in some alkali–borate glasses is shown in Figure 8.4. Two charge-
er absorption bands around 27 000 and 39 000 cm^{-1} are obtained, and
been assigned to $T_1 \rightarrow {}^3T_2$ and ${}^2T_2 \rightarrow {}^3T_2$ transitions respectively by
sberg and Helmholz [15], whereas the first transition has been assigned to
${}^2E$ by Liehr and Ballhausen [16].
described in Chapter 8, the absorption spectra of Cr^{6+} in higher
i–borate and alkali–silicate glasses are similar to those in alkaline aqueous
ions whereas in low-alkali–borate glasses (less than about 20 mol % alkali
e) the absorption spectra of Cr^{6+} closely resemble those in acidic aqueous
ions. Thus, Cr^{6+} absorption spectra can be used as indicators to study the
–base property of glass.
he other important transition-metal ions commercially used to produce near-
aviolet charge-transfer absorption in oxide glasses are Ti^{4+}, V^{5+}, Cu^{2+} and
+.

321

Out of all these transition-metal ions, the charge-transfer band produced by Cr^{6+} in glass lies at the lowest energy. Thus Cr^{6+} in glass completely cuts off all the near-ultraviolet radiation and is extensively used to produce ultraviolet-absorbing commercial glasses. One of the cheapest sources of chromium is chromite, $FeO \cdot Cr_2O_3$; when this is used, in addition to the Cr^{6+} absorption, some extra ultraviolet absorption due to Fe^{3+} is also obtained.

Up to now we have been discussing charge-transfer bands where an electron in a molecular orbital, predominantly ligand in nature, is excited to an orbital predominantly metal in character. In addition, charge-transfer absorption bands resulting from the interaction of two valence states of the same or different metal ions in glass sometimes may occur. These are known as *cooperative charge-transfer bands* and their occurrence in crystalline materials is well known. For example, broad strong absorption bands around $9000 \, cm^{-1}$ in yttrium iron garnet doped with ferrous impurity, has been ascribed to charge-transfer within the $Fe^{2+}-O-Fe^{3+}$ group [17]. Formation of a strong yellow colour by melting an oxide glass with a mixture of CeO_2 and TiO_2 is well known. Some typical absorption spectra of these glasses are shown in Figure 9.24 where the broad band around $29\,000 \, cm^{-1}$ is a cooperative charge-transfer band due to the $Ce^{3+}-O-Ti^{4+}$ centre. In Figure 9.24, curves 1 and 3 represent optical absorption of a $Na_2O-B_2O_3-Al_2O_3-SiO_2$ glass doped with $1.0 \, wt\%$ CeO_2 and $1.5 \, wt\%$ TiO_2 respectively; curve 2 represents the optical absorption of the same base glass

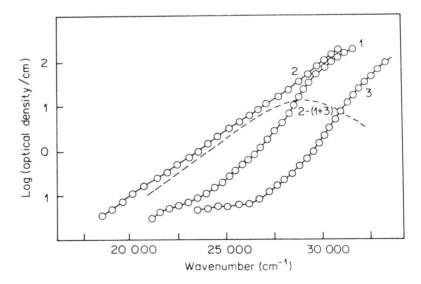

Fig. 9.24 Optical absorption of glasses and their differential absorption.
Glass (1) $12 \, Na_2O$, $20 \, B_2O_3$, $3 \, Al_2O_3$, $65 \, SiO_2$ glass $+ 1 \, wt.\%$ CeO_2
Glass (2) $12 \, Na_2O$, $20 \, B_2O_3$, $3 \, Al_2O_3$, $65 \, SiO_2$ glass $+ 1 \, wt.\%$ CeO_2
$+ 1.5 \, wt.\%$ TiO_2
Glass (3) $12 \, Na_2O$, $20 \, B_2O_3$, $3 \, Al_2O_3$, $65 \, SiO_2$ glass $+ 1.5 \, wt.\%$ TiO_2

ı a mixture of 1.0 wt % CeO_2 and 1.5 wt % TiO_2. The dotted curve with ıximum around 29 000 cm^{-1} is the differential absorption curve, and is ıperative charge-transfer absorption in the group $Ce^{3+}-O-Ti^{4+}$.

9.6 ANIONIC SUBSTITUTION IN GLASS

ʿide glasses transition-metal ions form coordination complexes with ligands. These oxygens may be of different donor capacity (depending ıs composition) and thus produce complexes of different symmetries. If ıelt contains ionic species other than oxides, such as halides, sulphides, ʿtc. coordination complexes can be formed with anionic substitution. t of this type of substitution will depend upon the activities of the ng species in the melt, the temperature of the melt, and the equilibrium free energy change) of the respective reactions. As a consequence of this ıstitution, the absorption spectra will change and the colour of the glass ʿred significantly. This type of ligand substitution reaction in aqueous ; well known. For example, when hydrochloric acid is added to aqueous sulphate solution, the pink colour of the solution changes to deep blue ırmation of $(CoCl_4)^{2-}$ species. Figure 9.25 shows cobalt (II) absorption ı a series of H_2O-HCl mixtures. A very similar change of absorp-

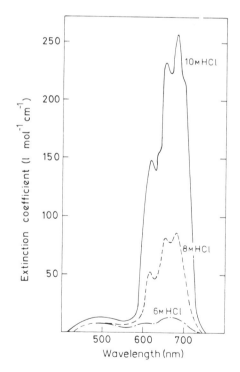

Fig. 9.25 Optical absorption of cobalt(II) in HCl–H$_2$O mixtures.

323

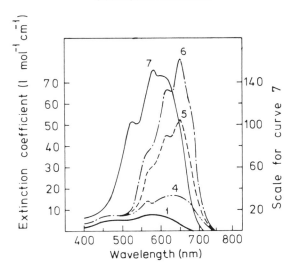

Fig. 9.26 Optical absorption of cobalt(II) in 11 Na$_2$O, 89 B$_2$O$_3$ glasses containing chloride.

(1) 0.00 g ion chloride/l glass
(4) 0.0114 g ion chloride/l glass
(5) 0.0252 g ion chloride/l glass
(6) 0.0317 g ion chloride/l glass
(7) Na$_2$O, 2B$_2$O$_3$ glass.

tion takes place when NaCl is added to some Na$_2$O–B$_2$O$_3$ glasses containing cobalt (II); some typical results are shown in Figure 9.26.

As is evident from the optical absorption, cobalt (II) assumes octahedral and tetrahedral symmetries in water and concentrated HCl respectively. When hydrochloric acid is added to a pink cobalt sulphate solution in water, the following equilibria are involved:

$$[Co(H_2O)_6]^{2+} + nHCl \rightleftharpoons [Co(H_2O)_{6-n}(Cl)_n]^{(n-2)-} + nH_3O^+ \quad (9.10)$$

$$\rightleftharpoons [Co(H_2O)_{4-m}(Cl)_m]^{(m-2)-} + mH_3O^+$$
$$+ 2H_2O \quad (9.11)$$

$$\rightleftharpoons [CoCl_4]^{2-} + 4H_3O^+ + 2H_2O \quad (9.12)$$

Formation of the species given in equations (9.11) and (9.12) gives rise to new absorption bands of much greater intensity and with band maxima at lower energies than those of the complexes with only oxygen coordination, on account of the relative weakness of the chloride ligand compared to water in the spectrochemical series. Figure 9.25 thus demonstrates the transformation of octahedral cobalt (II) to the tetrahedral symmetry. In aqueous solution when the chloride/cobalt ratio is made quite large, almost all the cobalt (II) is expected to be present as [CoCl$_4$]$^{2-}$. Thus the overall equilibrium may be simply written as:

$$[Co(H_2O)_6]^{2+} + 4Cl^- \rightleftharpoons [CoCl_4]^{2-} + 6H_2O \quad (9.13)$$

te solution in water the equilibrium constant, K, may be written as:

$$K = \frac{[[CoCl_4]^{2-}]}{[[Co(H_2O)_6]^{2+}]} \cdot \frac{1}{[Cl^-]^4}$$

] indicates activity of the species. Changing this to the concentration
m constant, K_c:

$$K_c = \frac{([CoCl_4]^{2-})}{([Co(H_2O)_6]^{2+})} \cdot \frac{1}{(Cl^-)^4}$$

) indicates concentration.

g $([CoCl_4]^{2-}) - \log([Co(H_2O)_6]^{2+}) = \log K_c + 4\log(Cl^-)$ (9.14)

chloride concentration is sufficient to convert only a small amount of
) from the octahedral to the tetrahedral configuration, equation (9.14)
vritten:

$$\log([CoCl_4]^{2-}) = \log K_c + \log T + 4\log(Cl^-) \qquad (9.15)$$

is the concentration of total cobalt (II) in the solution. If the chloride
ation is increased so that an appreciable part of the cobalt (II) is
d to the tetrahedral form, equation (9.15) will no longer hold good.
tinction coefficient of octahedral cobalt (II) in water is approximately 5
cm^{-1} at the peak of the absorbance curve while the extinction coefficient
edral cobalt (II) is approximately $550\,l\,mol^{-1}\,cm^{-1}$. Unfortunately the
on envelopes in these two symmetries overlap considerably. However, as
the concentration of the tetrahedral complex becomes comparable with
he octahedral complex, the area under the absorbance curve may be taken
oximately measuring the concentration of the tetrahedral complex (see
8). Thus, over an appropriate concentration range, equation (9.15)
that the logarithm of the area under the absorbance curve may be
d to give a straight line when plotted against the logarithm of the
ration of chloride. This logarithmic plot is found to be linear, from HCl
ration of 6 M (equivalent to log Cl$^-$ = 0.778) to above 10 M HCl, with a
4 as predicted by equation (9.15). As expected the linearity disappears at
concentrations of chloride.
gure 9.27 similar logarithmic plots are given for different Na_2O–B_2O_3
From these plots it may be seen that in the glasses which contain 20 mol %
Na_2O, slopes of approximately unity are obtained. Therefore, in soda-
xide glasses containing 20 per cent or less Na_2O on average, each
dral cobalt (II) combines with one chloride ion to produce a tetrahedral
II). In the more basic glasses containing 25 and 33 mol % Na_2O, there is no
of absorbance area with chloride concentration in the glass. As the
II) is in tetrahedral coordination in these high-soda glasses before any
e is added, the constancy of the absorbance area alone does not prove that
ride enters the complex. However, there is also very little evidence of any
f the wavelength of maximum absorbance; thus it may be concluded that
very little, chloride enters the tetrahedral complex.

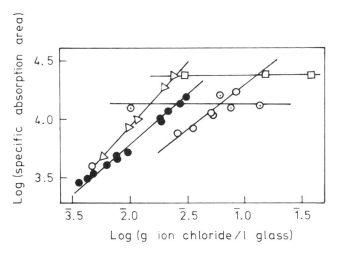

Fig. 9.27 Log (specific absorption area) of cobalt (II) in Na_2O–$NaCl$–B_2O_3 glasses plotted against log (g ion chloride/l glass).
  □ - Na_2O, $2\,B_2O_3$
  ⊙ - Na_2O, $3\,B_2O_3$
  ○ - Na_2O, $4\,B_2O_3$
  ● - $16\,Na_2O$, $84\,B_2O_3$
  △ - $11\,Na_2O$, $89\,B_2O_3$

From previous discussions it is evident that the formation of tetrahedral cobalt (II) with mixed oxide and chloride ligands in a melt will depend upon the activities of NaCl and cobalt (II) in the melt. There is a two-liquid region in the Na_2O–$NaCl$–B_2O_3 phase diagram. This diagram suggests that the activity of NaCl decreases as the soda content increases and, therefore, it would be wrong to ascribe the lack of reaction of chloride and the oxide-tetrahedral complex entirely to the state of binding of the cobalt (II) to its ligands. Further investigation is needed to resolve this point.

In low-alkali–borate glasses similar changes of coordination around cobalt (II) can be achieved by adding NaBr to the melt. Some typical results are shown in Figure 9.28. Addition of bromide intensifies the cobalt (II) absorption and modifies the absorption spectra considerably. There are two aspects of this modification:

(a) With increasing amounts of bromide, newer bands are developed around 15 500 and 6000 cm^{-1} in addition to bands around 18 250, 16 750 and 7400 cm^{-1} for the bromide-free glass (glass-A). These new bands are characteristic of the tetrahedral cobalt (II) complex which is also known to occur in high-alkali–soda–borate glasses, and in acidic Na_2O–$NaCl$–B_2O_3 glasses. However, the tetrahedral cobalt (II) absorption bands in glasses containing bromide are shifted to lower energies in comparison with oxide and oxide–chloride complexes. This is consistent with the spectrochemical series where oxide > chloride > bromide.

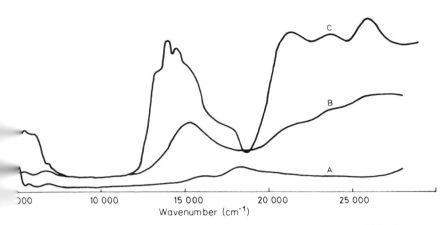

9.28 Optical absorption spectra of cobalt(II) and cobalt(III) in $(8-x)Na_2O$, $2x\,NaBr$, $92\,B_2O_3$ glasses.
(A) 0.000 mol NaBr/l glass
(B) 0.042 mol NaBr/l glass
(C) 0.179 mol NaBr/l glass

set of three new intense bands are developed around 21 500, 23 500 and 5 500 cm^{-1}; the intensity of all these bands increases similarly with increasing bromide content of the glass. These bands are due to cobalt(III) and it will be shown shortly that the cobalt(III) complex is most probably trans-$[CoO_4Br_2]^{m-}$. It should be pointed out that tetrahedral cobalt(II) complexes may also have absorption bands around 22 000 cm^{-1} corresponding to quartet → doublet transitions, but these bands being spin-forbidden are expected to be of low intensity and are unlikely to be those reported in Figure 9.28.

ss A is pink; the absorption spectrum being characteristic of octahedral alt(II). As NaBr is added, from glass B to C, two new major bands are eloped at 5875 and 14 750 cm^{-1}. These two bands may be assigned to $^4A_2$ (F) $^4T_1$ (F) and $^4A_2$ (F) → $^4T_1$ (P) transitions respectively. The other spin-allowed nsition $4A_2$ (F) → $^4T_2$ (F) could not be identified in the spectra due to very ong-OH absorption at 5000 cm^{-1} and lower energies.

The qualitative energy level diagram for cobalt (III), a $3d^6$ system, is shown in ure 9.11 (f) for octahedral symmetry. With cobalt (III) the $^1A_{1g}$ state originating in one of the high energy singlet states of the free ion drops very rapidly and osses the $^5T_{2g}$ state at a very low value of 10 Dq. Thus almost all cobalt(III) mplexes have diamagnetic ground states, and the situation is given by Figure 29 (b). The absorption spectra of the diamagnetic, octahedrally coordinated obalt(III) complex is made up of two strong bands corresponding

327

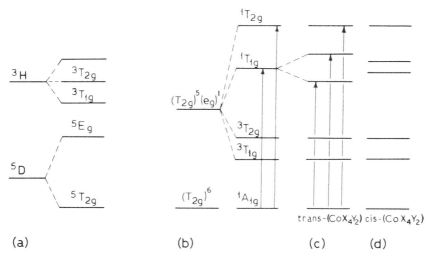

Fig. 9.29 Splitting scheme for a d^6 system in an octahedral field.
    (a) weak field
    (b) strong field–regular octahedron (CoX_6)
    (c) strong field – *trans*-(CoX_4Y_2)
    (d) strong field – *cis*-(CoX_4Y_2).

to the $^1A_{1g} \to {}^1T_{1g}$ and $^1A_{1g} \to {}^1T_{2g}$ transitions. In addition, some singlet → triplet bands could appear.

The substituted cobalt(III) complexes are very stable. When two of the six ligands are replaced by different ones, the lowering of the octahedral symmetry causes a splitting of the two upper states as shown in Figure 9.29 (c) and (d). It has been shown theoretically that the $^1T_{2g}$ state is not split observably, whereas the splitting of the $^1T_{1g}$ state should be at least twice as great for the *trans*-complex as for the *cis*-form [18]. Moreover, because the *cis*-form lacks a centre of symmetry, it may be expected to have a higher intensity of absorption. These simple predictions are nicely borne out in practice as can be seen in Figure 9.30 where the absorption spectra of *cis*- and *trans*-isomers of $[Co(en)_2F_2]^{1+}$ are shown.

In Figure 9.28 the absorption spectra of cobalt(III) resembles very closely that of the *trans*-form in shape and relative intensity. Three bands are obtained at 21 600, 23 500 and 25 750 cm^{-1}, the middle band being weaker than the other two. Now there are two possibilities for the cobalt(III) oxide–bromide complex in glass: $[CoO_4Br_2]$ or $[CoO_2Br_4]$. As all the glasses containing cobalt(III) and bromide also contained unknown and variable amounts of the cobalt(II)–oxide–bromide complex, the stoichiometry of the complex with respect to cobalt(III): bromide ratio cannot be established as in the case of cobalt(II)–oxide–chloride complex. Nevertheless it can be said, from the nature of the absorption spectra, that octahedral diamagnetic cobalt(III) complexes have been formed in these glasses with a mixture of oxide and bromide ligands, and at least two bromides are associated with the cobalt(III) centre at *trans*-positions.

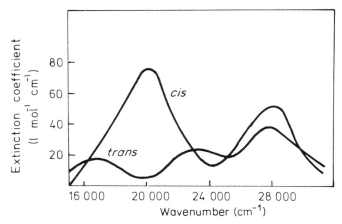

Fig. 9.30 Optical absorption spectra of $[Co(en)_2F_2]^+$.

...ar halide substitutions around iron(III), nickel(II), copper(II) etc. have ...en investigated and reported in the literature[31–36].

...example of anionic substitution having great industrial importance is ...r glass'. Amber glass is made commercially by melting a batch containing ...e' sand (that is, containing more iron than ordinary glass-making sand), ...n sulphate and carbon. Carbon reduces part of the sulphate to sulphide, ...is sulphide is then substituted for oxygen ligand around tetrahedral ferric ...Previously the origin of the amber colour was wrongly thought to be due to ...articles of carbon and probably of sulphur as well. Thus in some quarters ...now it is known as 'carbon–sulphur' amber glass. The optical absorption ...rum of the typical amber glass is shown in Figure 9.31; absorption spectra of ...us and ferric ions with exclusive oxygen coordination in the same glass are ...ncluded in the figure for comparison. The amber chromophore in the glass ...uces two distinct bands around 23 500 and 33 900 cm^{-1}; both the bands are ...intense, the intensity of the high-energy band being about three times that of ...ow-energy one. From quantitative spectroscopic (optical and e.s.r.) studies it ...been shown that the amber chromophore consists of a ferric ion coordinated ...three oxygen and one sulphide along with the appropriate number of alkali ...lkaline earth ions to neutralize the charge on the complex group[19]. ...extinction coefficient of the 23 500 cm^{-1} band has been estimated as ...0 l mol^{-1} cm^{-1}.

...he amber colour centre involves both an oxidized species (ferric iron) and a ...uced species (sulphide sulphur) in glass; the standard free energies for the ...tems FeO–Fe$_2$O$_3$ and Na$_2$S–Na$_2$SO$_4$ are shown in Figure 9.32. Thus, there is ...ly a limited range of oxygen partial pressure where both species can coexist in ...ss in reasonable proportions. For pure oxide systems at 1400°C, this ...rresponds to $pO_2 \sim 4.88 \times 10^{-5}$ to 1.78×10^{-6} atmosphere. According to ...own and Douglas[20], this range of pO_2 is approximately from 10^{-8} to 10^{-10} ...mospheres in a 30 Na$_2$O, 70 SiO$_2$ glass at 1400°C. It is needless to say that this ...itical range of oxygen pressure will vary with changing glass composition and

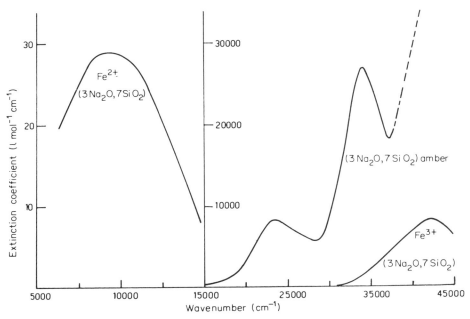

Fig. 9.31 Optical absorption of iron (II) and iron (III) in 3 Na$_2$O, 7 SiO$_2$ glass and in an amber glass of the same composition.

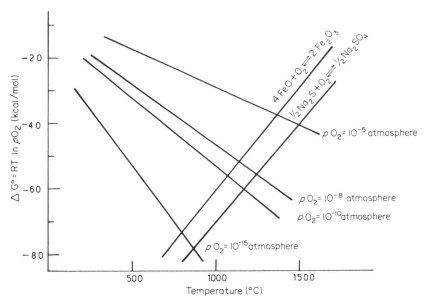

Fig. 9.32 Free energy–temperature diagram for the FeO–Fe$_2$O$_3$ and Na$_2$S–Na$_2$SO$_4$ systems.

e. In the 30 Na_2O, 70 SiO_2 glass at 1400°C, when pO_2 is less than
osphere, only about 2.8 per cent of total iron will be present as ferric;
$_2$ greater than 10^{-8} atmospheres will oxidize almost all the sulphide in
Thus, exceeding either of the limiting values effectively removes one of
al components of the amber colour centre.
strial practice the amber colour is controlled in most instances by
e effective oxygen pressure of the melt, often by adjusting the amount of
the batch. From the oxidation–reduction equations as written below, it
n that the ratio sulphate/sulphide is much more sensitive to changing
the ratio ferric/ferrous in glass.

$$[FeO]_{glass} + \tfrac{1}{4}O_2 \rightleftharpoons [\tfrac{1}{2}Fe_2O_3]_{glass}, \frac{[1/2Fe_2O_3]_{glass}}{[FeO]_{glass}} \propto pO_2^{1/4}$$

$$[S]^{2-}_{glass} + 2O_2 \rightleftharpoons [SO]^{2-}_{glass}, \frac{[SO_4]_{glass}}{[S^{2-}]_{glass}} \propto pO_2^{2}$$

hin the critical range of oxygen pressure, increasing pO_2 makes the glass
hile decreased pO_2 makes it darker, predominantly due to sulphide level
ns of the glass.
are other considerations in amber glass melting and forming which
restrictions on the pO_2 control range. Ferrous iron in glass produces
near-infrared absorption (Figure 9.31); its presence causes problems of
at transfer during melting, refining, conditioning and forming of glass.
his point of view it might seem advantageous to lower the total iron level of
ss. The proper intensity of the amber colour could be maintained by
the melt more reducing to increase the sulphide level. However, glasses
with severe reduction are very difficult to refine, and excessive sulphide
glass produces reboil. Since high iron levels impose a heat transfer
m, and high sulphide levels impose a stability problem, these two
erations restrict the pO_2 range which may be used to control amber colour
ustry.
manner very similar to iron–sulphur amber glass, when an oxygen around
hedrally coordinated ferric iron is replaced by a Se^{2-}, black colour results.
al absorption spectra of iron–selenium black glass are shown in Figure 9.33.
de being more electronegative than sulphide, the ligand–metal charge-
er bands occur at lower energy; in this particular case, almost at the middle
e visible range, and thus making the glass almost black. The intensity of
dual bands in Figure 9.33 has not yet been quantitatively estimated [21],
ver, these are expected to be as strong, if not more so than the iron–sulphur
r bands, and due to their low energy a perceptible colour results even with
small concentration of this chromophore in glass. Sometimes this poses
lems during decolourization of glass with selenium.

9.7 PHOTOSENSITIVE GLASSES

1 Solarization

en a glass is melted with two suitable redox oxides, exposure to solar radiation
y cause some of the constituent redox ions to undergo photo-oxidation or

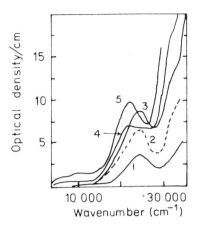

curve	weight %	
	Fe_2O_3	Se
(1)	–	0.50
(2)	0.02	0.50
(3)	0.04	0.50
(4)	0.10	0.50
(5)	0.50	0.50

Fig. 9.33 Optical absorption spectra of $3\,Na_2O$, $7\,SiO_2$ black glasses containing iron and selenium.

reduction, resulting in changes in the absorption spectrum of the glass. This process is known as solarization. In a glass containing Mn^{2+} and As^{5+} ions, for example, the solar energy is believed to be used to remove an electron from a Mn^{2+} ion which is oxidized to the Mn^{3+} state. The photo-electrons are readily accepted by an As^{5+} ion which becomes reduced to the As^{3+} state. As a result, the weakly coloured Mn^{2+} ion is transformed to the pink Mn^{3+} state. Since the photo-oxidized Mn^{3+} is in an environment previously occupied by Mn^{2+}, the absorption spectrum of photo-oxidized Mn^{3+} has been found to be different from that of normal Mn^{3+} ion in the same glass [22].

From a commercial point of view, solarization is an undesirable phenomenon. Glass always contains traces of iron as impurities and when a glass, decolourized with manganese, is exposed to solar radiation electron transfer between Mn^{2+} and Fe^{3+} takes place as follows:

$$Fe^{3+} + Mn^{2+} + h\nu \rightleftharpoons Fe^{2+} + Mn^{3+} \text{ (pink)}$$

Thus after solarization the near-ultraviolet transmission will increase (due to decrease of Fe^{3+}) and the glass will appear pink. Various pairs of redox oxides

332

f producing solarization in glass have been studied; some of the y studied couples are:

Mn–As, Fe–As, Ce–As, Ce–V, Ce–Cu, Ce–Ag etc.

omenon of solarization can be made reversible; when a piece of solarized se glass is heated to about 500°C, the pink colour bleaches and the on spectrum becomes similar to that of unexposed glass.

hotosensitive glass

nsitive glasses are melted with noble metal ions, particularly copper, gold. When exposed to short-wave radiation (ultraviolet), the noble ns are reduced to atoms, agglomerating into bigger particles when the neated so that a visible coloured image is formed on the areas exposed to n. The latent image which is produced in the photosensitive glass by active on becomes visible only after heat treatment. The origin of the image in ensitive glasses is based upon coagulation of particles and differs from ation by being a non-reversible process. With solarization, discolouration s by short-wave radiation is caused by transition of electrons of some ions er orbits. In contradiction to photosensitive glasses, no formation and ′ growth of crystal nuclei due to higher temperatures takes place in ed glass, but the electrons return to their original orbits and the ouration ceases.

mportant branch of photosensitive glass is photoplastic glasses. In the heat ient of photosensitive glass by gradual heating to the nucleation rature, metal atoms aggregate into larger, colloid-dispersed aggregates, colour the glass by their characteristic colour. If crystalline aggregates of a ient size are formed, they may act as heterogeneous nuclei for silicate allization from the glass. It is thus possible to obtain turbid images, and if the is chosen so as to have a composition such that crystalline phases, which are pitated by this heterogeneous nucleation, dissolve at substantially different (for instance in hydrofluoric acid), then the 'developed' crystalline image can emoved by etching, and we obtain a plastic (relief) image, or a copy may be ined with a high precision of shape and dimensions.

hotosensitive glasses have a basic composition similar to common silicate ses; they must not contain PbO, Tl_2O etc. as major components for these les have strong ultraviolet absorptions thus making the glass opaque to short- e radiation. Glasses containing higher B_2O_3 or P_2O_5 contents are not able because these glasses are very often too reducing to keep the noble tals, particularly gold and silver, in the ionic form. For making good otosensitive glass three constituents are essential: photosensitive metals, sitizers, and thermoreducing agents.

Photosensitive metals

ese 'metals' must be such that they are capable of existing in glass in the ionic rm and at the same time easily reducible in the glassy matrix. Gold, silver, and pper (noble metals) are suitable for this purpose. Gold and silver when melted nder mildly oxidizing conditions produce Au^{3+} and Ag^+ in glass, copper glasses

333

must be melted under mildly reducing conditions to get rid of blue Cu^{2+}; if copper glasses are melted under strongly reducing conditions, Cu^+ will be reduced to Cu and thus the photosensitivity will decrease. For photosensitive glasses containing copper there appears to be a critical reducing condition; if the reduction is less, the glass will be blue containing Cu^{2+} and will not be photosensitive. If reduction is too high, Cu^+ will be reduced to metallic copper, thus decreasing the photosensitivity. (For more detail see section 9.8 on copper ruby glasses). Usually 0.001 to 0.05 wt % gold, 0.001 to 0.3 wt % silver or 0.05 to 1.0 wt % copper is used. The temperature and melting conditions are adjusted so as to obtain the maximum possible amount of the desired metallic ion. Palladium (0.001 to 0.20 wt %) is sometimes used in combination with silver or gold to change the tint of the colour.

(b) Sensitizers

Sensitizers are substances which absorb ultraviolet radiation over a wider range of wavelength than that of the proper photosensitive substances, thereby oxidizing themselves and liberating an electron which has a very advantageous effect upon the activation of photosensitive elements. This increases the sensitivity of the glass to ultraviolet radiation. The most important of these substances is CeO_2 which is usually added in amounts up to 0.05 wt %. It should be noted that larger amounts of CeO_2, like PbO, decrease the penetrating power of ultraviolet radiation and thus are to be avoided.

(c) Thermoreducing agents

It is believed that thermoreducing agents facilitate and regulate the growth of crystal nuclei to particles of colloidal size during the heat treatment of photosensitive glasses. The most common are tin and antimony oxides and these are used in amounts up to 0.1 wt %. Considering that they also absorb ultraviolet radiation, they may reduce the depth of the exposed image if greater amounts are added. Furthermore, it has been found that at higher concentrations, they often cause discolouration in unexposed areas.

Some constituents in the glass interfere with the photosensitivity even if small quantities are present. The most important oxides causing interference are Fe_2O_3 (which is present in every glass as impurity) and TiO_2. By increasing the iron contents the sensitivity of the glass to u.v. radiation decreases rapidly and, with 0.03 wt % iron, only slight discolourations and brown-red tints are obtained. Arsenic has also been reported to have adverse effects on the photosensitivity of glass. Photosensitive glasses have properties similar to those of photographic materials from which they differ mainly by the fact that the exposure has to be carried out by short-wave radiation. Mercury-arc lamps are suitable; with a 150 W mercury arc lamp, the time of exposure at distances of 10 to 30 cm from the source varies from a few seconds to one hour, according to the image intensity required.

The latent image which is produced in the photosensitive glass by ultraviolet radiation becomes visible only after heat treatment. Photosensitive glasses containing gold are totally colourless after exposure and show no colour change. Silver-containing glasses usually have a slightly yellow tint and with copper-containing glasses a blue-grey tint is obtained. During heating to 400° C, however,

of colour disappears. The picture itself is formed only when the glass
o a desired temperature which is above the upper annealing range
ding to a viscosity of $\eta = 10^{13.4}$ poises) and below the softening point
ding to viscosity of $\eta = 10^{7.65}$ poises). The higher the temperature
t treatment the shorter the time necessary for production of the image.
temperature of about $150°$ C below the Littleton's softening point and a
een 15 and 30 minutes is used.
oplastic glasses, to facilitate the heterogeneous nucleation and separ-
he crystalline phase on the nuclei of photosensitive metals, the colour
produced first, and then the temperature is raised so that nucleation of
lline phase may take place. The temperature is then further raised, so
eterogeneous nuclei thus formed grow more rapidly to dimensions such
ausing visible opacity. These are then etched with a mixture of sulphuric
ofluoric acid. The relative concentrations of these two acids and the total
are adjusted to suit the individual glass composition.

otochromic glasses

t years inorganic glasses have been developed that show reversible colour
cal density changes; a property known as photochromism. Reversible
are activated by u.v. radiation, and the optical density returns to its
state when the source is removed. This reversal process is not only
e to heat (thermal fading), but also to visible light of relatively long
ngths (optical bleaching). The degree of optical density change depends on
ctral distribution and the energy density of the total light incident on the
chromic glass. Optimum activation occurs at wavelengths between 320 nm
20 nm. Optimum bleaching occurs at wavelengths between 550 nm and
n. The thermal fading rate increases at higher temperatures and is relatively
or some photochromic glasses at room temperature.
ow seems to be accepted that the photochromic properties of these glasses
from small silver halide crystals that are precipitated from the homo-
us glassy matrix during glass formation. Photographic emulsions also
in silver halide crystallites that are embedded in a suitable host matrix. The
n of incident photons causes the silver to separate from the halogens; these
to diffuse away from the original crystal sites if the composition of the host
ix permits such diffusion. In conventional photographic material the
sed halogens are no longer available for recombination when the irradiation
es, and a permanent image is formed after suitable chemical treatment.
tochromic glass on the other hand, has a host matrix almost impervious to
gen diffusion, and the released halogens are confined to the immediate
nity of the crystallites. Thus, gradual recombination can take place after
ation of the activation irradiation.
he range of activation wavelengths depends on the chemical composition of
glass. The optimum activation wavelength is around 350 nm for glass
taining only chloride. If bromide (from 350 to 550 nm) or iodide (up to
) nm) is added, the optimum shifts to longer wavelengths.
Some typical compositions for these silver halide photochromic glasses are
en in Table 9.11. All the glasses in this table are transparent in the unexposed
te, darkening to a grey or reddish grey when illuminated. At high con-

Table 9.11

Compositions of some typical photochromic silver halide glasses

Constituent	Glass 1	Glass 2	Glass 3	Glass 4	Glass 5	Glass 6	Glass 7
SiO_2	60.1%	62.8%	59.2%	59.2%	60.1%	52.4%	51.0%
Na_2O	10.0	10.0	10.9	14.9	10.0	1.8	1.7
Al_2O_3	9.5	10.0	9.4	9.4	9.5	6.9	6.8
B_2O_3	20.0	15.9	20.0	16.0	20.0	20.0	19.5
Li_2O	–	–	–	–	–	2.6	2.5
PbO	–	–	–	–	–	4.8	4.7
BaO	–	–	–	–	–	8.2	8.0
ZrO_2	–	–	–	–	–	2.1	4.6
Ag	0.40	0.38	0.50	0.50	0.40	0.31	0.30
Br	0.17	–	–	0.60	0.17	0.23	0.11
Cl	0.10	1.7	0.39	–	0.10	0.66	0.69
F	0.84	2.5	1.45	1.45	0.84	–	–
CuO	–	0.016	0.016	0.015	0.016	0.016	0.016

Note: Compositions are given in weight percent; halogens are given as weight percent additions to that of the base glass.

centrations of the halide crystallites, or following heat treatments which produce large average particle size, the glasses are translucent or opaque. The upper limit for the transparent glasses is usually about 0.7 wt% silver. Other metals, in the form of polyvalent oxides, including arsenic, antimony, tin, lead, copper and cadmium, increase the sensitivity and the photochromic absorbance. The mechanism of extra sensitization by these polyvalent ions is not clear.

In general, glasses with particles less than about 50 Å in diameter are not photochromic. As the time or temperature of heat treatment for any one glass is increased, the average number of particles is reduced, and their size is increased. Above about 300 Å the particles scatter light, and the resultant glass is opal. For particles of average diameter 100 Å present in concentration say 0.2 per cent in the glass, there will be about 4×10^{15} particles per cm^3 with average spacing 600 Å between particles.

It should be noted that most of the photochromic glasses contain a mixture of halides. During precipitation from the glass, silver may be precipitated either as a single halide complex or as a mixed halide complex; in extreme cases silver may be coordinated with a mixture of halides and oxygen of glass and correspondingly the photochromic sensitivity will change. However, no clear understanding of this area of the problem is yet available.

9.8 COPPER RUBY GLASSES

Ruby glasses are a special type of coloured glass; here the colour of glass originates from precipitation of very small crystalline particles with which the glass melt has become supersaturated at the heat treatment temperature. In a ruby glass the precipitated phase may be a single metal as in gold (Au) ruby or a simple compound as in chrome (Cr_2O_3) ruby or it may be a mixed crystal as in the

cadmium–sulphoselenide glasses where the precipitated phase is
be a $[CdS_{(1-x)}Se_x]$ mixed crystal. Since the physical chemistry
all the ruby glasses is more or less the same, only copper ruby
will be discussed.

production of copper ruby glass a batch containing copper oxide
ith some reducing agent (SnO is commonly used) is melted in reducing
. The melt initially shows the blue colour characteristic of copper (II),
melting proceeds the colour changes to a pale yellow straw colour, and
ge the glass is fabricated. By subsequent heat treatment, commonly
striking, at a temperature somewhere between the annealing and the
temperature, the ruby colour is developed. If, during melting, the
is allowed to proceed beyond the critical stage, the ultimate ruby colour
wn and appears dull. On the other hand if the copper is insufficiently
and any trace of the blue colour remains, the ruby colour cannot be
d.

by colour has been thought to be due to the precipitation from the glass
netallic copper and cuprous oxide. Atmaram and Prasad [23] have made
studies of copper ruby glasses; they stressed the importance of the critical
reduction in the production of good ruby glass and stated that the
ng chromophore of ruby glass is not metallic copper but cuprous oxide.
ise, they stated, it would be expected that the greater the degree of
on the better the ruby. To substantiate their view they measured
copic properties such as chemical durability and viscosity of ruby glasses
ing large amounts of copper, about 4–5 wt % total. The measurement of
macroscopic properties, especially for glasses containing much larger
ts of copper than is usual for ruby glasses, cannot give direct proof of their
nesis, but they also showed that there is great similarity between the
ption spectrum of ruby glass and that of colloidal cuprous oxide in an
us medium. On the other hand Rawson [24] has shown that according to
ght scattering theory of Mie, for small particle sizes (up to 100 Å diameter)
bsorption spectrum of copper ruby is consistent with metallic copper
les. Unfortunately the optical constants, n and k for cuprous oxide are not
n, and no comparison like that for metallic copper can be made.
e formation of the ruby colour is basically a problem of chemical
nodynamics in the glass melt and the kinetics of the precipitation of phases in
h the melt is supersaturated. During heat treatment there is very little
ability of any entry or egress of oxygen from the glass and, as rubies can be
e without any additional reducing agent being put in the glass batch, the
ibility of reaction with the reducing agent at the heat treatment temperature
be eliminated. Moreover, as the characteristic d–d band of copper (II) around
00 cm^{-1} is not developed in a ruby glass, the possibility of the Cannizaro
ction of the type

$$2Cu^+ \rightleftharpoons Cu^{2+} + Cu^0$$

suggested by Weyl [25] and others, can also be disregarded.
Banerjee and Paul [26] have studied the copper (I) – copper (0) equilibrium in
ow melting borate glass. A 30 Na$_2$O, 70 B$_2$O$_3$ glass was equilibrated at different
nperatures by melting in a solid metallic copper crucible at different oxygen

337

pressures in the range $pO_2 = 10^{-12}$ to 10^{-17} atmosphere. All these glasses were free from blue colour and contained no detectable amount of copper (II). On heat treatment the best ruby was formed in glasses equilibrated with a pO_2 of 10^{-14} atmosphere; with a higher oxygen pressure, say $pO_2 = 10^{-12}$ atmosphere, no ruby was formed, and at a lower oxygen pressure ($pO_2 \leqslant 10^{-15}$ atmosphere) the glass progressively produced weaker rubies.

The copper (I) – Copper (0) equilibrium in glass may be written as:

$$4 \langle Cu \rangle + (O_2) \rightleftharpoons 2 \langle Cu_2O \rangle \qquad (9.16)$$

$$2 \langle Cu_2O \rangle \rightleftharpoons 2 [Cu_2O]_{glass} \qquad (9.17)$$

$$2 [Cu_2O]_{glass} \rightleftharpoons 4 [copper (I)]_{glass} \qquad (9.18)$$

Here $\langle i \rangle$ and $[i]$ indicate the species, i, in the solid and in the solution state respectively.

The relation between the standard free energy of reaction (9.16) and absolute temperature, T is:

$$\Delta G^0 = -81\,000 - 7.84T \log T + 59.0T \text{ cal/mol} \qquad (9.19)$$

Thus, $^a\langle Cu_2O \rangle$ in equilibrium with solid metallic copper at different temperatures and oxygen pressures can be calculated; and these will be equal to $^a[Cu_2O]_{glass}$. Figure 9.34 shows a plot of the experimentally determined wt % Cu_2O in glass against $^a\langle Cu_2O \rangle$ calculated from equation (9.16). It is clear that the relationship strongly deviates from linearity and a small change of tempera-

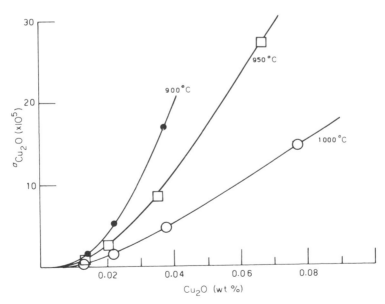

Fig. 9.34 Activity–concentration relationship of Cu_2O dissolved in a sodium–borate melt at different temperatures.

profound influence on it. From spectroscopic measurements (absorp-
nission), it is known that the cuprous ion exists in glass as a monomeric
mplex rather than the dimeric Cu_2O species. A plot of wt% Cu_2O
$/_o < Cu_2O >$ produces a series of linear relationships, one for each
re and all the lines pass through the origin. This clearly suggests that
nergy of oxidation of copper in glass will be strongly influenced by
9.18) and will be significantly different from that of reaction (9.16) and

$$4\langle Cu \rangle + (O_2) \underset{K}{\rightleftharpoons} 2[Cu_2O]_{glass} \qquad (9.20)$$

$$K = \frac{[{}^a Cu_2 O_{glass}]^2}{[{}^a \langle Cu \rangle]^4 \cdot pO_2} \qquad (9.21)$$

resent experimental set up, $^a \langle Cu \rangle = 1$. Thus

$$K = \frac{[{}^a Cu_2 O_{glass}]^2}{pO_2} = \frac{[{}^c Cu_2 O_{glass}]^2 \cdot [{}^\gamma Cu_2 O_{glass}]^2}{pO_2} \qquad (9.22)$$

Cu_2O_{glass} is the concentration of Cu_2O (expressed as metallic copper) in
n glass, and $^\gamma Cu_2 O_{glass}$ is the activity coefficient of Cu_2O in glass. From
n (9.22) and the relationship $\Delta G^\circ = -RT \ln K$, $[{}^\gamma Cu_2 O_{glass}]$ was
ted at different temperatures and this is shown plotted in Figure 9.35 as

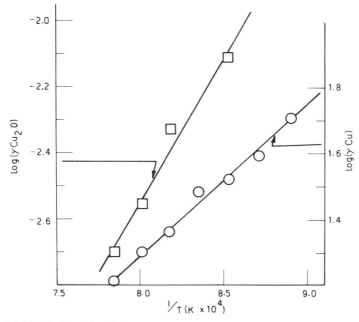

Fig. 9.35 Variation of activity coefficient, γ, of Cu_2O (glass) and Cu (glass) with
temperature.

339

$\log [^\gamma Cu_2 O_{glass}]$ against reciprocal temperature. Similarly $[^\gamma Cu_{glass}]$ was calculated from the solubility of copper(0) in the glass at different temperature with the relationship:

$$1 = {}^a\langle Cu \rangle = [^c Cu_{glass}] [^\gamma Cu_{glass}] \tag{9.23}$$

and these are also plotted in Figure 9.35 as $\log [^\gamma Cu_{glass}]$ against reciprocal temperature; in both the cases linear relationships have been observed.

When a glass melt is equilibrated at $1000°$ C with a definite and known partial pressure of oxygen, this gives an invariant point on the free-energy-temperature diagram (Ellingham diagram). Such a point for $pO_2 = 10^{-14}$ atmosphere at $1000°C$ is shown in Figure 9.36 at A. Figure 9.36 also contains the standrad free energy line for the $\langle Cu \rangle - \langle Cu_2 O \rangle$ system, i.e. where both copper and cuprous oxide are at unit activity; this at $1000°C$ corresponds to $pO_2 = 10^{-6.33}$ atmosphere (point 'B'). When the oxygen pressure is reduced to 10^{-14} atmosphere, since metallic copper cannot be further reduced, ${}^a\langle Cu \rangle$ remains unity, and the activity of $Cu_2 O$ decreases from 1 to about 1.5×10^{-4}, or in other words the free energy point 'B' is shifted to 'A'. If the system at A is now cut off from external atmosphere and allowed to cool, as was done during this experiment, the system will follow the line AC, provided the activity coefficients of both $[Cu_{glass}]$ and $[Cu_2 O_{glass}]$ remain unchanged with the change of temperature, or change in such a way so as to keep the ratio: $[^\gamma Cu_2 O_{glass}]^2 / [^\gamma Cu_{glass}]^4$ unchanged. However, as may be seen from Figure 9.35, both $[^\gamma Cu_{glass}]$ and $[^\gamma Cu_2 O_{glass}]$ change critically with temperature. It should be noted that $[^\gamma Cu_{glass}]$ is high (of the order of 10–30) and increases with decrease of temperature. Thus, a glass which has been equilibrated at $1000°C$ with metallic copper will continuously precipitate $\langle Cu \rangle$ on cooling to maintain the relationship of equation (9.23).

It should also be noted that under equilibration conditions $[^\gamma Cu_2 O_{glass}]$ is very small (of the order of 10^{-3} around $1000°C$), however, it also increases rapidly with decreasing temperature. In this experiment, since $[^a Cu_{glass}]$ is always unity at lower temperatures, the free energy of the $\langle Cu \rangle - [Cu_2 O]$ system in glass is controlled by the variation of $[^\gamma Cu_2 O_{glass}]$ with temperature. So long as the product $[^\gamma Cu_2 O_{glass}] \cdot [^c Cu_2 O_{glass}]$ is less than unity the experimental free energy will be more negative than the standard state line, CB; when the product becomes unity, the experimental point will fall on the standard state line, and on further cooling the glass will precipitate both copper and cuprous oxide, their relative concentrations in the precipitated phase being determined by the slopes of the lines in Figure 9.35 and the total amount of copper contained in the glass.

When equilibrated with a furnace atmosphere having $pO_2 = 10^{-14}$ atmosphere, a $30 Na_2 O$, $70 B_2 O_3$ glass at $1000°C$ was found to dissolve 0.06 wt % metallic copper and 0.07 wt % cuprous oxide (calculated as metallic copper); thus the total copper solubility is 0.13 wt %. On cooling this saturated glass to lower temperatures, $[^\gamma Cu_2 O_{glass}]$ and $[^\gamma Cu_{glass}]$ increase and the free energy of the system follows the path $A \to D \to C$. From the temperature D downwards, $\langle Cu_2 O \rangle$ in addition to $\langle Cu \rangle$ will precipitate from the glass.

Let us now equilibrate a similar glass containing 0.13 wt % total copper at $1000°C$ and at two other oxygen pressures, say $pO_2 = 10^{-16}$ and 10^{-12}

e; the former is more reducing than the optimum saturation with 0.13 copper, and the latter is more oxidizing.

ase of $pO_2 = 10^{-16}$ atmosphere, the glass will dissolve 0.06 wt % pper (for the solubility of metallic copper does not involve any change on state, and so is independent of pO_2); the $^a[Cu_2O_{glass}]$ will reduce 2×10^{-5} to 1.472×10^{-5}, and the glass will hold only 0.02 wt % Cu_2O. wt % Cu_2O will be reduced to metallic copper and, being super-will come out of solution in the glass; the new free energy point will be ed by the point E in Figure 9.36. On cooling the activity coefficients of d cuprous oxide will increase and the free energy of the system will e path E → F → C. It is to be noted that, by melting at a lower oxygen than the optimum, the glass retained less Cu_2O dissolved in it, and the tion of $\langle Cu_2O \rangle$ from the melt started at a lower temperature, F relative s important to point out that systems equilibrated at $pO_2 = 10^{-14}$ and tmospheres will precipitate the same amount of metallic copper on although precipitation of Cu_2O will be much smaller in the case of 10^{-16} ere.

e second case, when equilibrated with $pO_2 = 10^{-12}$ atmosphere, the of Cu_2O in contact with solid metallic copper can be calculated from n (9.16) to be 147.2×10^{-5}, i.e. the solubility of Cu_2O in glass will increase This will result in partial oxidation of dissolved metallic copper in the orming cuprous oxide with a little cupric oxide. In fact by appropriate dynamic calculation it can be shown that with $pO_2 = 10^{-12}$ atmosphere 1000°C, the 0.13 wt % total copper will be distributed in a 30 Na_2O, $)_3$ glass as follows:

$$Cu = 0.0289 \text{ wt }\%, \qquad Cu_2O = 0.1000 \text{ wt }\% \text{ and}$$
$$CuO = 0.0012 \text{ wt }\%$$

the resultant glass will be unsaturated with respect to both metallic copper uprous oxide; the free energy point for such a system is represented by the G in Figure 9.36. On cooling such a system the activity coefficients of both and (Cu_2O) will increase and the free energy of the system will follow G → H → C; at the temperature H, the glass becomes saturated with 0.0289 wt % llic copper; from the temperature I, Cu_2O in addition to metallic copper also s precipitating.

ne precipitation of Cu_2O and Cu from the glass melt on cooling cannot inue indefinitely. Eventually the viscosity of the glass increases so much that system cools in a metastable condition as shown by the vertical dotted line in ure 9.36. The exact *fictive* or *frozen-in* temperature is not definitely known but xpected to be higher than the upper annealing temperature of the glass. For mple, in the 30 Na_2O, 70 B_2O_3 glass, the upper annealing temperature is und 500°C; the optimum striking temperature has been experimentally found be 580°C; and no ruby colour was developed on heat treatment at 540°C; even er 24 hours.

From the above considerations it is clear that when a glass is melted at a definite nperature with a certain fixed amount of copper in it, thermodynamically these a unique oxygen pressure at which the glass will be just saturated. On cooling is glass the maximum amount of Cu_2O will precipitate from the system. By

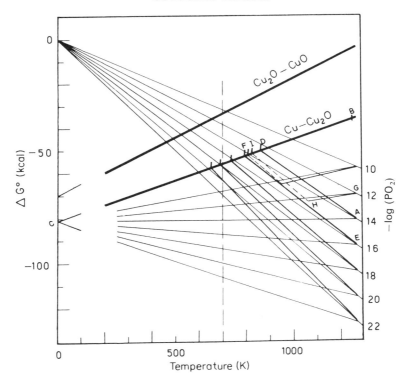

Fig. 9.36 Thermodynamic cooling path of Cu_2O (glass)–Cu (glass) system.

going either to lower or higher oxygen pressures, the amount of precipitated Cu_2O will decrease and the quality of copper ruby is also known to deteriorate.

Thus after considering the different possibilities concerning the precipitation of Cu(solid) and Cu_2O(solid) from the point of view of equilibrium thermodynamics, it may be concluded that Cu_2O(solid) is primarily responsible for the formation of good ruby colour in glass.

9.9 MEASUREMENT OF COLOUR, COLOUR DIAGRAM AND TRISTIMULUS VALUES

The term *colour* is normally used in three distinctly different senses: the chemist employs it as a generic term for dyes, pigments, and similar materials. The physicist regards the term as a description of certain phenomena in the field of optics. Physiologists and psychologists, on the other hand, are interested primarily in understanding the nature of the visual process, and use the term on occasions to denote a sensation in the consciousness of a human observer.

The visual stimulation that results when looking at a coloured surface depends upon the character of the light by which the surface is illuminated. Daylight, the traditional source by which samples are examined, consists of a mixtures of all the

.s of the visible spectrum in nearly equal proportions. A filter has been which, when used with a tungsten lamp operated at the proper ːe, provides a source that is a close approximation to average daylight. ard is known as CIE Illuminant C. The relative spectral distribution of radiated per unit time by CIE Illuminant C is shown in Figure 9.37.

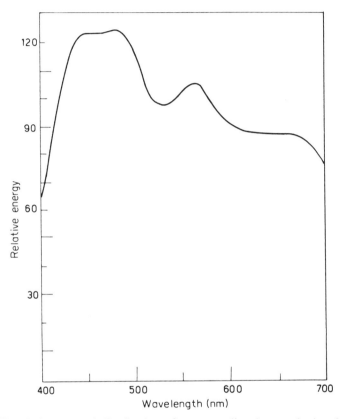

g. 9.37 Relative spectral distribution of energy radiated per unit time by CIE illuminant C.

is impossible to expose an observer to light of known spectral quality and ː ct him to describe the sensation that it produces. However, a normal observer duplicate the effect of any colour stimulus by mixing the light from three ːnary sources, Red, Green and Blue. The experimental technique is as follows: ːhe observer looks into an optical instrument containing a suitable photo- ːric field; (ii) the light whose colour is to be matched is introduced into one half ːhe field, and the light from the three primary sources introduced in controlled ːounts into the other half; (iii) by manipulation of the controls, a setting can be ːnd where an exact colour match between the two halves of the field is obtained.

343

By calibrating the controls, the amount of each primary can be recorded. The unknown colour can then be specified by three numbers, X, Y, Z. These are known as the *tristimulus* values, each number representing the amount of one of the primary stimuli. The tristimulus values that were adopted by the Commission International de l'Eclairage (CIE) for the various spectrum colours are given in Table 9.12.

Table 9.12 Tristimulus values of the spectrum colours

Wavelength (nm)	X	Y	Z
400	0.0143	0.0004	0.0679
410	0.0435	0.0012	0.2074
420	0.1344	0.0040	0.6456
430	0.2839	0.0116	1.3856
440	0.3483	0.0230	1.7471
450	0.3362	0.0380	1.7721
460	0.2908	0.0600	1.6692
470	0.1954	0.0910	1.2876
480	0.0956	0.1390	0.8130
490	0.0320	0.2080	0.4652
500	0.0049	0.3230	0.2720
510	0.0093	0.5030	0.1582
520	0.0633	0.7100	0.0782
530	0.1655	0.8620	0.0422
540	0.2904	0.9540	0.0203
550	0.4334	0.9950	0.0087
560	0.5945	0.9950	0.0039
570	0.7621	0.9520	0.0021
580	0.9163	0.8700	0.0017
590	1.0263	0.7570	0.0011
600	1.0622	0.6310	0.0008
610	1.0026	0.5030	0.0003
620	0.8544	0.3810	0.0002
630	0.6424	0.2650	0.0000
640	0.4479	0.1750	0.0000
650	0.2835	0.1070	0.0000
660	0.1649	0.0610	0.0000
670	0.0874	0.0320	0.0000
680	0.0468	0.0170	0.0000
690	0.0227	0.0082	0.0000
700	0.0114	0.0041	0.0000

uation of the quality of a colour (*chromaticity*) is accomplished by
ree new quantities as follows:

$$x = X/(X + Y + Z), \; y = Y/(X + Y + Z), \; z = Z/(X + Y + Z)$$

atities are called *trichromatic coefficients*. It should be noted that only
se quantities are independent since

$$x + y + z = 1$$

specify the chromaticity of a sample, it is necessary to give the values of
of these three quantities (*x* and *y* have generally been selected).
9.38 shows the trichromatic coefficients for certain spectrum colours;
e been calculated from the tristimulus values given in Table 9.12. Thus,
e of radiation of 600 nm, $X = 1.0662$, $Y = 0.6310$ and $Z = 0.0008$. From

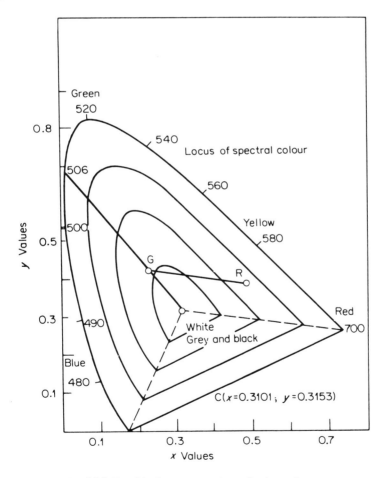

Fig. 9.38 Graphical representation of colour diagram.

345

these it can be calculated that the trichromatic coefficients are $x = 0.6270$ and $y = 0.3725$. The tristimulus of illuminant C can be determined by an integration process to be $X = 98.04$, $Y = 100.00$, and $Z = 118.12$. The corresponding trichromatic coefficients, $x = 0.3101$ and $y = 0.3163$, locate the position of illuminant C. A graphical representation such as Figure 9.38 is called a *colour diagram*.

The tristimulus values for either a transparent or an opaque material may be computed by multiplying at each wavelength the relative amount of energy in the source by the reflection or transmission factor of the sample. This product gives the relative amount of energy at this wavelength entering the eye of the observer. This product is then multiplied by the trichromatic coefficients of the spectrum colour, $\bar{x}$, $\bar{y}$ and $\bar{z}$ at that wavelength, in turn.

As an example let us determine the trichromatic coefficients of a red signal glass with illuminant A of CIE. The computational details are given in Table 9.13. The

Table 9.13
Sample calculation of a tristimulus value of a filter
by the weighted ordinate method

Wave length (nm)	T	$E_A \bar{x}$	$TE_A \bar{x}$
380			
570	0.002	81.69	0.2
580	0.100	104.85	10.5
590	0.500	124.97	62.5
600	0.775	137.04	106.2
610	0.850	136.73	116.2
620	0.870	122.72	106.8
630	0.876	96.90	84.9
640	0.880	68.20	60.0
650	0.883	44.66	39.4
660	0.887	26.81	23.8
670	0.890	15.63	13.9
680	0.894	8.67	7.8
690	0.895	4.35	3.9
700	0.897	2.25	2.0
710	0.897	1.19	1.1
720	0.897	0.61	0.5
730	0.897	0.31	0.3
740	0.897	0.15	0.1
750	0.897	0.08	0.1
760	0.897	0.04	0.0
770	0.897	0.02	0.0
			Sum = 640.2

n gives the wavelength of the light at intervals of 10 nm; the second
es the transmission factor as determined by a spectrophotometer; the
in gives the values of $E_A \bar{x}$ for illuminant A; and the fourth column is
t of the second and third columns. The sum of the entries in the fourth
measure of X. Repeating this operation with $E_A \bar{y}$ and $E_A \bar{z}$ substituted
gives $X = 640.2$, $Y = 321.3$, and $Z = 0.0$. From these values, the
ic coefficients are $x = 0.6658$ and $y = 0.3342$.

al transmission can now be found by determining the ratio of the value
value of Y for a hypothetical glass whose transmission factor is 1.00 at
elength. The value of Y obtained for such a hypothetical glass is found
.0, whereas the value obtained for the red glass is 321.3. The ratio
.0 = 0.2978 or 29.78 per cent.

ortunate that the trichromatic coefficients do not indicate the nature of
difference when a difference exists. In a simple case, if two colours are
ed by $X = 40$, $Y = 50$, $Z = 30$ and $X' = 20$, $Y' = 25$ and $Z' = 15$
ly, it can be reasoned that the two colours are alike in quality or
city, but one is twice as bright as the other in the sense that it reflects
much light.

aximum visibility for a normal human eye occurs at 555 nm and the
becomes almost zero at 400 and 700 nm. The primaries of the CIE
city diagram are so chosen that y corresponds to 550 nm (green), and
umerical value is a direct measure of the *brightness* of the sample on the
iagram.

ure 9.38, suppose that a certain red is located at R and a certain green at
rdless of the proportions in which these colours are mixed, the resultant
vill always lie on the line joining R and G. Because of this additive
of the diagram, it will be seen that all real colours must lie within the area
by the outer envelope and the two broken lines, since every real colour
considered to be a mixture of its spectral components in various
ions. For example, the green G is shown to be a mixture of illuminant C
ctrum light having wavelength of 506 nm. This wavelength is known as the
t *wavelength*. Since this green lies on a line which terminates at a pure
m colour at one end and at the illuminant point at the other end, the
is evidently not so pure a green as the corresponding spectrum colour. In
e, the distance of the sample point from the illuminant point is 20 per cent
distance of the spectrum locus from the illuminant point. The sample is
re said to have a *purity* of 20 per cent.

portion of the diagram lying within the triangle below the dashed lines
nts the purples or magentas. It is evident that purple cannot be obtained
ing white light with a single spectrum colour.

REFERENCES

raddon, D. P. in (1961), *An Introduction to Co-ordination Chemistry*, Pergamon
ress, p. 1.
Iantzsch, A. and Werner, A. (1890), *Ber. Deutsch. Chem. Ges.*, **23**, 11.
Werner, A. (1891), *Vierteljahresschrift Naturforsch. Ges.* Zürich, **36**, 129.
Werner, A. (1893), *Z. Anorg. Allg. Chem.*, **3**, 267.

[5] Racah, G. *Phys. Rev.*, (1942; 1943), **62**, 438; **63**, 367.
[6] Figgis, B. N. (1966), in *Introduction to Ligand Fields*, Interscience.
[7] Bates, T. and Douglas, R. W. (1959), *Trans. Soc. Glass Technol*, **43**, 289.
[8] Paul, A. (1970), *Phys. Chem. Glasses*, **11**, 168.
[9] Cotton, F. A. and Wilkinson, C. (1966), *Advanced Inorganic Chemistry*, Interscience, London.
[10] Juza, R., Seidel, H. and Tiedemann, T. (1966), *Angew. Chem. (Int.)*, **5**, 85.
[11] Lever, A. B. P. (1974), *J. Chem. Education*, **51**, 612.
[12] Ford, P., De, F. P., Gaunder, R. and Taube, H. (1968), *J. Amer. Chem. Soc.*, **90**, 1187.
[13] Jorgensen, C. K. (1963), *Inorganic Complexes*, Academic Press, London, p. 5.
[14] Steele, F. N. and Douglas, R. W. (1965), *Phys. Chem. Glasses*, **6**, 246.
[15] Wolfsberg, M. and Helmholz, L. (1952), *J. Chem. Phys.*, **20**, 837.
[16] Liehr, A. D. and Ballhausen, C. J. (1958), *J. Mol. Spec.*, **2**, 342.
[17] Wood, D. L. and Remeika, J. P. (1966), *J. Appl. Phys.*, **37**, 1232.
[18] Ballhausen, C. J. (1962), *Introduction to Ligand Field Theory*, McGraw Hill, London, p. 107.
[19] Douglas, R. W. and Zaman, M. S. (1969), *Phys. Chem. Glasses*, **10**, 125.
[20] Brown, D. and Douglas, R. W. (1965), *Glass Technol.*, **6**, 190.
[21] Paul, A. (1973), *Phys. Chem. Glasses*, **14**, 96.
[22] Kumar, S. and Sen, P. (1960), *Phys. Chem. Glasses*, **1**, 175.
[23] Atmaram and Prasad, S. N. (1962), *Advances in Glass Technology*, Plenum Press, New York, p. 256.
[24] Rawson, H. (1965), *Phys. Chem. Glasses*, **6**, 81.
[25] Weyl, W. A. (1967), *Coloured Glasses*, Society of Glass Technology, p. 420.
[26] Banerjee, S. and Paul, A. (1974), *J. Amer. Ceram. Soc.*, **57**, 286.
[27] Bingham, K. and Parke, S. (1965), *Phys. Chem. Glasses.* **6**, 224.
[28] Edwards, R. J., Paul, A. and Douglas, R. W. (1972), *Phys. Chem. Glasses*, **13**, 137.
[29] Paul, A. and Douglas, R. W. (1968), *Phys. Chem. Glasses*, **9**, 27.
[30] Rowe, M. D., McCaffery, A. J., Gale, R., and Copsey, D. N., (1972), *Inorg. Chem.*, **11**, 3090.
[31] Paul, A. (1970), *Phys. Chem. Glasses*, **11**, 159.
[32] Paul, A. and Tiwari, A. N. (1974), *J. Mater. Sci.*, **9**, 1037.
[33] Paul, A. Sen, S. C. and Srivastava, D. (1973), *J. Mater. Sci.*, **8**, 1110.
[34] Paul, A. (1970), *Phys. Chem. Glasses*, **11**, 46.
[35] Paul, A. and Gomalka, S. (1975), *Phys. Chem. Glasses*, **16**, 57.
[36] Paul, A. (1972), *Phys. Chem. Glasses*, **13**, 144.

CHAPTER 10

Polymeric Nature of Glass Melts

fraction patterns from SiO_2 and B_2O_3 in the glassy and molten states similar. In fused silica the basic building units are SiO_4 tetrahedra are corners with one another; glassy B_2O_3 is built up of BO_3 triangles corners. The radial distribution functions (RDF) for fused silica, boric oxide, and three glasses of molar compositions .66 SiO_2, 34 Na_2O.66 B_2O_3 and 43 CaO.57 P_2O_5 are shown in 0.1(a) and (b). In the case of fused silica the first peak occurs at 1.62 Å rresponds closely to the silicon–oxygen distance in many silicate crystals. e area under this first peak, a coordination number of 4.3 ± 10 per cent toms about each silicon has been estimated [1], so establishing that the rahedra which exist in the crystalline form of silica persist in the glass. In ' for silica the atomic density falls to almost zero at 2.2 Å, indicating that e between silicon and oxygen atoms is not atomically populated. Thus atoms within a sphere of 2.2 Å radius are permanently associated with icon atom, because if they were to move with any significant frequency this boundary, the time-average atomic density (RDF) could not be zero. completely prohibited distance does not occur with molten salts (where id is predominantly ionic), and arises because there are strong covalent between silicon and oxygen. The RDF curve for silica immediately 2 Å consists of two overlapping peaks – one for the distance between two s (2.65 Å) and another for the distance between two silicons (3.2 Å). n a basic metal oxide like Na_2O is added to a network-forming oxide, the bridges between groups are broken. In the case of silica this may be nted by the equation:

$$\equiv Si - O - Si \equiv + Na_2O \rightleftharpoons 2 \equiv Si - O^- + 2\,Na^+$$

eaction introduces negative charges on the unshared corners where the nt bonds between tetrahedra are broken, and the cations (Na^+ in this case) o be located in the vicinity of these negative charges.

Na_2O–SiO_2 glass in Figure 10.1(a) shows the first peak as fused silica 0 Å) with a similar zero atomic density at about 2.0 Å. The area under this gives a coordination number of about four. This is clear evidence that 1–oxygen tetrahedra, with their covalent silicon–oxygen bonds, are not yed by the addition of Na_2O. The next peak occurs at about 2.4 Å, which be resolved into two overlapping peaks as indicated by the broken lines. of these peaks corresponds to an O–O distance of 2.65 Å, the same as in fused , and the other to an Na–O distance of 2.35 Å. The areas for the Na–O peaks

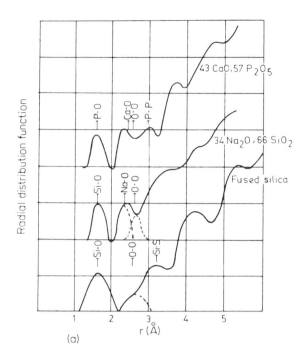

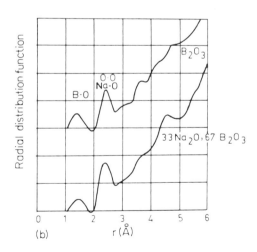

Fig. 10.1 Radial distribution functions for glasses obtained by X-ray diffraction.
(a) silicate and phosphate glasses
(b) borate glasses
The short vertical lines indicate the interatomic distances found in the corresponding crystals.

found to vary with the ratio of Na_2O to SiO_2 and the coordination oxygen about each sodium has been reported to fall from nearly 8 at contents to between 5 and 6 where the mol fraction of Na_2O exceeds

treous B_2O_3 at room temperature, the first peak on the RDF curve 1.39 Å, and the area under it shows a coordination number of 3.1. The of Na_2O shifts the peak to greater distances and increases the ion number. Some typical results are given in Table 10.1. The boron _s change from triangular to tetrahedral coordination. The structure akin to that of silica, but with sodium ions packed into the network to he charges.

Table 10.1

Boron–oxygen distances and boron coordination in sodium borate glasses

Mol% Na_2O	B–O distance (Å)	Average coordination number
0	1.30	3.1
11.4	1.37	3.2
22.5	1.42	3.7
33.3	1.48	3.9

(after Waren and Biscoe, [1])

RDF curve for $CaO–P_2O_5$ is very similar to that of $Na_2O–SiO_2$ glass. The eak, corresponding to the P–O distance, occurs at 1.57 Å and its area tes a coordination number of 4.2 for oxygen around the phosphorus. Since s a zero atomic density at 2.0 Å, we may infer that the phosphate glasses are up of PO_4 tetrahedra linked together by P–O–P bridges.

: first crystalline structures which can be formed when metal oxides are d to silica have the general formula $R_2O \cdot 2\,SiO_2$. They are stable only when ations are large and univalent – i.e. Na^+, K^+, Rb^+ and Cs^+. In this osition each tetrahedron has one unshared corner, so that sheets of silicate ledra connected at three corners can be built up in a regular arrangement. cations then lie between the sheets. The addition of higher proportions of l oxide causes progressive breakdown of the silica network and leads to rent kinds of silicates as listed in Table 10.2. Similar considerations apply to crystalline phosphates and borates as well; some typical examples of borate ems are also included in Table 10.2.

analogy to crystalline silicates it may be expected that a glass made from an osilicate or a pyrosilicate would consist solely of the SiO_4^{4-} or $Si_2O_6^{6-}$ ns which occur in the corresponding crystalline solids, and that glasses with apositions intermediate between two crystal compositions might contain only types of ion, such as SiO_4^{4-} and $Si_2O_7^{6-}$. In recent years, however, erimental evidence has been put forward which strongly suggests that glass is olymeric substance and consists of a wide distribution of anions of different es and chain lengths. The evidence for these wide distributions comes mainly m the chromatographic work on simple phosphate glasses [3]. It has been

Table 10.2
Some stable crystalline silicates and borates

Formula	Anions
R_2O, $2SiO_2$	Endless sheets
R_2O, SiO_2	Endless chains
$3R_2O$, $2SiO_2$	$(Si_2O_7)^{6-}$
$2R_2O$, SiO_2	$(SiO_4)^{4-}$

SiO_2 (three dimensional network)

$2R_2O$, $4SiO_2$ (three-linked cages)

R_2O, $2SiO_2$ (sheets)
Micas

R_2O, SiO_2 (rings)
Beryl

R_2O, SiO_2 (chains)
Wallastonite

$3R_2O$, $2SiO_2$ (dimer)
Hemimorphite

Formula	Anions

2R$_2$O, SiO$_2$ (monomer)
Olivine

B$_2$O$_3$ (complex polymer)

K$_2$O, 5B$_2$O$_3$ (B$_5$O$_8$)$^-$ group

Cs$_2$O, 3B$_2$O$_3$ (B$_3$O$_5$)$^-$ group

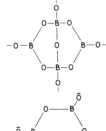

Li$_2$O, 2B$_2$O$_3$ (B$_4$O$_7$)$^{2-}$ group

Na$_2$O, B$_2$O$_3$ (B$_3$O$_6$)$^{3-}$ group

2CaO, B$_2$O$_3$ (linear chain)

2CoO, B$_2$O$_3$ (dimer)

found that if a crystalline sodium phosphate is dissolved in water under carefully controlled conditions, it gives a solution containing the one type of phosphate chain which was present in the original crystal. There is virtually no hydrolysis and no breakdown of the chains. If a sodium phosphate glass of similar chemical composition is dissolved in water under identical conditions this produces a distribution of chains differing in lengths which can be easily and quantitatively separated by simple paper chromatography. It has been found that alkali phosphates richer in metal oxide than the metaphosphate ($Na_2O \cdot P_2O_5$) give rise to anions which consist primarily of long linear chains, and no chain branching occurs [4].

Linear chain   Branching chain

Figure 10.2(a) shows the distribution of different phosphorus chains (as ion fractions) in $5\,Na_2O$, $3P_2O_5$ and $5\,ZnO \cdot 3\,P_2O_5$ glasses. Figure 10.2(b) shows the ion fractions expected from Flory distributions [5]. It should be pointed out that

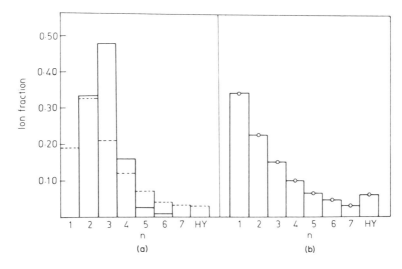

Fig. 10.2 The distribution of phosphorus chains of lengths n expressed as ion fractions.
(a) —— sodium–phosphate, and - - - - - zinc–phosphate glasses
(b) ion fractions expected from Flory distributions.

354

phosphates of this particular molar composition consist of only one
n $P_3O_{10}^{5-}$, whereas in glass chains both shorter and longer than 3
m Figure 10.2 it is clear that the experimentally observed distributions
/er than the Flory distributions, and the distribution in $5\,Na_2O\cdot3P_2O_5$
:r than that in $5\,ZnO\cdot3P_2O_5$ glass.
ilibria between the different kinds of phosphate chains may be written
eral form:

$$[P_nO_{2n+1}]^{(n+2)-} \rightleftharpoons [P_{n+1}O_{2n+4}]^{(n+3)-} + [P_{n-1}O_{2n-2}]^{(n+1)-} \qquad (10.1)$$

quilibrium constant

$$K_n = \frac{[P_{n+1}][P_{n-1}]}{[P_n]^2} \qquad (10.2)$$

: noted that forming or altering the distributions by the above reaction
alter the numbers of chains, so that the average number of phosphorus
:r chain, $\bar{n}$, depends entirely on the metal oxide to phosphorus pentoxide
the glass. However, if a reaction of the following type occurs:

$$2[PO_4]^{3-} = [P_2O_7]^{4-} + O^{2-} \qquad (10.3)$$

ill increase with increasing O^{2-} content of the glass. Fortunately in metal
hosphate melts O^{2-} concentration becomes significant at $\bar{n} = 1.5$ and
' the chromatographic work done so far has been with $\bar{n} \geqslant 2$.
e were to have a random distribution of chain lengths in the glass, one
ave the distribution calculated by Flory [5]. For this distribution it can be
that

$$X_n = \frac{1}{n} \cdot \frac{(\bar{n}-1)^{n-1}}{\bar{n}} \qquad (10.4)$$

X_n is the mol fraction of ions containing n phosphorus atoms
$_nO_{2n+1}]^{(n-2)-}$ chain.
ation (10.2) can be rewritten as

$$K_n = \frac{X_{n+1} \cdot X_{n-1}}{X_n^2} \qquad (10.5)$$

ituting the values of X from equation (10.4), K can be shown to be equal to
for all values of n and $\bar{n}$, for the Flory or random distribution. The values of
)r a number of compositions of different phosphate glasses have been
lated from the experimentally measured distributions. Typical results are
n in Figure 10.3. It is apparent that within the limits of error K_n tends to
' as n increases.
e fact that most values of K_n are less than unity indicates that ΔG is positive
eaction (10.1), whereas it would be zero for Flory distribution.
om a vast amount of work that has been done on different alkali and alkaline
l phosphates, it has become clear that the values of K_n, and the widths of the
ributions, become smaller, the more negative the free energy of mixing of the
al oxide with P_2O_5. Heats of formation of some binary phosphates are shown
igure 10.4.

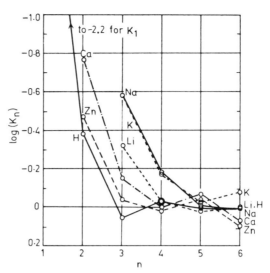

Fig. 10.3 Values of K_n for phosphates of K, Na, Li, Ca, Zn and H.

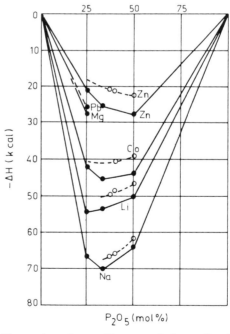

Fig. 10.4 Heats of formation of crystalline, ●, and glassy, ○, phosphates, per mol of crystalline oxides ($MO + P_2O_5$) at 35°C. The broken lines are heats of formation calculated for single chain glasses (after Meadowcroft and Richardson).

ergy of mixing of a metal oxide with P_2O_5 is temperature dependent; asing temperature the free energy of mixing becomes more positive and itly the distribution becomes wider. The question therefore arises as to se temperature to which the chain distributions measured at room are apply. Furthermore, it is to be borne in mind that even with a well schedule of cooling a glass-forming melt, the fictive temperature may ely depending upon the composition of the glass. Thus, a straightforward n about chain distributions in melts from measurements of anionic ions at room temperature is questionable. Unlike phosphate glasses, ory quantitative data regarding polymer distribution in silicate systems is ited, but there is qualitative evidence that chain length distributions in , where the ratio of metal oxide : silica ranges from 2 to 1 : 1, may resemble ' phosphates. This comes from the measured distributions of linear ethyl cates and methyl polysiloxanes (silicone fluids) which consists of oxygen chains with organic side groups [6]. These polymers can be ed by gas–liquid chromatography and the values of K_n can be calculated. typical values are given in Figure 10.5 along with binary sodium and gen phosphates for comparison. There is a striking resemblance among the with the values of K_n rising to unity as n increases.

main difficulty in studying the polymer distribution of silicate glasses is ery low solubility (without breaking the structural backbone in hydrogen de or alkali hydroxide solutions), and the strong tendency of silicic acids to nse, gradually eliminating water and changing into an insoluble silica gel. tly a silanation method [7,8] has been reported which, under carefully olled conditions, appears to avoid side – as well as polymerization and

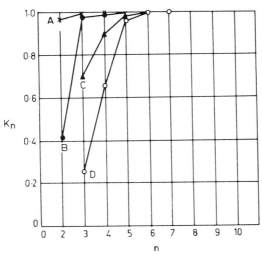

Fig. 10.5 K_n in organic silicate polymers and in inorganic phosphate polymers as a function of n.

(A) methyl polysiloxane   (C) ethyl polysilicate
(B) $H_2O + P_2O_5$ (D) $Na_2O + P_2O_5$

357

depolymerization – reactions. This method relies on cation substitution in silicates with Me_3Si–O–groups (silanation reaction) which can be represented as follows:

$$Si(OH)_4 + 4\,Me_3Si\text{–}OH \rightarrow Si(O\text{–}SiMe_3)_4 + 4\,H_2O$$

$$Si_2O(OH)_6 + 6\,Me_3Si\text{–}OH \rightarrow Si_2O(O\text{–}SiMe_3)_6 + 6\,H_2O$$

$$Si_nO_{n-1}(OH)_{2n+2} + (2n+2)\,Me_3Si\text{–}OH \rightarrow Si_nO_{n-1}(O\text{–}SiMe_3)_{2n+2}$$
$$+ (2n+2)\,H_2O$$

For silanation reaction Lentz[7] suggested a solvent consisting of a mixture of hydrochloric acid, hexamethyldisiloxane and water. Hydrochloric acid produces silicic acids and trimethylsilyl by reacting with metal-silicates and hexamethyldisiloxane respectively, the trimethylsilyl formed in solution reacts immediately with silicic acids thus preventing polymerization. Götz and Masson[8] found that maintenance of distribution in crystalline silicates is surer if a mixture of trimethylsilylchloride, hexamethyldisiloxane and isopropyl alcohol is used.

Trimethylsilyl derivative species of silicates are volatile and can be separated by conventional gas chromatography[9]. At various temperatures different derivatives volatalize, giving distinct peaks on the chromatogram. A typical chromatogram of $Cu_6(Si_6O_{18})\cdot 6H_2O$ is shown in Figure 10.6. The polymer corresponding to a certain peak can be identified by collecting the given fraction and analysing it by a mass spectrometer.

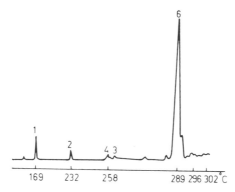

Fig. 10.6 Gas chromatogram of $Cu_6(Si_6O_{18})\cdot 6H_2O$ obtained by the silanation method.

Polymer distribution in binary Na_2O–SiO_2 glasses, as determined by the silanation technique is shown in Figure 10.7. It is clear that, like the phosphate system, silicate glasses containing high soda contain a mixture of polymers of different size, and when the Na_2O/SiO_2 ratio approaches unity the hypolymers (polymeric groups with a high molecular weight) become prevalent.

Gotz and Masson[8] have studied the polymer distribution in several natural silicates whose structure was known, by X-ray analysis; good agreement between X-ray and chromatographic results was observed. Polymer distribution in lead

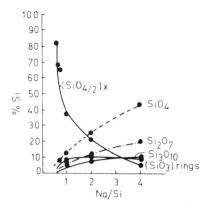

Fig. 10.7 Polymer distribution in sodium–silicate glasses.

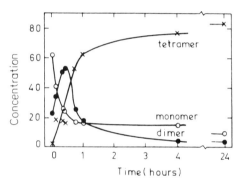

Fig. 10.8 Concentration of various species in lead orthosilicate as
a function of the thermal treatment.

osilicate glasses, heat treated at 500° C for various lengths of time, is shown in
re 10.8. It is clear that heat-treatment at 500° C, which involves phase
sformation in this particular glass, has a profound influence on the polymer
ribution. On heat-treatment the concentration of the monomer decreases
ceably. Initially, the concentration of the dimer increases, reaching a
imum in about 30 minutes and then decreasing continuously with time of
t-treatment. The concentration of the tetramer, which is about 1 per cent in
chilled glass, reaches 60 per cent after one hour of heat-treatment. The
ractograms showed that the heat treated glasses gradually crystallized; the
stals include, surprisingly, cyclic tetramers in very high concentration. This
ult clearly indicates that significant modifications in structure take place, at
st in this system, during crystallization.
Further recent advances in the study of polymer distribution in silicate systems
ay be found in references [15–18].
Following the success with phosphate glasses, a number of attempts have been

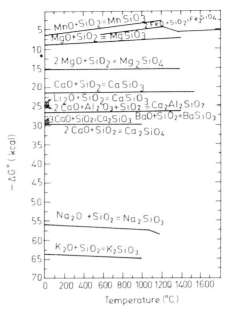

Fig. 10.9 Free energies of formation of crystalline silicates from their component oxides.

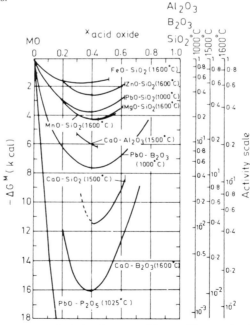

Fig. 10.10 Free energies of mixing for some binary silicate, aluminate and borate melts (after Jeffes).

late metal oxide activities in binary silicate melts to a single parameter
etal oxide. Unfortunately, most of the metal oxides so far studied are of
cal importance, like CaO, FeO, MnO, PbO, etc. where the free energies
on are much less negative than those of the binary alkali silicates which
amount importance to conventional glass technology. Free energies of
1 of some crystalline silicates, and free energies of mixing of some typical
and borates, are shown in Figures 10.9 and 10.10 respectively. Partial
e energies can be read from the curves of Figure 10.10 by the tangent-
method as described in Chapter 4. In Figure 10.10 the activity scales
n be used with this method for three different temperatures are marked
ght hand side of the figure. Typical activity curves in binary silicates of
$MO-SiO_2$ are illustrated in Figure 10.11. All these activity curves bear a
1 feature – substantial positive deviations from Raoultian behaviour for
mol fractions in excess of 0.60 and sharp falls in silica activities between
1 0.33 mol fraction.

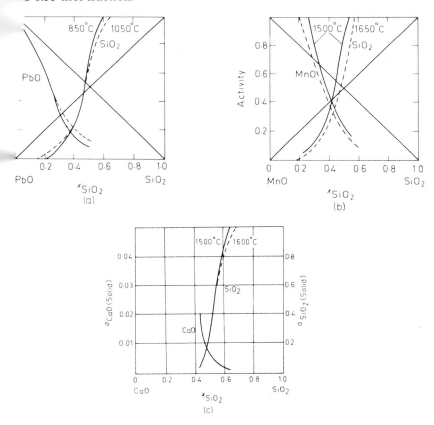

Fig. 10.11 Activity curves for binary silicate melts containing (a) PbO (b) MnO, and
(c) CaO (activities of CaO and SiO_2 are relative to solids).

361

One of the most useful approaches for estimating size distribution, and thereby calculating the activity of component oxides in silicate melts, is due to Masson [10]. This is based on the idea that silicate melts involve a series of polymerization equilibria:

$$2[SiO_4]^{4-} \rightleftharpoons [Si_2O_7]^{6-} + O^{2-} \tag{10.6}$$

$$2[Si_2O_7]^{6-} \rightleftharpoons [Si_3O_{10}]^{8-} + [SiO_4]^{4-} \tag{10.7}$$

or

$$[SiO_4]^{4-} + [Si_2O_7]^{6-} \rightleftharpoons [Si_3O_{10}]^{8-} + O^{2-} \tag{10.8}$$

The equilibrium constants of reactions (10.6), (10.7) and (10.8) will be referred to as K_1, K_2 and K_{21} respectively. It can be seen that K_{21} is equal to $K_1 K_2$. Since free energies of formation of silicates are much less negative than those of phosphates, Masson assumed that all K_n are equal to unity when $n \geqslant 2$. He further assumed that K_1 in a silicate system does not vary with composition, and on the basis of these two assumptions he has developed two models for silicate melts; in one of these the anions may have branching chains and in the other only linear chains, as has been found for phosphates. With the further assumption that the activity of the metal oxide is given by the Tempkin equation:

$$a_{MO} = x_{M^{2+}} \cdot x_{O^{2-}} \tag{10.9}$$

one can get

$$\frac{1}{x_{SiO_2}} = 2 + \frac{1}{1 - a_{MO}} - \frac{3}{1 - a_{MO} + (3a_{MO}/K_1)} \tag{10.10}$$

for branching chains, and

$$\frac{1}{x_{SiO_2}} = 3 - K_1 + \frac{a_{MO}}{1 - a_{MO}} + \frac{K_1(K_1 - 1)}{K_1 + [a_{MO}/(1 - a_{MO})]} \tag{10.11}$$

for the linear chains.

According to Masson, one can closely reproduce the activity curves for the metal oxides in binary silicates, by choosing a value for K_1 in equation (10.10) for each binary case as shown in Figure 10.12. Masson's treatment appears satisfactory at least for cases where the minimum in the free energy of mixing curves does not fall below about -12 kcal and the temperatures are reasonably high ($\geqslant 1100°$ C).

From equation (10.6) it appears that K_1 may be linked to the free energies of formation (from liquid MO and liquid SiO_2) of the liquid pyro- and orthosilicates:

$$RT \ln K_1 = 2\Delta G_{M_2SiO_4} - \Delta G_{M_3Si_2O_7} \tag{10.12}$$

But the ΔG values under present consideration are for the single chain (monodisperse) melts and are not the $\Delta G°$ values for the equilibrium melts which are multichain or polydisperse. At present it is not possible to derive one from the other; however, rough values of K_1 can only be obtained by using $\Delta G°$ values in equation (10.12).

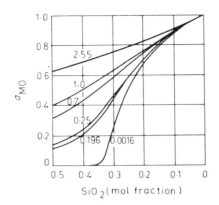

0.12 Calculated curves of activity of MO in binary M–silicates according to equation (10). Values of K_1 against each curve. From top to bottom the curves apply to SnO (1100° C), FeO (1300° C), MnO (1600° C), PbO (1000° C), and CaO (1600° C).

REFERENCES

arren, B. E. and Biscoe, J. (1938), *J. Amer. Ceram. Soc.*, **21**, 259.

chardson, F. D. (1974), *Physical Chemistry of Melts in Metallurgy*, Academic Press, ondon.

estman, A. E. R. and Gartaganis, P. A. (1957), *J. Amer. Chem. Soc.*, **40**, 293.

edowcroft, T. R. and Richardson, F. D. (1965), *Trans. Farad. Soc.*, **61**, 54.

lory, P. J. (1936; 1942), *J. Amer. Chem. Soc.*, **58**, 1877; **64**, 2205.

effes, J. H. E. (1975), *Silicate Industries*, **12**, 325.

entz, C. W. (1964), *Inorg. Chem.*, **3**, 574.

Gotz, J. and Masson, C. R. (1970; 1971), *J. Chem. Soc.*, A., 2683; A., 686.

Wu, F. F. H., Gotz, J., Jamieson, W. D. and Masson, C. R. (1970), *J. Chromatog.*, **48**, 515.

Masson, C. R. (1965), *Proc. Roy. Soc.*, **A287**, 201.

Richardson, F. D., Jeffes, J. H. E. and Withers, G. (1950), *J. Iron Steel Inst.*, **166**, 213.

Sridhar, R. and Jeffes, J. H. E. (1967), *Trans. Inst. Min. Metall.*, **76**, c. 44.

Mehta, S. R. and Richardson, F. D. (1965), *J. Iron Steel Inst.*, **203**, 524.

Sharma, R. A. and Richardson, F. D. (1962), *J. Iron Steel Inst.*, **200**, 373.

Gaskell, D. R. (1973), *Metallurgical Trans.*, **4**, 185.

Baes, C. F. (1970), *J. Solid State Chem.*, **1**, 159.

Pretnar, B. (1968), *Ber. der. Bunsengesllschaff, Bd.*, **72**, 773.

Kapoor, M. L., Mehrotra, G. M. and Frohberg, M. G. (1974), *Arch. Eisenhuttenwes*, **45**, 213.

Index